STUDENT SOLUTIONS M/

# BRIEF APPLIED CALCULUS

**DANIEL CLEGG**
PALOMAR COLLEGE

BROOKS/COLE
CENGAGE Learning™

Australia · Brazil · Japan · Korea · Mexico · Singapore · Spain · United Kingdom · United States

## BROOKS/COLE
CENGAGE Learning

For product information and technology assistance, contact us at
**Cengage Learning Customer & Sales Support,**
**1-800-354-9706**

For permission to use material from this text or product, submit all requests online at **www.cengage.com/permissions.** Further permissions questions can be e-mailed to **permissionrequest@cengage.com.**

ISBN-13: 978-0-534-42387-2
ISBN-10: 0-534-42387-6

**Brooks/Cole**
20 Davis Drive
Belmont, CA 94002-3098
USA

Cengage Learning is a leading provider of customized learning solutions with office locations around the globe, including Singapore, the United Kingdom, Australia, Mexico, Brazil, and Japan. Locate your local office at **www.cengage.com/global**

Cengage Learning products are represented in Canada by Nelson Education, Ltd.

To learn more about Brooks/Cole, visit **www.cengage.com/brookscole**

Purchase any of our products at your local college store or at our preferred online store **www.cengagebrain.com**

Printed in the United States of America
1 2 3 4 5 6 7 16 15 14 13 12

# Preface

This Student Solutions Manual contains detailed solutions to selected exercises in the text *Brief Applied Calculus* by James Stewart and Daniel Clegg. Specifically, it includes solutions to odd-numbered exercises in each chapter section and review section. It also includes all solutions to the Prepare Yourself exercises.

Each solution is presented in the context of the corresponding section of the text. In general, solutions to the initial exercises involving a new concept illustrate that concept in more detail; this knowledge is then utilized in subsequent solutions. Thus, while the intermediate steps of a solution are given, you may need to refer back to earlier exercises in the section or prior sections for additional explanation of the concepts involved. Note that, in many cases, different routes to an answer may exist which are equally valid; also, answers can be expressed in different but equivalent forms. Thus, the goal of this manual is not to give the definitive solution to each exercise, but rather to assist you as a student in understanding the concepts of the text and learning how to apply them to the challenge of solving a problem.

I would like to thank James Stewart for offering suggestions and Kathi Townes of TECH-arts for typesetting and producing this manual as well as creating the illustrations. Gina Sanders prepared solutions for comparison of accuracy and style in addition to proofreading manuscript; her assistance and suggestions were very helpful and much appreciated. I would also like to thank Jon Booze for initially preparing many solutions. Finally, I thank Richard Stratton and Laura Wheel of Brooks/Cole, Cengage Learning for their trust, assistance, and patience.

DANIEL CLEGG

Palomar College

# Contents

# 7 Functions of Several Variables 193

# Appendixes 213

# ☐ DIAGNOSTIC TEST

**1.** (a) $(-3)^4$ means $(-3)(-3)(-3)(-3) = 81$.

(b) $-3^4$ means $-(3 \cdot 3 \cdot 3 \cdot 3) = -81$.

(c) $3^{-4} = \dfrac{1}{3^4} = \dfrac{1}{81}$

(d) $\dfrac{5^{23}}{5^{21}} = 5^{23-21} = 5^2 = 25$

(e) $\left(\frac{2}{3}\right)^{-2} = \dfrac{1}{(2/3)^2} = \left(\frac{3}{2}\right)^2 = \frac{9}{4}$

(f) $16^{-3/4} = \dfrac{1}{16^{3/4}} = \dfrac{1}{\left(\sqrt[4]{16}\right)^3} = \dfrac{1}{2^3} = \dfrac{1}{8}$

**2.** (a) Note that $\sqrt{200} = \sqrt{100 \cdot 2} = 10\sqrt{2}$ and $\sqrt{32} = \sqrt{16 \cdot 2} = 4\sqrt{2}$. Thus $\sqrt{200} - \sqrt{32} = 10\sqrt{2} - 4\sqrt{2} = 6\sqrt{2}$.

(b) $(3a^3b^3)(4ab^2)^2 = 3a^3b^3 \cdot (4)^2(a)^2(b^2)^2 = 3a^3b^3 \cdot 16a^2b^4 = 3 \cdot 16 \cdot a^{3+2} \cdot b^{3+4} = 48a^5b^7$

(c) $\left(\dfrac{3x^{3/2}y^3}{x^2y^{-1/2}}\right)^{-2} = \left[3x^{(3/2)-2}y^{3-(-1/2)}\right]^{-2} = \left(3x^{-1/2}y^{7/2}\right)^{-2} = (3)^{-2}(x^{-1/2})^{-2}(y^{7/2})^{-2}$

$$= 3^{-2}x^1y^{-7} = \dfrac{x}{3^2y^7} = \dfrac{x}{9y^7}$$

*Or:*  $\left(\dfrac{3x^{3/2}y^3}{x^2y^{-1/2}}\right)^{-2} = \left(\dfrac{x^2y^{-1/2}}{3x^{3/2}y^3}\right)^2 = \dfrac{(x^2y^{-1/2})^2}{(3x^{3/2}y^3)^2} = \dfrac{x^4y^{-1}}{9x^3y^6} = \dfrac{x^4}{9x^3y^6y} = \dfrac{x}{9y^7}$

**3.** (a) $3(x+6) + 4(2x-5) = 3x + 18 + 8x - 20 = 11x - 2$

(b) $(x+3)(4x-5) = 4x^2 - 5x + 12x - 15 = 4x^2 + 7x - 15$

(c) $\left(\sqrt{a} + \sqrt{b}\right)\left(\sqrt{a} - \sqrt{b}\right) = \left(\sqrt{a}\right)^2 - \sqrt{a}\sqrt{b} + \sqrt{a}\sqrt{b} - \left(\sqrt{b}\right)^2 = a - b$

*Or:*  Use the formula for the difference of two squares to see that $\left(\sqrt{a} + \sqrt{b}\right)\left(\sqrt{a} - \sqrt{b}\right) = \left(\sqrt{a}\right)^2 - \left(\sqrt{b}\right)^2 = a - b.$

(d) $(2x+3)^2 = (2x+3)(2x+3) = 4x^2 + 6x + 6x + 9 = 4x^2 + 12x + 9.$

*Note:* A quicker way to expand this binomial is to use the formula $(a+b)^2 = a^2 + 2ab + b^2$ with $a = 2x$ and $b = 3$:
$(2x+3)^2 = (2x)^2 + 2(2x)(3) + 3^2 = 4x^2 + 12x + 9$

(e) $(x+2)^3 = (x+2)(x+2)^2 = (x+2)(x^2 + 4x + 4) = x^3 + 4x^2 + 4x + 2x^2 + 8x + 8$
$= x^3 + 6x^2 + 12x + 8$

**4.** (a) Using the difference of two squares formula, $a^2 - b^2 = (a+b)(a-b)$, we have
$4x^2 - 25 = (2x)^2 - 5^2 = (2x+5)(2x-5).$

(b) Factoring by trial and error, we get $2x^2 + 5x - 12 = (2x - 3)(x + 4).$

(c) Using factoring by grouping and the difference of two squares formula, we have

$$x^3 - 3x^2 - 4x + 12 = x^2(x-3) - 4(x-3) = (x^2-4)(x-3) = (x+2)(x-2)(x-3).$$

(d) $x^4 + 27x = x(x^3 + 27) = x(x+3)(x^2 - 3x + 9)$

This last expression was obtained using the sum of two cubes formula, $x^3 + y^3 = (x+y)(x^2 - xy + y^2)$ with $y = 3$.

[See Reference Page 1 in the textbook.]

(e) $x^3y - 4xy = xy(x^2 - 4) = xy(x+2)(x-2)$

**5.** (a) $\dfrac{x^2 + 3x + 2}{x^2 - x - 2} = \dfrac{(x+1)(x+2)}{(x+1)(x-2)} = \dfrac{x+2}{x-2}$

(b) $\dfrac{2x^2 - x - 1}{x^2 - 9} \cdot \dfrac{x+3}{2x+1} = \dfrac{(2x+1)(x-1)}{(x-3)(x+3)} \cdot \dfrac{x+3}{2x+1} = \dfrac{x-1}{x-3}$

(c) $\dfrac{x^2}{x^2 - 4} - \dfrac{x+1}{x+2} = \dfrac{x^2}{(x-2)(x+2)} - \dfrac{x+1}{x+2} = \dfrac{x^2}{(x-2)(x+2)} - \dfrac{x+1}{x+2} \cdot \dfrac{x-2}{x-2} = \dfrac{x^2 - (x+1)(x-2)}{(x-2)(x+2)}$

$$= \dfrac{x^2 - (x^2 - x - 2)}{(x+2)(x-2)} = \dfrac{x^2 - x^2 + x + 2}{(x+2)(x-2)} = \dfrac{x+2}{(x+2)(x-2)} = \dfrac{1}{x-2}$$

(d) $\dfrac{\dfrac{y}{x} - \dfrac{x}{y}}{\dfrac{1}{y} - \dfrac{1}{x}} = \dfrac{\dfrac{y}{x} - \dfrac{x}{y}}{\dfrac{1}{y} - \dfrac{1}{x}} \cdot \dfrac{xy}{xy} = \dfrac{\dfrac{y}{x} \cdot xy - \dfrac{x}{y} \cdot xy}{\dfrac{1}{y} \cdot xy - \dfrac{1}{x} \cdot xy} = \dfrac{y^2 - x^2}{x - y} = \dfrac{(y+x)(y-x)}{-(y-x)} = \dfrac{y+x}{-1} = -(x+y) \text{ or } -x - y$

**6.** (a) $\dfrac{\sqrt{10}}{\sqrt{5} - 2} = \dfrac{\sqrt{10}}{\sqrt{5} - 2} \cdot \dfrac{\sqrt{5} + 2}{\sqrt{5} + 2} = \dfrac{\sqrt{50} + 2\sqrt{10}}{(\sqrt{5})^2 - 2^2} = \dfrac{5\sqrt{2} + 2\sqrt{10}}{5 - 4} = 5\sqrt{2} + 2\sqrt{10}$

(b) $\dfrac{\sqrt{4+h} - 2}{h} = \dfrac{\sqrt{4+h} - 2}{h} \cdot \dfrac{\sqrt{4+h} + 2}{\sqrt{4+h} + 2} = \dfrac{4 + h - 4}{h(\sqrt{4+h} + 2)} = \dfrac{h}{h(\sqrt{4+h} + 2)} = \dfrac{1}{\sqrt{4+h} + 2}$

**7.** (a) $x + 5 = 14 - \frac{1}{2}x \ \Rightarrow \ x + \frac{1}{2}x = 14 - 5 \ \Rightarrow \ \frac{3}{2}x = 9 \ \Rightarrow \ x = \frac{2}{3} \cdot 9 \ \Rightarrow \ x = 6$

(b) $\dfrac{2x}{x+1} = \dfrac{2x-1}{x} \ \Rightarrow \ 2x^2 = (2x-1)(x+1) \quad \text{[cross multiply]} \ \Rightarrow \ 2x^2 = 2x^2 + x - 1 \ \Rightarrow$

$0 = x - 1 \ \Rightarrow \ x = 1$

(c) $x^2 - x - 12 = 0 \ \Rightarrow \ (x+3)(x-4) = 0 \ \Rightarrow \ x + 3 = 0 \text{ or } x - 4 = 0 \ \Rightarrow \ x = -3 \text{ or } x = 4$

(d) By the quadratic formula, $2x^2 + 4x + 1 = 0 \ \Rightarrow$

$$x = \dfrac{-4 \pm \sqrt{4^2 - 4(2)(1)}}{2(2)} = \dfrac{-4 \pm \sqrt{8}}{4} = \dfrac{-4 \pm 2\sqrt{2}}{4} = \dfrac{-4}{4} \pm \dfrac{2\sqrt{2}}{4} = -1 \pm \dfrac{\sqrt{2}}{2} = -1 \pm \tfrac{1}{2}\sqrt{2}.$$

(e) $x^4 - 3x^2 + 2 = 0 \ \Rightarrow \ (x^2 - 1)(x^2 - 2) = 0 \ \Rightarrow \ (x+1)(x-1)(x^2 - 2) = 0 \ \Rightarrow \ x + 1 = 0 \text{ or } x - 1 = 0$

or $x^2 - 2 = 0$; thus $x = -1$ or $x = 1$ or $x^2 = 2 \ \Rightarrow \ x = \pm\sqrt{2}.$

(f) Multiplying through $2x(4-x)^{-1/2} - 3\sqrt{4-x} = 0$ by $(4-x)^{1/2}$ gives $2x - 3(4-x) = 0 \ \Rightarrow$

$2x - 12 + 3x = 0 \ \Rightarrow \ 5x - 12 = 0 \ \Rightarrow \ 5x = 12 \ \Rightarrow \ x = \frac{12}{5}.$

8. (a) $-4 < 5 - 3x \le 17 \quad \Rightarrow \quad -9 < -3x \le 12 \quad \Rightarrow \quad 3 > x \ge -4$ or $-4 \le x < 3$.

In interval notation, the solution is $[-4, 3)$.

(b) $x^2 < 2x + 8 \quad \Rightarrow \quad x^2 - 2x - 8 < 0 \quad \Rightarrow \quad (x+2)(x-4) < 0$. Now $(x+2)(x-4) = 0$ when $x = -2$ or $x = 4$, and $(x+2)(x-4)$ can change sign only at the critical values $x = -2$ and $x = 4$. Thus the possible intervals of solution are $(-\infty, -2)$, $(-2, 4)$, and $(4, \infty)$. By choosing a single test value from each interval, we see that $(x+2)(x-4) < 0$ only when $-2 < x < 4$. In interval notation, the solution to the inequality is $(-2, 4)$.

(c) First $x(x-1)(x+2) = 0$ when $x = 0$, $x = 1$, or $x = -2$. Thus the possible intervals of solution to the inequality $x(x-1)(x+2) > 0$ are $(-\infty, -2)$, $(-2, 0)$, $(0, 1)$ and $(1, \infty)$. By choosing a single test value from each interval, we see that both intervals $(-2, 0)$ and $(1, \infty)$ satisfy the inequality. Thus, the solution is the union of these two intervals: $(-2, 0) \cup (1, \infty)$.

9. (a) False. In order for the statement to be true, it must hold for all real numbers, so, to show that the statement is false, pick $p = 1$ and $q = 2$ and observe that $(1+2)^2 \ne 1^2 + 2^2$. In general, $(p+q)^2 = p^2 + 2pq + q^2$.

(b) True as long as $a$ and $b$ are nonnegative real numbers. To see this, think in terms of the laws of exponents:
$$\sqrt{ab} = (ab)^{1/2} = a^{1/2}b^{1/2} = \sqrt{a}\,\sqrt{b}.$$

(c) False. To see this, let $a = 1$ and $b = 2$, then $\sqrt{1^2 + 2^2} \ne 1 + 2$.

(d) False. To see this, let $T = 1$ and $C = 2$, then $\dfrac{1 + 1(2)}{2} \ne 1 + 1$. In general, $\dfrac{1 + TC}{C} = \dfrac{1}{C} + \dfrac{TC}{C} = \dfrac{1}{C} + T$.

(e) False. To see this, let $x = 2$ and $y = 3$, then $\dfrac{1}{2-3} \ne \dfrac{1}{2} - \dfrac{1}{3}$.

(f) True since $\dfrac{1/x}{a/x - b/x} \cdot \dfrac{x}{x} = \dfrac{1}{a - b}$, as long as $x \ne 0$ and $a - b \ne 0$.

10. (a) Using the point $(2, -5)$ and $m = -3$ in the point-slope equation of a line, $y - y_1 = m(x - x_1)$, we get
$$y - (-5) = -3(x - 2) \quad \Rightarrow \quad y + 5 = -3x + 6 \quad \Rightarrow \quad y = -3x + 1.$$

(b) A line parallel to the $x$-axis must be horizontal and thus have a slope of 0. Since the line passes through the point $(2, -5)$, the $y$-coordinate of every point on the line is $-5$, so the equation is $y = -5$.

(c) A line parallel to the $y$-axis is vertical with undefined slope. So the $x$-coordinate of every point on the line is 2 and so the equation is $x = 2$.

(d) Note that $2x - 4y = 3 \quad \Rightarrow \quad -4y = -2x + 3 \quad \Rightarrow \quad y = \frac{1}{2}x - \frac{3}{4}$. This line has slope $m = \frac{1}{2}$, and the line we are looking for is parallel, so its slope is also $m = \frac{1}{2}$. Thus an equation of the line is $y - (-5) = \frac{1}{2}(x - 2) \quad \Rightarrow$
$y + 5 = \frac{1}{2}x - 1 \quad \Rightarrow \quad y = \frac{1}{2}x - 6$.

11. The radius of the circle is the distance from the point $(-1, 4)$ to the point $(3, -2)$:
$r = \sqrt{[3 - (-1)]^2 + (-2 - 4)^2} = \sqrt{4^2 + (-6)^2} = \sqrt{52}$. The standard equation of a circle with center $(h, k)$ and radius $r$ is $(x - h)^2 + (y - k)^2 = r^2$, so an equation of the circle here is $[x - (-1)]^2 + (y - 4)^2 = (\sqrt{52})^2 \quad \Leftrightarrow$
$(x + 1)^2 + (y - 4)^2 = 52$.

**12.** (a) The slope between the points $(-7, 4)$ and $(5, -12)$ is $m = \dfrac{y_2 - y_1}{x_2 - x_1} = \dfrac{-12 - 4}{5 - (-7)} = \dfrac{-16}{12} = -\dfrac{4}{3}$.

(b) Using the point-slope equation of a line with the point $(-7, 4)$ and slope $m = -\frac{4}{3}$, an equation is

$y - 4 = -\frac{4}{3}[x - (-7)] \;\Rightarrow\; y - 4 = -\frac{4}{3}x - \frac{28}{3} \;\Rightarrow\; 3y - 12 = -4x - 28 \;\Rightarrow\; 4x + 3y = -16$ or

$4x + 3y + 16 = 0$. [In slope-intercept form, the equation is $y = -\frac{4}{3}x - \frac{16}{3}$.]

When $y = 0$, we have $4x + 16 = 0 \;\Rightarrow\; x = -4$, so the $x$-intercept is $-4$. When $x = 0$ we have

$3y + 16 = 0 \;\Rightarrow\; y = -\frac{16}{3}$, so the $y$-intercept is $-\frac{16}{3}$.

(c) The distance between points $A(-7, 4)$ and $B(5, -12)$ is

$$\sqrt{[5 - (-7)]^2 + (-12 - 4)^2} = \sqrt{12^2 + (-16)^2} = \sqrt{144 + 256} = \sqrt{400} = 20.$$

**13.** (a) The graph of $y = 1 - \frac{1}{2}x$ is a line with slope $-\frac{1}{2}$ and $y$-intercept 1.

(b) We first graph $y = 1 - \frac{1}{2}x$ as a dotted line. Since $y < 1 - \frac{1}{2}x$, the points in the

region lie *below* this line.

(c) The graph of $y = x^2 - 1$ is a parabola opening upward with vertex $(0, -1)$.

(d) The graph of the equation $x^2 + y^2 = 4$ is a circle centered at the origin with

radius 2.

(e) Graph the corresponding horizontal lines (given by the equations $y = -1$ and

$y = 3$) as solid lines. The inequality $y \geq -1$ describes the points $(x, y)$ that lie

on or *above* the line $y = -1$. The inequality $y \leq 3$ describes the points $(x, y)$

that lie on or *below* the line $y = 3$. So the pair of inequalities $-1 \leq y \leq 3$

describes the points that lie on or *between* the lines $y = -1$ and $y = 3$.

# 1   FUNCTIONS AND MODELS

## 1.1   Functions and Their Representations

In exercises requiring estimations or approximations, your answers may vary slightly from the answers given here.

**1.**   (a) $f(10)$ is the price of a bag of potting soil that weighs 10 pounds, so $f(10) = \$0.40 \times 10 = \$4.00$.

    (b) The domain of $f$ is $\{4, 10, 50\}$ and the range is the outputs corresponding to the domain values: $\{1.60, 4.00, 20.00\}$.

**3.**   The equation $P(8) = 64.3$ means that on January 1, 2008, the population of the city was 64,300. $P(4.5)$ represents the population on July 1, 2004 (4.5 years after January 1, 2000).

**5.**   The equation $F(65) = 24.7$ says that the average fuel economy of the car is 24.7 miles per gallon when the car is being driven at 65 mi/h.

**7.**   (a) The point $(-1, -2)$ is on the graph of $f$, so $f(-1) = -2$.

    (b) When $x = 2$, $y$ is about 2.8, so $f(2) \approx 2.8$.

    (c) $f(x) = 2$ is equivalent to $y = 2$. When $y = 2$, we have $x = -3$ and $x = 1$.

    (d) When $y = 0$, it appears that $x \approx -2.5$ and $x \approx 0.3$.

    (e) From the graph, $f(x)$ is defined when $-3 \le x \le 3$, so the domain is $[-3, 3]$. Looking at the $y$-values on the graph, $f$ takes on all values from $-2$ to 3, so the range is $[-2, 3]$.

**9.**   From Figure 1 in the text, the lowest point occurs at about $(t, a) = (12, -85)$. The highest point occurs at about $(17, 115)$. Thus, the range of the vertical ground acceleration is $-85 \le a \le 115$. Written in interval notation, we get $[-85, 115]$.

**11.**   The person's weight increased to about 160 pounds at age 20 and stayed fairly steady for 10 years. The person's weight dropped to about 120 pounds for the next 5 years, then increased rapidly to about 170 pounds. The next 30 years saw a gradual increase to 190 pounds. Possible reasons for the drop in weight at 30 years of age: diet, exercise, illness.

**13.**   The water will cool down almost to freezing as the ice melts. Then, when the ice has melted, the water will slowly warm up to room temperature.

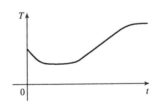

**15.**   Of course, this graph depends strongly on the geographical location!

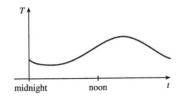

**17.** As the price increases, the amount sold decreases.

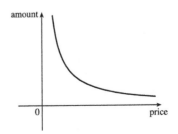

**19.**

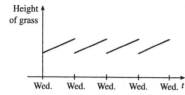

**21.** (a)

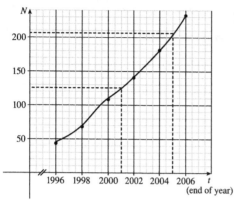

(b) From the graph, we estimate the number of US cell phone subscribers to be about 126 million in 2001 and 207 million in 2005.

**23.** $f(x) = 3x^2 - x + 2$.

$f(2) = 3(2)^2 - 2 + 2 = 12 - 2 + 2 = 12$.

$f(-2) = 3(-2)^2 - (-2) + 2 = 12 + 2 + 2 = 16$.

$f(a) = 3a^2 - a + 2$.

$f(-a) = 3(-a)^2 - (-a) + 2 = 3a^2 + a + 2$.

$f(a + 1) = 3(a + 1)^2 - (a + 1) + 2 = 3(a^2 + 2a + 1) - a - 1 + 2 = 3a^2 + 6a + 3 - a + 1 = 3a^2 + 5a + 4$.

$2f(a) = 2 \cdot f(a) = 2(3a^2 - a + 2) = 6a^2 - 2a + 4$.

$f(2a) = 3(2a)^2 - (2a) + 2 = 3(4a^2) - 2a + 2 = 12a^2 - 2a + 2$.

$f(a^2) = 3(a^2)^2 - (a^2) + 2 = 3(a^4) - a^2 + 2 = 3a^4 - a^2 + 2$.

$[f(a)]^2 = [3a^2 - a + 2]^2 = (3a^2 - a + 2)(3a^2 - a + 2)$
$= 9a^4 - 3a^3 + 6a^2 - 3a^3 + a^2 - 2a + 6a^2 - 2a + 4 = 9a^4 - 6a^3 + 13a^2 - 4a + 4$.

$f(a + h) = 3(a + h)^2 - (a + h) + 2 = 3(a^2 + 2ah + h^2) - a - h + 2 = 3a^2 + 6ah + 3h^2 - a - h + 2$.

**25.** $f(x) = x^2 + 1$, so $f(4 + h) = (4 + h)^2 + 1 = 16 + 8h + h^2 + 1 = 17 + 8h + h^2$, and

$$\frac{f(4 + h) - f(4)}{h} = \frac{(17 + 8h + h^2) - 17}{h} = \frac{8h + h^2}{h} = \frac{h(8 + h)}{h} = 8 + h.$$

**27.** $f(x) = 4 + 3x - x^2$, so $f(3 + h) = 4 + 3(3 + h) - (3 + h)^2 = 4 + 9 + 3h - (9 + 6h + h^2) = 4 - 3h - h^2$, and

$$\frac{f(3 + h) - f(3)}{h} = \frac{(4 - 3h - h^2) - 4}{h} = \frac{-3h - h^2}{h} = \frac{h(-3 - h)}{h} = -3 - h.$$

**29.** $\dfrac{f(x) - f(a)}{x - a} = \dfrac{\dfrac{1}{x} - \dfrac{1}{a}}{x - a} = \dfrac{\dfrac{a - x}{ax}}{x - a} = \dfrac{a - x}{ax(x - a)} = \dfrac{-1(x - a)}{ax(x - a)} = -\dfrac{1}{ax}$

**31.** $f(x) = x/(3x - 1)$ is defined for all $x$ except when $3x - 1 = 0 \iff x = \frac{1}{3}$, so the domain is

$\left\{ x \mid x \neq \frac{1}{3} \right\} = \left( -\infty, \frac{1}{3} \right) \cup \left( \frac{1}{3}, \infty \right)$.

**33.** $f(t) = \sqrt{2t + 6}$ is defined when $2t + 6 \geq 0 \iff t \geq -3$, so the domain is $\{ t \mid t \geq -3 \} = [-3, \infty)$.

**35.** No, the scatter plot is not the graph of a function because we can draw a vertical line (at $x = 5$ or $x = 12$) that intersects two points.

**37.** No, the curve is not the graph of a function because a vertical line can be drawn that intersects the curve more than once. Hence, the curve fails the Vertical Line Test.

**39.** Yes, the curve is the graph of a function because it passes the Vertical Line Test. The domain is $[-3, 2]$ and the range is $[-3, -2) \cup [-1, 3]$.

**41.** $f(x) = \begin{cases} x + 2 & \text{if } x < 0 \\ 1 - x & \text{if } x \geq 0 \end{cases}$

$f(-3) = -3 + 2 = -1$, $f(0) = 1 - 0 = 1$, and $f(2) = 1 - 2 = -1$.

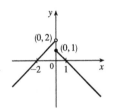

**43.** $f(x) = \begin{cases} x + 1 & \text{if } x \leq -1 \\ x^2 & \text{if } x > -1 \end{cases}$

$f(-3) = -3 + 1 = -2$, $f(0) = 0^2 = 0$, and $f(2) = 2^2 = 4$.

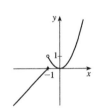

**45.** Let the length and width of the rectangle be $L$ and $W$. Then the perimeter is $2L + 2W = 20$ and the area is $A = LW$.

Solving the first equation for $W$ in terms of $L$ gives $W = \dfrac{20 - 2L}{2} = 10 - L$. Thus, $A(L) = L(10 - L) = 10L - L^2$. Since lengths are positive, the domain of $A$ is $0 < L < 10$. If we further restrict $L$ to be larger than $W$, then $5 < L < 10$ would be the domain.

**47.** Let each side of the base of the box have length $x$, and let the height of the box be $h$. Since the volume is 2, we know that $2 = hx^2$, so that $h = 2/x^2$, and the surface area is $S = x^2 + 4xh$. Thus, $S(x) = x^2 + 4x(2/x^2) = x^2 + (8/x)$, with domain $x > 0$.

**49.** The height of the box is $x$ and the length and width are $L = 20 - 2x$, $W = 12 - 2x$. Then $V = LWx$ and so

$V(x) = (20 - 2x)(12 - 2x)(x) = (240 - 64x + 4x^2)(x) = 4x^3 - 64x^2 + 240x$. The sides $L$, $W$, and $x$ must be positive.

Thus, $L > 0 \iff 20 - 2x > 0 \iff 20 > 2x \iff x < 10$;

$W > 0 \iff 12 - 2x > 0 \iff 12 > 2x \iff x < 6$; and $x > 0$. Combining these restrictions gives us the domain $0 < x < 6$.

**51.** (a)

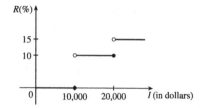

(b) On $14,000, tax is assessed on $4000, and 10% of $4000 is

$0.10 \times \$4000 = \$400.$

On $26,000, tax is assessed on $16,000, and

$10\%(\$10,000) + 15\%(\$6000) = \$1000 + \$900 = \$1900.$

(c) There is no tax for $0 \leq I \leq 10,000$, so the graph of $T$ is a horizontal line segment at $T = 0$ there. For $10,000 < I \leq 20,000$ the tax is 10%; as in part (b), there is $1000 tax assessed on $20,000 of income, so the graph of $T$ is a line segment from $(10,000, 0)$ to $(20,000, 1000)$. The tax rate then increases to 15%; the tax on $30,000 is $2500, so the graph of $T$ for $I > 20,000$ is a (half) line beginning at $(20,000, 1000)$ and passing through $(30,000, 2500)$.

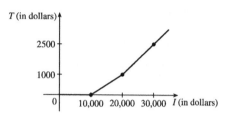

**53.** $f$ is an odd function because its graph is symmetric about the origin. $g$ is an even function because its graph is symmetric with respect to the $y$-axis.

**55.** (a) Because an even function is symmetric with respect to the $y$-axis, and the point $(5, 3)$ is on the graph of this even function, the point $(-5, 3)$ must also be on its graph.

(b) Because an odd function is symmetric with respect to the origin, and the point $(5, 3)$ is on the graph of this odd function, the point $(-5, -3)$ must also be on its graph.

**57.** $f(x) = \dfrac{x}{x^2 + 1}$.

$f(-x) = \dfrac{-x}{(-x)^2 + 1} = \dfrac{-x}{x^2 + 1} = -\dfrac{x}{x^2 + 1} = -f(x).$

So $f$ is an odd function.

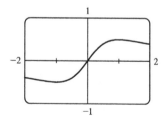

**59.** $f(x) = \dfrac{x}{x + 1}$, so $f(-x) = \dfrac{-x}{-x + 1} = \dfrac{x}{x - 1}$.

Since this is neither $f(x)$ nor $-f(x)$, the function $f$ is neither even nor odd.

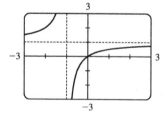

**61.** $f(x) = 1 + 3x^2 - x^4$.

$f(-x) = 1 + 3(-x)^2 - (-x)^4 = 1 + 3x^2 - x^4 = f(x).$

So $f$ is an even function.

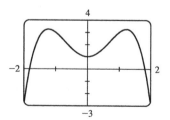

**63.** (i) If $f$ and $g$ are both even functions, then $f(-x) = f(x)$ and $g(-x) = g(x)$. Now

$h(-x) = f(-x) + g(-x) = f(x) + g(x) = h(x)$, so $h$ is an *even* function.

(ii) If $f$ and $g$ are both odd functions, then $f(-x) = -f(x)$ and $g(-x) = -g(x)$. Now

$h(-x) = f(-x) + g(-x) = -f(x) + [-g(x)] = -[f(x) + g(x)] = -h(x)$, so $h$ is an *odd* function.

(iii) If $f$ is an even function and $g$ is an odd function, then $h(-x) = f(-x) + g(-x) = f(x) + [-g(x)] = f(x) - g(x)$,

which is neither $h(x)$ nor $-h(x)$, so $h$ is *neither* even nor odd. (Exception: if $f$ is the zero function, then $h$ will be *odd*. If $g$ is the zero function, then $h$ will be *even*.)

## 1.2   Combining and Transforming Functions

**1.** $g(t)$ is the total number of students, both male and female, that attended a math class at the university on day $t$ of this year.

**3.** $f(n)$ is the value, in dollars, of gold held in the bank's vault at the end of the $n$th day of this year.

**5.** $C(x)$ is the number of bushels of corn divided by the number of acres of land, so it represents the average number of bushels of corn yielded per acre during year $x$.

**7.** (a) $f(t) = S(t) + C(t) = (42 + 1.8t) + (16.4 + 0.6t) = 58.4 + 2.4t$

(b) $f(4) = 58.4 + 2.4(4) = 68$. During 2004, the employee earned a total of $68,000.

**9.** (a) $A(x) = f(x) + g(x) = (x^2 - 5x) + (3x + 12) = x^2 - 2x + 12$

(b) $B(x) = f(x) - g(x) = (x^2 - 5x) - (3x + 12) = x^2 - 5x - 3x - 12 = x^2 - 8x - 12$

(c) $C(x) = f(x)g(x) = (x^2 - 5x)(3x + 12) = 3x^3 + 12x^2 - 15x^2 - 60x = 3x^3 - 3x^2 - 60x$

(d) $D(x) = f(x)/g(x) = (x^2 - 5x)/(3x + 12)$

**11.** $g(3) = 4(3) - 2 = 10$, so $A(3) = f(g(3)) = f(10) = (10)^2 + 1 = 101$.

$f(3) = (3)^2 + 1 = 10$, so $B(3) = g(f(3)) = g(10) = 4(10) - 2 = 38$.

**13.** $N(3) = 3(3) + 7 = 16$, so $C(3) = M(N(3)) = M(16) = 16 + \sqrt{16} = 16 + 4 = 20$.

$M(4) = 4 + \sqrt{4} = 4 + 2 = 6$, so $D(4) = N(M(4)) = N(6) = 3(6) + 7 = 25$.

**15.** $p(x) = f(g(x)) = f(2x + 1) = (2x + 1)^2 - 1 = (4x^2 + 4x + 1) - 1 = 4x^2 + 4x$;

$q(x) = g(f(x)) = g(x^2 - 1) = 2(x^2 - 1) + 1 = 2x^2 - 2 + 1 = 2x^2 - 1$.

**17.** $p(x) = f(g(x)) = f(1 - \sqrt{x}) = (1 - \sqrt{x})^3 + 2(1 - \sqrt{x}) = (1 - \sqrt{x})^3 + 2 - 2\sqrt{x}$;

$q(x) = g(f(x)) = g(x^3 + 2x) = 1 - \sqrt{x^3 + 2x}$.

**19.** $p(x) = f(g(x)) = f(x + 2) = x + 2 + \dfrac{1}{x + 2}$; $q(x) = g(f(x)) = g\left(x + \dfrac{1}{x}\right) = x + \dfrac{1}{x} + 2$.

**21.** We input the year $t$ into the inner function $N$. The output of $N$, the number of surfboards produced, is then used as the input for the outer function $P$, which gives the profit earned. Thus $f(t) = P(N(t))$ is the profit, in thousands of dollars, the manufacturer earns during year $t$.

**23.** The output of $f$ (a percentage of commuters) is not an appropriate input for $g$, where the input is time. So $g(f(t))$ is not meaningful. However, if we input the time $t$ into the function $g$, we get the average price per gallon of gasoline. This output can in turn be used as the input into the function $f$, which gives the average percentage of commuters who carpool. Thus $f(g(t))$ is meaningful, and it gives the average percentage of commuters who carpooled $t$ months after January 1, 2011.

**25.** (a) $A(m) = P(f(m)) = P(0.5m + 3\sqrt{m}) = 14.7 + 0.433(0.5m + 3\sqrt{m}) = 14.7 + 0.2165m + 1.299\sqrt{m}$. $A(m)$ is the pressure, in PSI, that Paul experiences $m$ minutes after he starts his dive.

(b) $A(25) = 14.7 + 0.2165(25) + 1.299\sqrt{25} = 26.6075$; Paul experiences a pressure of about 26.6 PSI, 25 minutes after the start of his dive.

**27.** (a) $g(2) = 5$, because the point $(2, 5)$ is on the graph of $g$. Thus, $f(g(2)) = f(5) = 4$, because the point $(5, 4)$ is on the graph of $f$.

(b) $g(f(0)) = g(0) = 3$

(c) $f(g(0)) = f(3) = 0$

(d) $f(f(4)) = f(2) = -2$

**29.** One option is to consider $x^2 + 1$ the inside function and call it $g$. The outside function takes the 10th power of the input, so let $f(x) = x^{10}$. Then $h(x) = f(g(x)) = f(x^2 + 1) = (x^2 + 1)^{10}$.

**31.** Let $g(x) = 2x^2 + 5$. Then if $f(x) = \sqrt{x}$, we have $h(x) = f(g(x)) = f(2x^2 + 5) = \sqrt{2x^2 + 5}$.

**33.** (a) If the graph of $f$ is shifted 4 units upward, its equation becomes $y = f(x) + 4$.

(b) If the graph of $f$ is shifted 4 units downward, its equation becomes $y = f(x) - 4$.

(c) If the graph of $f$ is shifted 4 units to the right, its equation becomes $y = f(x - 4)$.

(d) If the graph of $f$ is shifted 4 units to the left, its equation becomes $y = f(x + 4)$.

(e) If the graph of $f$ is reflected about the $x$-axis, its equation becomes $y = -f(x)$.

(f) If the graph of $f$ is reflected about the $y$-axis, its equation becomes $y = f(-x)$.

(g) If the graph of $f$ is stretched vertically by a factor of 3, its equation becomes $y = 3f(x)$.

(h) If the graph of $f$ is compressed vertically by a factor of 3, its equation becomes $y = \frac{1}{3}f(x)$.

**35.** (a) Graph 3: The graph of $f$ is shifted 4 units to the right and has equation $y = f(x - 4)$.

(b) Graph 1: The graph of $f$ is shifted 3 units upward and has equation $y = f(x) + 3$.

(c) Graph 4: The graph of $f$ is compressed vertically by a factor of 3 and has equation $y = \frac{1}{3}f(x)$.

(d) Graph 5: The graph of $f$ is shifted 4 units to the left and reflected about the $x$-axis. Its equation is $y = -f(x + 4)$.

(e) Graph 2: The graph of $f$ is shifted 6 units to the left and stretched vertically by a factor of 2. Its equation is $y = 2f(x + 6)$.

**37.** (a) To graph $y = f(2x)$ we compress the graph of $f$ horizontally by a factor of 2.

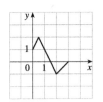

The point $(4, -1)$ on the graph of $f$ corresponds to the point

$\left(\frac{1}{2} \cdot 4, -1\right) = (2, -1)$.

(b) To graph $y = f\left(\frac{1}{2}x\right)$ we stretch the graph of $f$ horizontally by a factor of 2.

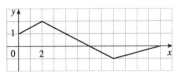

The point $(4, -1)$ on the graph of $f$ corresponds to the point $(2 \cdot 4, -1) = (8, -1)$.

(c) To graph $y = f(-x)$ we reflect the graph of $f$ about the $y$-axis.

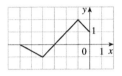

The point $(4, -1)$ on the graph of $f$ corresponds to the point

$(-1 \cdot 4, -1) = (-4, -1)$.

(d) To graph $y = -f(-x)$ we reflect the graph of $f$ about the $y$-axis, then about the $x$-axis.

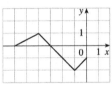

The point $(4, -1)$ on the graph of $f$ corresponds to the point $(-1 \cdot 4, -1 \cdot -1) = (-4, 1)$.

**39.** To graph $y = \sqrt{x + 3}$ we shift the graph of $y = \sqrt{x}$ left 3 units.

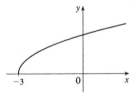

**41.** To graph $y = -\sqrt{x - 1}$ we shift the graph of $y = \sqrt{x}$ to the right 1 unit and then reflect the graph about the $x$-axis.

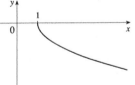

**43.** To graph $y = -x^2 + 2$ we reflect the graph of $y = x^2$ about the $x$-axis and then shift the graph upward 2 units.

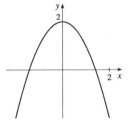

**45.** To graph $f(x) = \frac{1}{4}x^2 - 3$ we compress the graph of $y = x^2$ vertically by a factor of 4 and then shift the graph downward 3 units.

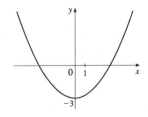

**47.** (a) We first reflect the graph of $y = \sqrt{x}$ about the $x$-axis: $y = -\sqrt{x}$. Then shift the resulting graph upward 4 units:

$y = -\sqrt{x} + 4$.

(b) We shift the graph of $y = \sqrt{x}$ to the right 4 units and then down 1 unit: $y = \sqrt{x - 4} - 1$.

**49.** (a) The output of $g$ is always 15 less than the output of $f$ (for the same input), so the second reservoir's water level is always 15 feet lower than that of the first reservoir.

(b) The output of $g$ for a given input is the same as the output of $f$ for an input that is 2 units smaller (the graph of $g$ is the graph of $f$ shifted 2 units to the right), so the second reservoir's depth always matches the depth of the first reservoir from two months prior.

(c) The output of $g$ for a given input is the same as the output of $f$ for an input that is 2 units greater (the graph of $g$ is the graph of $f$ shifted 2 units to the left), so the *first* reservoir's depth always matches the depth of the *second* reservoir from two months prior.

(d) The output of $g$ is always 0.8 times the output of $f$ (for the same input), so the second reservoir's depth is always 0.8 times, or 80% of, that of the first reservoir.

**51.** (a) The output of $B$ is always 1.3 times the output of $A$ (for the same input), so the number of songs sold by the rival service is always 1.3 times (30% more than) the number sold by the original service.

(b) The output of $B$ is always 23 more than the output of $A$ (for the same input), so the number of songs sold by the rival service is always 23,000 more than the number sold by the original service.

(c) The output of $B$ for a given input is 5 more than the output of $A$ for an input that is 1 unit less (the graph of $B$ is the graph of $A$ shifted 1 unit to the right and 5 units upward), so the rival service sells 5000 more songs each month than the original service sold the previous month.

**53.** (a)

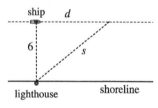

From the figure, we have a right triangle with legs 6 and $d$, and hypotenuse $s$. By the Pythagorean Theorem, $d^2 + 6^2 = s^2 \Rightarrow$ $s = f(d) = \sqrt{d^2 + 36}$.

(b) Using the relationship *distance = rate × time*, we get $d = (30 \text{ km/h})(t \text{ hours}) = 30t$ (in km). Thus, $d = g(t) = 30t$.

(c) $f(g(t)) = f(30t) = \sqrt{(30t)^2 + 36} = \sqrt{900t^2 + 36}$. This function represents the distance between the lighthouse and the ship as a function of the time elapsed since noon.

**55.** (a) Using the relationship *distance = rate × time* with the radius $r$ as the distance, we have $r(t) = 60t$ (in cm).

(b) The area of a circle is $A = \pi r^2$, so $A(r(t)) = \pi(60t)^2 = 3600\pi t^2$. This formula gives the area (in cm$^2$) enclosed by the circular ripple at any time $t$.

**57.** $p(x) = f(g(h(x))) = f(g(x + 3)) = f((x + 3)^2 + 2)$
$= f(x^2 + 6x + 11) = \sqrt{(x^2 + 6x + 11) - 1} = \sqrt{x^2 + 6x + 10}$

**59.** Stretch the graph of $f$ horizontally by a factor of 2, reflect the graph about the $x$-axis, and then shift the graph upward 3 units:

$y = -f(\frac{1}{2}x) + 3.$

**61.** Reflect the graph of $f$ about the $x$-axis, then shift the graph down 1 unit:   $y = -f(x) - 1.$

**63.** Compress the graph of $f$ horizontally by a factor of 2 and shift the graph downward 1 unit:   $y = f(2x) - 1$.

**65.** The graph of $y = f(x) = \sqrt{3x - x^2}$ has been shifted 4 units to the left, reflected about the $x$-axis, and shifted downward 1 unit. Thus, a function describing the graph is

$$y = \underbrace{-1 \cdot}_{\substack{\text{reflect} \\ \text{about } x\text{-axis}}} \quad \underbrace{f \; (x+4)}_{\substack{\text{shift} \\ \text{4 units left}}} \quad \underbrace{- \; 1}_{\substack{\text{shift} \\ \text{1 unit down}}}$$

This function can be written as

$$y = -f(x+4) - 1 = -\sqrt{3(x+4) - (x+4)^2} - 1 = -\sqrt{3x + 12 - (x^2 + 8x + 16)} - 1 = -\sqrt{-x^2 - 5x - 4} - 1$$

**67.** If $f(x) = m_1 x + b_1$ and $g(x) = m_2 x + b_2$, then

$h(x) = f(g(x)) = f(m_2 x + b_2) = m_1(m_2 x + b_2) + b_1 = m_1 m_2 x + m_1 b_2 + b_1.$

So $h$ is a linear function with slope $m_1 m_2$.

## 1.3   Linear Models and Rates of Change

**1.** The slope $m$ through the points $(3, 7)$ and $(5, 10)$ is $\dfrac{\Delta y}{\Delta x} = \dfrac{y_2 - y_1}{x_2 - x_1} = \dfrac{10 - 7}{5 - 3} = \dfrac{3}{2}.$

**3.** $m = \dfrac{y_2 - y_1}{x_2 - x_1} = \dfrac{2240 - 1860}{26 - 45} = \dfrac{380}{-19} = -20.$

**5.** By the slope-intercept form of the equation of a line, an equation of the line is $y = 3x - 2.$

**7.** By the point-slope form of the equation of a line, an equation of the line through $(2, -3)$ with slope 6 is $y - (-3) = 6(x - 2)$, which can be written as $y + 3 = 6x - 12$   or   $y = 6x - 15.$

**9.** The slope of the line through $(2, 1)$ and $(1, 6)$ is $m = \dfrac{6 - 1}{1 - 2} = -5$, so an equation of the line is $y - 1 = -5(x - 2)$   or

$y = -5x + 11.$

**11.** The slope of the line through $(4, 84)$ and $(13, -312)$ is $m = \dfrac{-312 - 84}{13 - 4} = \dfrac{-396}{9} = -44$, so an equation of the line is

$y - 84 = -44(x - 4)$ or $y = -44x + 260.$

**13.** Since the line passes through $(1, 0)$ and $(0, -3)$, its slope is $m = \dfrac{-3 - 0}{0 - 1} = 3.$ The $y$-intercept is $-3$, so an equation is

$y = 3x - 3.$

**15.** The slope is $-\frac{1}{5}$, so to find a second point on the line we move down 1 and right 5 from $(-2, 6)$, arriving at $(3, 5)$.

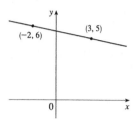

**17.** Write the equation $2x + 5y = 15$ in slope-intercept form $(y = mx + b)$: $5y = -2x + 15$ $\Leftrightarrow$ $y = -\frac{2}{5}x + 3$. Then the slope is $m = -\frac{2}{5}$, and the $y$-intercept is $b = 3$.

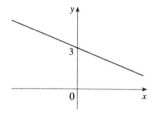

**19.** $-5x + 6y = 42$ $\Leftrightarrow$ $6y = 5x + 42$ $\Leftrightarrow$ $y = \frac{5}{6}x + 7$, so the slope is $m = \frac{5}{6}$ and the $y$-intercept is $b = 7$.

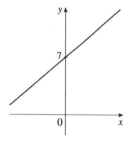

**21.** $f(x) = -2x + 14$ is in slope-intercept form, so we see that the slope is $m = -2$. The $y$-intercept is $f(0) = 14$, and since $f(x) = 0$ when $-2x + 14 = 0$ $\Leftrightarrow$ $-2x = -14$ $\Leftrightarrow$ $x = 7$, the $x$-intercept is 7.

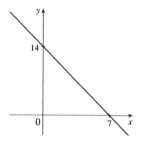

**23.** The slope of $A(t) = 0.2t - 4$ is $m = 0.2$. Since $A(0) = -4$ and $A(t) = 0$ when $0.2t - 4 = 0$ $\Leftrightarrow$ $t = \frac{4}{0.2} = 20$, the $t$-intercept is 20 and the $y$-intercept is $-4$.

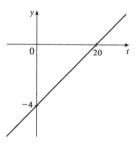

**25.** $h$ is a linear function where $h(7) = 329$ and $h(11) = 553$, so the slope of its graph is

$$m = \frac{h(11) - h(7)}{11 - 7} = \frac{553 - 329}{11 - 7} = \frac{224}{4} = 56.$$ An equation of the line is $y - 329 = 56(x - 7)$ or $y = 56x - 63$, thus $h(x) = 56x - 63$.

**27.** The slope of $L$ is $\frac{\Delta L}{\Delta t} = \frac{L(12) - L(8)}{12 - 8} = \frac{8.36 - 5.32}{12 - 8} = \frac{3.04}{4} = 0.76$, and the units are millions of viewers per week. Thus the program gains 0.76 million additional viewers each week.

**29.** The slope of $V$ is $\frac{\Delta V}{\Delta t} = \frac{-16{,}500}{5} = -3300$, and the units are dollars per year. Thus the value of the machine will decrease by \$3300 each year.

**31.** The rate of change of $T$ is 0.26, the coefficient of $p$ (and the slope of the graph), and is measured in thousands of dollars in tax per thousand dollars of profit. It measures the amount of tax the company pays ($260) for each thousand dollars of profit.

**33.** (a) $D = 200$, so $c = 0.0417D(a+1) = 0.0417(200)(a+1) = 8.34a + 8.34$. The slope is 8.34, which represents the change in mg of the dosage for a child for each change of 1 year in age.

   (b) For a newborn, $a = 0$, so $c = 8.34$ mg.

**35.** (a)

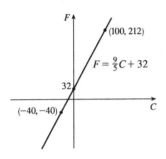

   (b) The slope of $\frac{9}{5}$ means that $F$ increases $\frac{9}{5}$ degrees for each increase of $1°$C. (Equivalently, $F$ increases by 9 when $C$ increases by 5 and $F$ decreases by 9 when $C$ decreases by 5.) The $F$-intercept of 32 is the Fahrenheit temperature corresponding to a Celsius temperature of 0.

**37.** (a) The slope is $\dfrac{\Delta T}{\Delta N} = \dfrac{80 - 70}{173 - 113} = \dfrac{10}{60} = \dfrac{1}{6}$. So a linear equation is $T - 80 = \frac{1}{6}(N - 173)$ $\Leftrightarrow$

     $T - 80 = \frac{1}{6}N - \frac{173}{6}$ $\Leftrightarrow$ $T = \frac{1}{6}N + \frac{307}{6}$ $\left[\frac{307}{6} \approx 51.2\right]$.

   (b) The slope of $\frac{1}{6}$ means that the temperature in Fahrenheit degrees increases one-sixth as rapidly as the number of cricket chirps per minute. Said differently, each increase of 6 cricket chirps per minute corresponds to an increase of $1°$F.

   (c) When $N = 150$, the temperature is given approximately by $T = \frac{1}{6}(150) + \frac{307}{6} \approx 76\,°$F.

**39.** (a) Let $P$ be the pressure and $d$ the depth. The rate of change of $P$ is $\dfrac{\Delta P}{\Delta d} = \dfrac{4.34}{10} = 0.434$ (lb/in$^2$) per foot. When $d = 0$ we have $P = 15$, so using the slope-intercept form of a line, $P = 0.434d + 15$.

   (b) When $P = 100$, then $100 = 0.434d + 15$ $\Leftrightarrow$ $0.434d = 85$ $\Leftrightarrow$ $d = \frac{85}{0.434} \approx 195.85$ feet. Thus, the pressure is $100$ lb/in$^2$ at a depth of approximately 196 feet.

**41.** (a)

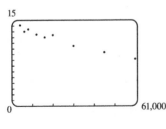

A linear model does seem appropriate.

   (b) Using the points $(8000, 13.4)$ and $(60,000, 8.2)$, the slope is

     $\dfrac{8.2 - 13.4}{60,000 - 8000} = \dfrac{-5.2}{52,000} = -0.0001$, and an equation of the line is

     $y - 13.4 = -0.0001\,(x - 8000)$ or, equivalently, $y = -0.0001x + 14.2$.

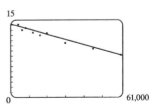

   (c) When $x = 90,000$, the model predicts that $y = -0.0001(90,000) + 14.2 = 5.2$, so 5.2 people per 100 population would suffer from a peptic ulcer. Because 90,000 is outside the given data, this is an example of extrapolation.

   (d) When $x = 200,000$, $y$ is negative, so the model does not apply.

**43.** (a)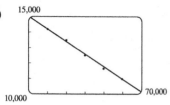

It appears that a line through the first and last data points is a good fit. Then the slope is

$$\frac{10{,}970 - 14{,}245}{60{,}000 - 20{,}000} = \frac{-3275}{40{,}000} = -0.081875,$$

and an equation of the line is $y - 14{,}245 = -0.081875(x - 20{,}000)$   or   $y = -0.081875x + 15{,}882.5$. Thus a linear model is $V(x) = -0.081875x + 15{,}882.5$, where $x$ is the mileage of the car and $V(x)$ is the car's value (in dollars).

(b) $V(12{,}000) = -0.081875(12{,}000) + 15{,}882.5 = 14{,}900$, so we estimate the value to be $\$14{,}900$.

(c) Solving $V(x) = 0$, we have $-0.081875x + 15{,}882.5 = 0$   $\Leftrightarrow$   $x = \dfrac{15{,}882.5}{0.081875} \approx 193{,}985$. Thus the model predicts

that the vehicle is worth $\$0$ after being driven 193,985 miles. This is not realistic, as the vehicle should always have a value greater than $\$0$, even if just for scrap or salvage.

**45.** (a) Using a graphing calculator or other computing device, we obtain the least squares regression line

$y \approx -0.00009979x + 13.951$.

The following commands and screens illustrate how to find the least squares regression line on a TI-84 Plus.

Enter the data into list one (L1) and list two (L2). Press $\boxed{\text{STAT}}\,\boxed{1}$ to enter the editor.

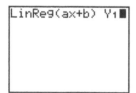 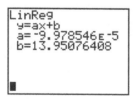 

Find the regression line and store it in $Y_1$. Press $\boxed{\text{2nd}}\,\boxed{\text{QUIT}}\,\boxed{\text{STAT}}\,\boxed{\blacktriangleright}\,\boxed{4}\,\boxed{\text{VARS}}\,\boxed{\blacktriangleright}\,\boxed{1}\,\boxed{1}\,\boxed{\text{ENTER}}$.

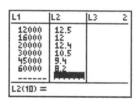

Note from the last figure that the regression line has been stored in $Y_1$ and that Plot1 has been turned on (Plot1 is highlighted). You can turn on Plot1 from the Y= menu by placing the cursor on Plot1 and pressing $\boxed{\text{ENTER}}$ or by pressing $\boxed{\text{2nd}}\,\boxed{\text{STAT PLOT}}\,\boxed{1}\,\boxed{\text{ENTER}}$.

Now press ZOOM 9 to produce a graph of the data and the regression line. Note that choice 9 of the ZOOM menu automatically selects a window that displays all of the data.

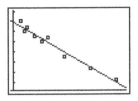

(b) When $x = 25{,}000$, $y \approx 11.456$; or about 11.5 per 100 population.

(c) When $x = 80{,}000$, $y \approx 5.968$; or about a 6% chance.

**47.** (a) A linear model seems appropriate over the time interval considered.

(b) Using a computing device, we obtain the regression line
$$y \approx 0.0265x - 46.8759.$$

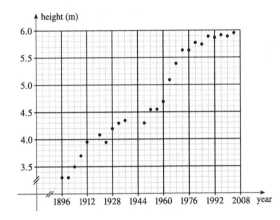

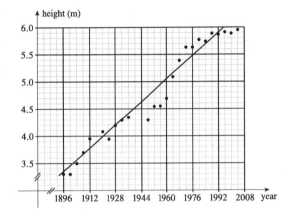

(c) For $x = 2008$, the linear model predicts a winning height of 6.27 m, considerably higher than the actual winning height of 5.96 m.

(d) It is *not* reasonable to use the model to predict the winning height at the 2100 Olympics because 2100 is too far from the 1896–2004 range on which the model is based.

**49.** (a) All members have slope 3, so graphs of members of the family are parallel lines.

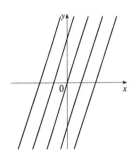

(b) All members have a $y$-intercept of 3.

(c) $f(x) = 3x + 3$

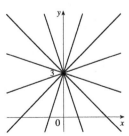

**51.** Although the graph of $f$ appears steeper than the graph of $g$, the scales of the $y$-axes are different. The slope of the graph of $g$ is larger than the slope of the graph of $f$, so $g$ has the greater rate of change.

**53.** From a scatter plot of the data (where $t = 0$ corresponds to 1995) we observe two different trends. From 1995 to 1999 the data appear to be roughly linear with a small negative slope, and from 2000 to 2003 we have a linear relationship with a positive steep slope.

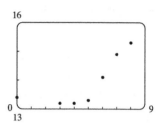

A line through the points $(0, 13.4)$ and $(4, 13.2)$ models the data from 1995 to 1999 well. The slope is $\dfrac{13.2 - 13.4}{4 - 0} = \dfrac{-0.2}{4} = -0.05$, and an equation of the line is $y = -0.05t + 13.4$. For 2000 to 2003, a line through the points $(5, 13.3)$ and $(8, 15.3)$ fits well. The slope is $\dfrac{15.3 - 13.3}{8 - 5} = \dfrac{2}{3}$, and an equation of the line is $y - 13.3 = \frac{2}{3}(t - 5)$ or $y \approx 0.667t + 9.97$. Thus a piecewise function that models the data is

$$H(t) = \begin{cases} -0.05t + 13.4 & \text{if } t < 5 \\ 0.667t + 9.97 & \text{if } t \geq 5 \end{cases}$$

where $t = 0$ corresponds to 1995.

## 1.4  Polynomial Models and Power Functions

**1.** (a) $g(w) = w^4$ is a polynomial of degree 4 and also a power function.

(b) $f(x) = \sqrt[5]{x} = x^{1/5}$ is a power function.

(c) $A(t) = -2t^7 + 3t - 1$ is a polynomial of degree 7.

(d) $r(x) = \dfrac{x^2 + 1}{x^3 + x}$ is a rational function because it is a ratio of polynomials.

**3.** (a) The graph of $y = x^2$ is a parabola (function $h$).

(b) $y = x^5$ is a power function; because the exponent is odd, its graph is similar to the graph of $y = x^3$ (function $f$).

(c) $y = \sqrt{x}$ is the square root function with domain $[0, \infty)$ (function $g$).

**5.** To graph $f(x) = (x + 2)^2 + 5$ we shift the graph of $y = x^2$ left 2 and upward 5. The vertex is $(-2, 5)$.

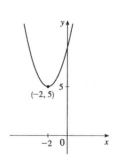

**7.** To graph $K(t) = -t^2 + 3$ we reflect the graph of $y = t^2$ about the $t$-axis and then shift the graph upward 3 units. The vertex is $(0, 3)$.

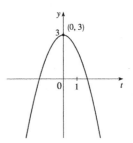

**9.** To graph $A(p) = \frac{1}{2}(p-1)^2 - 2$ we compress the graph of $y = p^2$ vertically by a factor of 2 and then shift the graph right 1 and down 2. The vertex is $(1, -2)$.

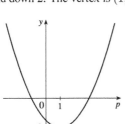

**11.** To graph $C(x) = -2(x-3)^2 - 4$ we stretch the graph of $y = x^2$ vertically by a factor of 2 and reflect it about the $x$-axis. Then we shift the graph right 3 and down 4. The vertex is $(3, -4)$.

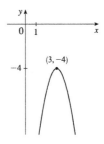

**13.** The vertex of the parabola is $(1, -2)$, so the quadratic function is $f(x) = a(x-1)^2 - 2$. (We shift the graph of $y = ax^2$ right 1 and down 2.) Because the point $(0, -3)$ is on the graph, we know $f(0) = -3$ and we use this information to determine $a$:

$a(0-1)^2 - 2 = -3 \quad \Rightarrow \quad a - 2 = -3 \quad \Rightarrow \quad a = -1$. Thus the function is $f(x) = -(x-1)^2 - 2$ or equivalently

$f(x) = -(x^2 - 2x + 1) - 2 = -x^2 + 2x - 1 - 2 = -x^2 + 2x - 3$.

**15.** The vertex of the parabola is $(0, 22)$, so an equation is $y = ax^2 + 22$. The point $(8, 6)$ is on the parabola, so we substitute 8 for $x$ and 6 for $y$ to find $a$. $6 = a(8)^2 + 22 \quad \Rightarrow \quad 64a = -16 \quad \Rightarrow \quad a = \frac{-16}{64} = -\frac{1}{4}$, so an equation for $f$ is

$f(x) = -\frac{1}{4}x^2 + 22$.

**17.** The vertex of the parabola is $(55, 1840)$, so an equation is $y = a(x-55)^2 + 1840$. The point $(0, 2203)$ is on the parabola, so we substitute 0 for $x$ and 2203 for $y$ to find $a$. $2203 = a(0-55)^2 + 1840 \quad \Rightarrow$

$2203 = 3025a + 1840 \quad \Rightarrow \quad a = \frac{363}{3025} = 0.12$, so an equation for $f$ is $f(x) = 0.12(x-55)^2 + 1840$ or equivalently

$f(x) = 0.12(x^2 - 110x + 3025) + 1840 = 0.12x^2 - 13.2x + 2203$.

**19.** (a) For convenience we let $t = 0$ correspond to July 1, 2000.

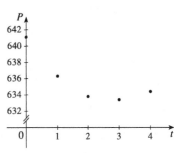

(b) From the scatter plot it appears that we get a good quadratic model

by choosing the point $(3, 633.4)$ as the vertex of a parabola, so its

equation is $P = a(t - 3)^2 + 633.4$. To determine $a$, we estimate that

the graph should pass through the point $(0, 641.1)$. Substituting, we

get $641.1 = a(0 - 3)^2 + 633.4 \Rightarrow 7.7 = 9a \Rightarrow$

$a = \frac{7.7}{9} \approx 0.856$. So a quadratic model for the data is

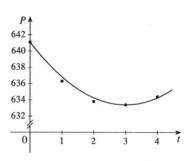

$P(t) = 0.856(t - 3)^2 + 633.4$. We see from the graph that the

model is a reasonably good fit.

(c) July 1, 2011, corresponds to $t = 11$, and $P(11) = 0.856(11 - 3)^2 + 633.4 \approx 688.2$, so we estimate that the population

then to be about 688,200.

**21.** (a) We draw a scatter plot, using $t = 0$ to represent the year 1995.

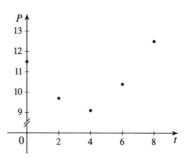

(b) We get a parabola that fits the data fairly well by using the point $(4, 9.1)$ as the vertex and choosing $(8, 12.5)$ as another

point on the graph. Thus an equation of the model is $P = a(t - 4)^2 + 9.1$, and we substitute $t = 8$, $P = 12.5$ to

determine $a$: $12.5 = a(8 - 4)^2 + 9.1 \Rightarrow 3.4 = 16a \Rightarrow a = \frac{3.4}{16} = 0.2125$. So a quadratic model for the data is

$P(t) = 0.2125(t - 4)^2 + 9.1$.

(c) 2002 corresponds to $t = 7$ and $P(7) = 0.2125(7 - 4)^2 + 9.1 = 11.0125$, so we estimate that in 2002, about 11.0% of

persons under the age of 65 were covered by Medicaid.

**23.** The average cost (in dollars) per magazine is given by

$$a(x) = \frac{C(x)}{x} = \frac{131{,}000 + 0.41x}{x} = \frac{131{,}000}{x} + 0.41$$

We have $a(65{,}000) = \frac{131{,}000}{65{,}000} + 0.41 \approx 2.43$, so the average cost is about \$2.43 per magazine when 65,000 copies are

printed monthly.

**25.** (a) Revenue = number of files × price per file = $f(p)$ thousand files × $p$ dollars per file, so

$R(p) = \left(142 - 91.4\sqrt{p}\right)p = 142p - 91.4p\sqrt{p}$ thousand dollars.

(b) From the graph of the revenue function $R$, we estimate that the highest point

on the curve is approximately $(1.07, 50.8)$. Thus annual revenue is

maximized when the files are priced at \$1.07 each. The number of files sold

at this price is $f(1.07) = 142 - 91.4\sqrt{1.07} \approx 47.455$ thousand, or 47,455.

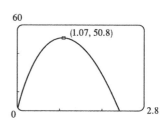

**27.** Looking at the graph, we estimate that $f$ is increasing on the interval $(-5, 8)$ and decreasing on the intervals $(-\infty, -5)$, $(8, \infty)$.

**29.** It appears that $f$ is increasing on approximately $(-\infty, -2)$, $(6, 9)$ and decreasing on $(-2, 6)$, $(9, \infty)$.

**31.** After graphing $f$ we see that the graph rises until the curve reaches the origin. The graph then falls until it reaches the point $(1.065, -3.02)$ (approximately) and then rises. Thus we estimate that $f$ is increasing on $(-\infty, 0)$, $(1.065, \infty)$ and decreasing on $(0, 1.065)$.

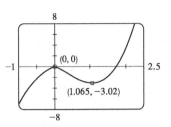

**33.** $A$ varies directly with $t$, so $A = kt$ for some constant $k$. If $A = 80$ when $t = 25$, then $80 = k \cdot 25$ $\Leftrightarrow$ $k = \frac{80}{25} = 3.2$ and $A = 3.2t$. When $t = 36$ we have $A = 3.2(36) = 115.2$.

**35.** $f(x)$ is proportional to $x^3$, so we can write $f(x) = kx^3$. We are given that $f(2) = 14.4$, so $8k = 14.4$ $\Leftrightarrow$ $k = \frac{14.4}{8} = 1.8$. Thus $f(x) = 1.8x^3$ and $f(5) = 1.8(5)^3 = 225$.

**37.** $T(p)$ is inversely proportional to $p$, so we can write $T(p) = k/p$. Since $T(8) = 2$, we have $k/8 = 2$ $\Rightarrow$ $k = 16$, so $T(p) = 16/p$ and $T(30) = \frac{16}{30} = \frac{8}{15}$.

**39.** (a) $D$ varies directly with $t$, so $D = kt$, and $D = 1.5$ when $t = 8$, so we have $1.5 = 8k$ $\Rightarrow$ $k = \frac{1.5}{8} = 0.1875$. Thus $D = 0.1875t$, where $t$ is measured in seconds and $D$ in miles.

(b) If $t = 14$, the storm is $D = 0.1875(14) = 2.625$ miles away.

**41.** $L = k/d^2$ and $L = 70$ when $d = 10$, so $70 = k/(10)^2$ $\Rightarrow$ $k = 7000$ and $L = 7000/d^2$. When $d = 100$, the loudness of the sound is $L = \frac{7000}{100^2} = 0.7$ dB.

**43.** If $f$ is the frequency of the string and $L$ is the length, then $f = k/L$. If we double the length of the string, the frequency becomes $\dfrac{k}{2L} = \dfrac{1}{2} \cdot \dfrac{k}{L} =$ half the original frequency.

**45.** We enter the data into a graphing calculator (or other computing device), using $t = 0$ to represent the year 2000. A quadratic regression model is $H(t) \approx 15.93t^2 - 5.79t + 1577$. The year 2008 corresponds to $t = 8$ and $H(8) = 2550.2$, so we estimate that the number of housing units completed in 2008 was 2550.2 thousand, or about 2,550,000.

**47.** We let $t = 0$ represent the year 1900 and enter the data into a graphing calculator, which gives a cubic regression model as approximately $P(t) = -0.000285t^3 + 0.522t^2 - 6.40t + 1721$. The estimated population in 1925 is $P(25) = -0.000285(25)^3 + 0.522(25)^2 - 6.40(25) + 1721 \approx 1883$ million.

**49.** When capital is 407, production is $P(407) = 52.5(407)^{1/4} \approx 235.8$. If capital is 20% higher, we have $x = 407 + 0.20(407) = 1.20(407) = 488.4$ and $P(488.4) = 52.5(488.4)^{1/4} \approx 246.8$. Since $\frac{246.8}{235.8} \approx 1.047$, this causes production to be about 1.047 times greater, or an increase of 4.7%.

**51.** Let $d$ be the distance from the light source to the object and let $I$ be the illumination. Then $I(d) = k/d^2$. If we begin with distance $d_1$, the illumination at half the distance is $I(\frac{1}{2}d_1) = k/(\frac{1}{2}d_1)^2 = k/\left(\frac{1}{4}d_1^2\right) = 4k/d_1^2 = 4I(d_1)$. Thus moving halfway to the lamp makes the illumination four times brighter.

**53.** (a) A graphing calculator gives a power model as approximately $N = 3.105A^{0.3080}$.

(b) When $A = 291$, $N = 3.105(291)^{0.3080} \approx 17.8$, so we would expect to find 18 species.

## 1.5 Exponential Models

**1.** (a) $f(x) = a^x$, $a > 0$      (b) $\mathbb{R}$        (c) $(0, \infty)$

(d) (i) See Figure 3(b)      (ii) See Figure 3(a)

**3.** All of these graphs approach the $x$-axis toward the left, all of them pass through the point $(0, 1)$, and all of them are increasing toward the right. The larger the base, the faster the function increases for $x > 0$.

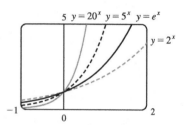

**5.** The functions with bases greater than 1 ($y = 3^x$ and $y = 10^x$) are increasing, while those with bases less than 1 $\left[y = \left(\frac{1}{3}\right)^x \text{ and } y = \left(\frac{1}{10}\right)^x\right]$ are decreasing. The graph of $y = \left(\frac{1}{3}\right)^x = 3^{-x}$ is the reflection of that of $y = 3^x$ about the $y$-axis, and the graph of $y = \left(\frac{1}{10}\right)^x = 10^{-x}$ is the reflection of that of $y = 10^x$ about the $y$-axis. The graph of $y = 10^x$ increases more quickly than that of $y = 3^x$ for $x > 0$.

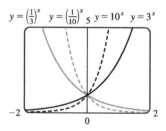

**7.** We start with the graph of $y = 4^x$ (Figure 2) and then shift 3 units downward. This shift doesn't affect the domain, but the range of $y = 4^x - 3$ is $(-3, \infty)$. There is a horizontal asymptote of $y = -3$.

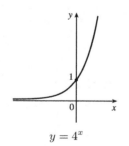

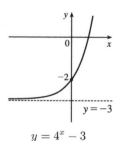

**9.** We start with the graph of $y = 2^x$ (Figure 2), reflect it about the $y$-axis, and then reflect it about the $x$-axis (or just rotate $180°$ to handle both reflections) to obtain the graph of $y = -2^{-x}$. In each graph, $y = 0$ is the horizontal asymptote.

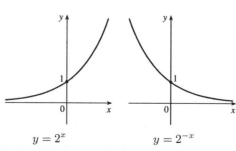

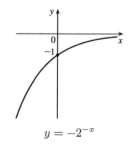

**11.** We start with the graph of $y = e^x$ (Figure 13), reflect the graph about the $y$-axis, and then vertically stretch by a factor of 3. There is a horizontal asymptote of $y = 0$, and the $y$-intercept is 3.

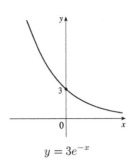

$$y = 3e^{-x}$$

**13.** (a) To find the equation of the graph that results from shifting the graph of $y = e^x$ downward 2 units, we subtract 2 from the original function to get $y = e^x - 2$.

(b) To find the equation of the graph that results from shifting the graph of $y = e^x$ to the right 2 units, we replace $x$ with $x - 2$ in the original function to get $y = e^{x-2}$.

(c) To find the equation of the graph that results from reflecting the graph of $y = e^x$ about the $x$-axis, we multiply the original function by $-1$ to get $y = -e^x$.

(d) To find the equation of the graph that results from reflecting the graph of $y = e^x$ about the $y$-axis, we replace $x$ with $-x$ in the original function to get $y = e^{-x}$.

(e) To find the equation of the graph that results from reflecting the graph of $y = e^x$ about the $x$-axis and then about the $y$-axis, we first multiply the original function by $-1$ (to get $y = -e^x$) and then replace $x$ with $-x$ in this equation to get $y = -e^{-x}$.

**15.** $x^3 x^5 = x^{3+5} = x^8$

**17.** $(u^4)^2 = u^{4 \cdot 2} = u^8$

**19.** $\left(\dfrac{p^3}{2}\right)^3 = \dfrac{(p^3)^3}{(2)^3} = \dfrac{p^{3 \cdot 3}}{8} = \dfrac{p^9}{8}$

**21.** $4^{2/3} = \sqrt[3]{4^2} = \sqrt[3]{16}$   or   $2\sqrt[3]{2}$

**23.** $e^{1/4} = \sqrt[4]{e}$

**25.** $P \cdot 3^{3x} = P \cdot (3^3)^x = P \cdot 27^x$

**27.** $500 \cdot (1.025)^{4t} = 500 \cdot \left[(1.025)^4\right]^t \approx 500 \cdot (1.1038)^t$

**29.** $4^{x+3} = 4^x \cdot 4^3 = 4^x \cdot 64 = 64 \cdot 4^x$

**31.** The outputs of $f$ increase at a constant percentage rate (each time the input increases by 1, the output doubles), so the function could be exponential. Since $f(0) = 5$, $f(1) = 5 \cdot 2$, $f(2) = 5 \cdot 2^2$, $f(3) = 5 \cdot 2^3$, and so on, a possible equation for $f$ is $f(x) = 5 \cdot 2^x$.

**33.** As the inputs increase by 1, the outputs decrease but not with a constant rate of change or a constant percentage change. So the function can't be linear or exponential.

**35.** (a) 15 hours represents 5 doubling periods (one doubling period is three hours), so the population will be $100 \cdot 2^5 = 3200$.

(b) In $t$ hours, there will be $t/3$ doubling periods. The initial population is 100, so the population $P$ at time $t$ is $P = 100 \cdot 2^{t/3}$.

(c) When $t = 20$ we have $P = 100 \cdot 2^{20/3} \approx 10{,}159$.

(d) We graph $P_1 = 100 \cdot 2^{t/3}$ and $P_2 = 50{,}000$. The two curves intersect at $t \approx 26.9$, so the population reaches 50,000 in about 26.9 hours.

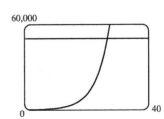

**37.** The point $(1, 6)$ is on the graph, so $f(1) = 6$ ⇒ $C \cdot a^1 = 6$ ⇒ $C = 6/a$. Also the point $(3, 24)$ is on the graph, so

$f(3) = 24$ ⇒ $C \cdot a^3 = 24$ and substituting, we have $(6/a) \cdot a^3 = 24$ ⇒ $6a^2 = 24$ ⇒ $a^2 = 4$ ⇒ $a = 2$

[since $a > 0$]. Then $C = \frac{6}{2} = 3$, and the function is $f(x) = 3 \cdot 2^x$.

**39.** If $f(x) = 5^x$, then $\dfrac{f(x+h) - f(x)}{h} = \dfrac{5^{x+h} - 5^x}{h} = \dfrac{5^x 5^h - 5^x}{h} = \dfrac{5^x (5^h - 1)}{h} = 5^x \left( \dfrac{5^h - 1}{h} \right).$

**41.** $2$ ft $= 24$ in, and $f(24) = 24^2$ in $= 576$ in $= 48$ ft.

$g(24) = 2^{24}$ in, or equivalently, $2^{24}$ in $\times$ $(1$ ft$/12$ in$)$ $\times$ $(1$ mi$/5280$ ft$)$ $\approx 265$ mi.

**43.** The graph of $g$ finally surpasses that of $f$ at $x \approx 35.8$.

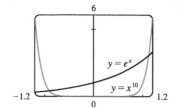 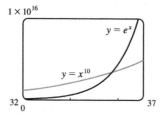

**45.** Let $t = 0$ correspond to 1950. A graphing calculator gives the exponential model $P(t) \approx 2614.086\,(1.01693)^t$. According to

the model, the population in 1993 is $P(43) \approx 2614.086\,(1.01693)^{43} \approx 5381$ million. The model predicts that the population

in 2020 will be $P(70) \approx 8466$ million.

**47.** (a) A graphing calculator gives an exponential model for the data as approximately $f(t) = 0.06698\,(1.3516)^t$, where $t = 0$

corresponds to the year 1980.

(b) Because exponential functions increase at a constant percentage rate, we can choose any value, say $f(0) = 0.06698$,

and double it, giving $0.13396$. Then solving $f(t) = 0.13396$, we get $0.06698\,(1.3516)^t = 0.13396$ ⇒

$(1.3516)^t = 2$ ⇒ $\ln (1.3516)^t = \ln 2$ ⇒ $t \ln (1.3516) = \ln 2$ ⇒ $t = \dfrac{\ln 2}{\ln (1.3516)} \approx 2.3$. Thus, according to

the model, the number of transistors doubles about every 2.3 years. Moore's Law predicts that the number doubles every

1.5 years.

(c) The model estimates that the number of transistors in 2004 is $f(24) = 0.06698\,(1.3516)^{24} \approx 92.5$ million. This is less

than $\frac{1}{6}$ of the actual value!

**49.** (a) January 1, 2007 corresponds to $t = 7$, and the model estimates the population then to be

$P(7) = \dfrac{23.7}{1 + 4.8 e^{-0.2(7)}} \approx 10.853$ thousand, or 10,853.

(b) The carrying capacity is $M = 23.7$ thousand.

(c)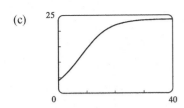

Looking at the graph, we estimate that $P(t) = \frac{1}{2}(23.7) = 11.85$ when $t \approx 7.8$ [graph $P(t)$ and the line $y = 11.85$ and find the intersection]. Thus the population reaches half its carrying capacity about 7.8 years after January 1, 2000 (in October 2007).

**51.** (a) Reflecting about the line $y = 3$ is equivalent to reflecting about the $x$-axis and then shifting upward $2 \cdot 3 = 6$ units, so an equation is $y = -2^x + 6$.

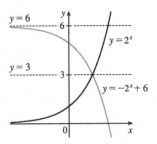

(b) Reflecting about the line $x = -4$ is equivalent to reflecting about the $y$-axis and then shifting left $2 \cdot 4 = 8$ units. Reflecting the graph of $y = 2^x$ about the $y$-axis gives the equation $y = 2^{-x}$. To shift left 8 units, we replace $x$ by $x + 8$: $y = 2^{-(x+8)}$. Notice that the point $(0, 1)$ on the original graph is reflected to $(-8, 1)$ on the second graph.

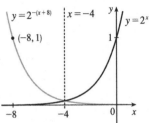

**53.**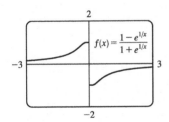

From the graph, it appears that $f$ is an odd function ($f$ is undefined for $x = 0$). To prove this, we must show that $f(-x) = -f(x)$.

$$f(-x) = \frac{1 - e^{1/(-x)}}{1 + e^{1/(-x)}} = \frac{1 - e^{(-1/x)}}{1 + e^{(-1/x)}} = \frac{1 - \frac{1}{e^{1/x}}}{1 + \frac{1}{e^{1/x}}} \cdot \frac{e^{1/x}}{e^{1/x}} = \frac{e^{1/x} - 1}{e^{1/x} + 1}$$

$$= -\frac{1 - e^{1/x}}{1 + e^{1/x}} = -f(x)$$

so $f$ is an odd function.

## 1.6 Logarithmic Functions

**1.** (a) It is defined as the inverse of the exponential function with base $a$, that is, $\log_a x = y \iff a^y = x$.

(b) $(0, \infty)$

(c) $\mathbb{R}$

**3.** (a) $\log_2 64 = 6$ since $2^6 = 64$.

(b) $\log_6 \frac{1}{36} = -2$ since $6^{-2} = \frac{1}{36}$.

**5.** (a) $\ln e^3 = 3$ by the first cancellation equation in (2).

(b) $e^{\ln 7} = 7$ by the second cancellation equation in (2).

**7.** $\ln 100 \approx 4.6052$

**9.** $\dfrac{\ln 28}{\ln 4} \approx 2.4037$

**11.** (a) $\log_8 4 = \frac{2}{3}$ ⇔ $8^{2/3} = 4$

(b) $\log_6 u = v$ ⇔ $6^v = u$

**13.** (a) $10^3 = 1000$ ⇔ $\log 1000 = 3$

(b) $y = 4^x$ ⇔ $\log_4 y = x$

**15.** Reflect the graph of $y = \ln x$ about the $x$-axis to obtain the graph of $y = -\ln x$.

**17.** Shift the graph of $y = \ln x$ left 1 and up 3. The $x$-intercept $(1, 0)$ on the original graph is shifted to $(0, 3)$.

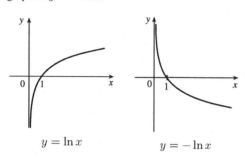

$y = \ln x$        $y = -\ln x$

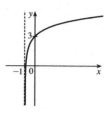

**19.** (a) To find the equation of the graph that results from shifting the graph of $y = \ln x$ 3 units upward, we add 3 to the original function to get $y = \ln x + 3$.

(b) To find the equation of the graph that results from shifting the graph of $y = \ln x$ 3 units to the left, we replace $x$ with $x + 3$ in the original function to get $y = \ln (x + 3)$.

(c) To find the equation of the graph that results from reflecting the graph of $y = \ln x$ about the $x$-axis, we multiply the original equation by $-1$ to get $y = -\ln x$.

(d) To find the equation of the graph that results from reflecting the graph of $y = \ln x$ about the $y$-axis, we replace $x$ with $-x$ in the original equation to get $y = \ln(-x)$.

**21.** 3 ft $= 36$ in, so we need to find $x$ such that $\ln x = 36$ ⇔ $x = e^{36} \approx 4.311 \times 10^{15}$ (4,311,000,000,000,000 inches). In miles, this is $4.311 \times 10^{15}$ in $\cdot \dfrac{1 \text{ ft}}{12 \text{ in}} \cdot \dfrac{1 \text{ mi}}{5280 \text{ ft}} \approx 6.8 \times 10^{10} = 68{,}000{,}000{,}000$ (68 billion) miles!

**23.** False. From Law 1 of logarithms (see page 67), we have $\ln c + \ln d = \ln(cd)$, not $\ln(c + d)$.

**25.** False. From Law 2 we have $\ln(u/3) = \ln u - \ln 3$.

**27.** $2\ln 4 - \ln 2 = \ln 4^2 - \ln 2 = \ln 16 - \ln 2 = \ln \frac{16}{2} = \ln 8$

**29.** $3\ln u - 2\ln 5 = \ln u^3 - \ln 5^2 = \ln u^3 - \ln 25 = \ln(u^3/25)$

**31.** (a) By Law 3 of logarithms, $\ln(x^3) = 3\ln x$ and the domain for both functions is $\{x \mid x > 0\} = (0, \infty)$, so the graphs are the same.

(b) Similarly, $\ln(x^2) = 2\ln x$, but the domain of $\ln(x^2)$ is $\{x \mid x \neq 0\} = (-\infty, 0) \cup (0, \infty)$ while the domain of $2\ln x$ is $\{x \mid x > 0\} = (0, \infty)$. The graphs are identical for $x > 0$, but for $x < 0$ the graph of $y = 2\ln x$ is empty [and the graph of $\ln(x^2)$ is not].

**33.** (a) $2\ln x = 1$ ⇔ $\ln x = \frac{1}{2}$ ⇔ $e^{1/2} = x$, so $x = \sqrt{e} \approx 1.6487$.

(b) $e^{-x} = 5 \iff \ln 5 = -x \iff x = -\ln 5 \approx -1.6094$

**35.** Taking the natural logarithm of both sides, we have $\ln(5^t) = \ln 20 \implies t \cdot \ln 5 = \ln 20 \implies t = \dfrac{\ln 20}{\ln 5} \approx 1.8614.$

**37.** $2^{x-5} = 3 \implies \ln(2^{x-5}) = \ln 3 \implies (x-5)\ln 2 = \ln 3 \implies x - 5 = \dfrac{\ln 3}{\ln 2} \implies x = 5 + \dfrac{\ln 3}{\ln 2} \approx 6.5850$

**39.** $8e^{3x} = 31 \iff e^{3x} = \frac{31}{8} \iff \ln \frac{31}{8} = 3x$  [by Equation 1]  $\iff x = \frac{1}{3}\ln\frac{31}{8} \approx 0.4515$

**41.** $6 \cdot (2^{x/7}) = 11.4 \iff 2^{x/7} = \frac{11.4}{6} = 1.9.$ Taking the natural logarithm of both sides, we get $\ln(2^{x/7}) = \ln 1.9 \implies$

$(x/7)\ln 2 = \ln 1.9 \implies \dfrac{x}{7} = \dfrac{\ln 1.9}{\ln 2} \implies x = \dfrac{7\ln 1.9}{\ln 2} \approx 6.4820.$

**43.** The population reaches one million when $P(t) = 437.2(1.036)^t = 1000 \iff (1.036)^t = \frac{1000}{437.2}.$ Taking the natural

logarithm of both sides gives $\ln(1.036)^t = \ln\frac{1000}{437.2} \implies t\ln 1.036 = \ln\frac{1000}{437.2} \implies t = \ln\left(\frac{1000}{437.2}\right)/\ln(1.036) \approx 23.39.$

Thus the population reaches one million about 23.4 years after the end of 1995 (May 2019).

**45.** We need to find the depth $x$ when $I = 5$:  $5 = 10e^{-0.008x} \iff e^{-0.008x} = \frac{5}{10} = \frac{1}{2} \iff \ln\frac{1}{2} = -0.008x \iff$

$x = -\frac{1}{0.008}\ln\frac{1}{2} \approx 86.64.$ Thus the light intensity drops to 5 lumens about 86.64 feet down.

**47.** When $n = 50{,}000$ we have $100 \cdot 2^{t/3} = 50{,}000 \implies 2^{t/3} = 500 \implies \ln(2^{t/3}) = \ln 500 \implies \dfrac{t}{3}\ln 2 = \ln 500 \implies$

$t = \dfrac{3\ln 500}{\ln 2} \approx 26.9$ hours.

**49.** (a) A graphing calculator gives a logarithmic model as approximately $Y(v) = -203.94 + 22.14\ln v$, where $v$ is the value of
the account in dollars and $Y(v)$ is the time measured in years.

(b) The time required is $Y(25{,}000) = -203.94 + 22.14\ln 25{,}000 \approx 20.3$ years.

**51.** We need to find the value of $t$ for which $f(t) = 0.30$:

$$\frac{0.41}{1 + 0.52e^{-0.4t}} = 0.3$$

$$\frac{0.41}{0.3} = 1 + 0.52e^{-0.4t}$$

$$0.52e^{-0.4t} = \frac{0.41}{0.3} - 1$$

$$e^{-0.4t} = \frac{1}{0.52}\left(\frac{0.41}{0.3} - 1\right)$$

$$-0.4t = \ln\left[\frac{1}{0.52}\left(\frac{0.41}{0.3} - 1\right)\right]$$

$$t = -\frac{1}{0.4}\ln\left[\frac{1}{0.52}\left(\frac{0.41}{0.3} - 1\right)\right] \approx 0.873$$

Thus 30% of households have seen the program about 0.87 years after January 1, 2005, or about mid-November 2005.

## 1  Review

**1.** (a) When $x = 2$, $y \approx 2.7$. Thus, $f(2) \approx 2.7.$

(b) $f(x) = 3 \implies x \approx 2.3, 5.6$

(c) The domain of $f$ is $-6 \le x \le 6$, or $[-6, 6]$.

(d) The range of $f$ is $-4 \le y \le 4$, or $[-4, 4]$.

**3.** (a)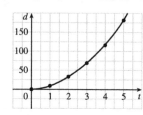

(b) From the graph, we see that the distance traveled after 4.5 seconds is about 150 feet.

**5.** $f(x) = \sqrt{4 - 10x}$ is defined when $4 - 10x \ge 0$ $\Leftrightarrow$ $x \le \frac{2}{5}$, so the domain is $\left\{ x \mid x \le \frac{2}{5} \right\} = \left(-\infty, \frac{2}{5}\right]$.

**7.** $y = 2^x + 1$ is defined for any real number $x$, so the domain is $\mathbb{R}$.

**9.** $p(x) = x^2 - 3x$.

$p(-2) = (-2)^2 - 3(-2) = 4 + 6 = 10$.

$p(x - 5) = (x - 5)^2 - 3(x - 5) = x^2 - 10x + 25 - 3x + 15 = x^2 - 13x + 40$.

$$\frac{p(a) - p(4)}{a - 4} = \frac{[a^2 - 3a] - [(4)^2 - 3(4)]}{a - 4} = \frac{a^2 - 3a - 4}{a - 4} = \frac{(a - 4)(a + 1)}{a - 4} = a + 1.$$

$$\frac{p(x + h) - p(x)}{h} = \frac{[(x + h)^2 - 3(x + h)] - [x^2 - 3x]}{h} = \frac{x^2 + 2xh + h^2 - 3x - 3h - x^2 + 3x}{h}$$

$$= \frac{2xh + h^2 - 3h}{h} = \frac{h(2x + h - 3)}{h} = 2x + h - 3.$$

**11.** (a) $f(x) = 2x^5 - 3x^2 + 2$, so

$f(-x) = 2(-x)^5 - 3(-x)^2 + 2 = -2x^5 - 3x^2 + 2$. Since this is neither $f(x)$ nor $-f(x)$, $f$ is neither even nor odd. Also, the graph of $f$ is not symmetric about the $y$-axis or about the origin.

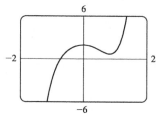

(b) $f(x) = x^3 - x^7$, so

$f(-x) = (-x)^3 - (-x)^7 = -x^3 + x^7 = -(x^3 - x^7) = -f(x)$, and thus $f$ is an odd function. Also, the graph of $f$ appears to be symmetric about the origin.

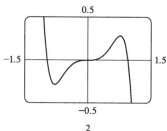

(c) $f(x) = e^{-x^2}$, so $f(-x) = e^{-(-x)^2} = e^{-x^2} = f(x)$, and thus $f$ is even.

The graph of $f$ appears to be symmetric about the $y$-axis.

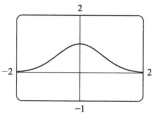

**13.** The value $h(t)$ is David's salary times the percentage contributed to a retirement account, so $h(t)$ measures the amount of money, in dollars, he contributed to the account during year $t$.

**15.** (a) $A(x) = f(x) - g(x) = (3x^2 + 4) - (2^x - 5) = 3x^2 + 4 - 2^x + 5 = 3x^2 - 2^x + 9$

(b) $B(x) = f(g(x)) = f(2^x - 5) = 3(2^x - 5)^2 + 4$

(c) $C(x) = g(f(x)) = g(3x^2 + 4) = 2^{3x^2 + 4} - 5$

**17.** (a) To obtain the graph of $y = f(x) + 8$, we shift the graph of $y = f(x)$ up 8 units.

(b) To obtain the graph of $y = f(x + 8)$, we shift the graph of $y = f(x)$ left 8 units.

(c) To obtain the graph of $y = 1 + 2f(x) = 2f(x) + 1$, we stretch the graph of $y = f(x)$ vertically by a factor of 2, and then shift the resulting graph 1 unit upward.

(d) To obtain the graph of $y = f(x - 2) - 2$, we shift the graph of $y = f(x)$ right 2 units, and then shift the resulting graph 2 units downward.

(e) To obtain the graph of $y = -f(x)$, we reflect the graph of $y = f(x)$ about the $x$-axis.

**19.** To graph $y = (x - 2)^2 - 3$ we shift the graph of $y = x^2$ to the right 2 and down 3.

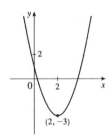

**21.** To graph $y = 2e^x + 3$ we stretch the graph of $y = e^x$ vertically by a factor of 2 and then shift the graph 3 units upward. There is a horizontal asymptote of $y = 3$, and the $y$-intercept is 5.

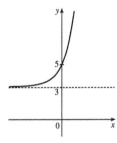

**23.** (a) The slope of $L$ is $\dfrac{\Delta L}{\Delta x} = \dfrac{L(18) - L(15)}{18 - 15} = \dfrac{245 - 281}{18 - 15} = \dfrac{-36}{3} = -12$ thousand units per dollar. This means that for each price increase of one dollar, 12,000 fewer units will be purchased.

(b) We have $m = -12$, and using the point-slope form of the equation of a line, $y - y_1 = m(x - x_1)$ ⇒

$y - 281 = -12(x - 15)$ ⇒ $y - 281 = -12x + 180$ ⇒ $y = -12x + 461$. Thus $L(x) = -12x + 461$.

**25.** We draw a scatter plot, using $t = 0$ to represent the year 1900.

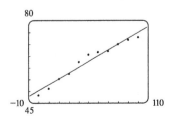

It appears that a line through the data points $(20, 55.2)$ and $(80, 70.0)$ is a good fit. Then the slope is

$m = \dfrac{70.0 - 55.2}{80 - 20} = \dfrac{14.8}{60} \approx 0.2467$ and an equation of the line is $y - 55.2 = 0.2467(t - 20)$ or $y \approx 0.2467t + 50.27$.

Thus a linear model is $L(t) = 0.2467t + 50.27$. The year 2015 corresponds to $t = 115$, so we predict the life span of a male born in 2015 to be about $L(115) = 0.2467(115) + 50.27 \approx 78.6$ years.

**27.** The graph is a parabola with vertex $(-8, 2)$, so an equation is $y = a(x + 8)^2 + 2$. The point $(2, 4)$ is on the parabola, so we substitute 2 for $x$ and 4 for $y$ to find $a$. $4 = a(2 + 8)^2 + 2$ $\Rightarrow$ $100a = 2$ $\Rightarrow$ $a = \frac{1}{50}$. Thus an equation for the function is $y = \frac{1}{50}(x + 8)^2 + 2$.

**29.** $A$ varies inversely with $x$, so $A = k/x$. If $A = 28$ when $x = 112$, then $28 = k/112$ $\Leftrightarrow$ $k = 3136$ and $A = 3136/x$.

**31.** It appears that $f$ is increasing on approximately $(-\infty, -8)$, $(5, \infty)$ and decreasing on $(-8, 5)$.

**33.** (a) $(3xy^4)^2 = (3)^2(x)^2(y^4)^2 = 9x^2y^8$

(b) $e^{2\ln 3} = e^{\ln(3^2)} = e^{\ln 9} = 9$ [Or: $e^{2\ln 3} = (e^{\ln 3})^2 = 3^2 = 9$]

(c) $\log_4 16 = 2$ because $4^2 = 16$.

**35.** $f(x) = C \cdot a^x$ and we are given that $f(0) = 8.3$ $\Rightarrow$ $C \cdot a^0 = 8.3$ $\Rightarrow$ $C = 8.3$. Also $f(4) = 20.9$ $\Rightarrow$

$8.3(a^4) = 20.9$ $\Rightarrow$ $a^4 = \frac{20.9}{8.3}$ $\Rightarrow$ $a = \sqrt[4]{\frac{20.9}{8.3}} \approx 1.2597$, so $f(x) \approx 8.3(1.2597)^x$.

**37.** (a) 20 years represents 4 tripling periods, so the population will be $4000 \cdot 3^4 = 324{,}000$.

(b) In $t$ hours there will be $t/5$ tripling periods. The initial population is 4000, so the population is $P(t) = 4000 \cdot 3^{t/5}$.

(c) The population reaches one million when $P(t) = 1{,}000{,}000$ $\Rightarrow$ $4000 \cdot 3^{t/5} = 1{,}000{,}000$ $\Rightarrow$ $3^{t/5} = 250$ $\Rightarrow$

$\ln(3^{t/5}) = \ln 250$ $\Rightarrow$ $\dfrac{t}{5}\ln 3 = \ln 250$ $\Rightarrow$ $t = \dfrac{5\ln 250}{\ln 3} \approx 25.13$, so in about 25.13 years.

**39.** A formula for the bacteria population after $t$ hours is $P = C \cdot 2^t$, where $C$ is the initial population. When the population triples, we have $3C = C \cdot 2^t$ $\Rightarrow$ $2^t = 3$ $\Rightarrow$ $\ln(2^t) = \ln 3$ $\Rightarrow$ $t\ln 2 = \ln 3$ $\Rightarrow$ $t = \dfrac{\ln 3}{\ln 2} \approx 1.585$. Thus the population triples after about 1.585 hours.

**41.** We enter the data into a graphing calculator (or other computing device) and use $t = 0$ to represent the year 2000. The calculator gives the regression line as approximately $y = 1.0607t + 21.96$, where $y$ is the hourly rate in dollars. The year 2012 corresponds to $t = 12$, and the model predicts the average hourly rate for registered nurses then to be $1.0607(12) + 21.96 \approx 34.69$ dollars.

# 2  THE DERIVATIVE

## 2.1  Measuring Change

**1.** $\dfrac{\Delta f}{\Delta x} = \dfrac{f(3) - f(1)}{3 - 1} = \dfrac{24 - 6}{3 - 1} = \dfrac{18}{2} = 9$

**3.** $\dfrac{\Delta A}{\Delta v} = \dfrac{A(13) - A(6)}{13 - 6} = \dfrac{4 - 3}{13 - 6} = \dfrac{1}{7}$

**5.** $\dfrac{\Delta P}{\Delta t} = \dfrac{P(84) - P(16)}{84 - 16} = \dfrac{(4.7\ln 84 + 1.8) - (4.7\ln 16 + 1.8)}{84 - 16} = \dfrac{4.7\ln 84 - 4.7\ln 16}{68} \approx 0.115$

**7.** $\dfrac{\Delta N}{\Delta w} = \dfrac{N(22) - N(16)}{22 - 16} = \dfrac{5e^{0.2(22)} - 5e^{0.2(16)}}{22 - 16} = \dfrac{5e^{4.4} - 5e^{3.2}}{6} \approx 47.432$

**9.** (a) From March 15 to May 17 is a difference of 63 days (From March 15 to April 15 is 31 days, then 30 days to May 15, and

two more days to May 17), so the average rate of change is $\dfrac{\Delta \text{ price}}{\Delta \text{ days}} = \dfrac{699.50 - 556.50}{63} = \dfrac{143}{63} \approx 2.27$ dollars per day.

Thus from March 15 to May 17, the price of gold increased at an average rate of $2.27 per day.

(b) From January 11 to February 8 is a difference of 28 days, and the average rate of change is

$\dfrac{\Delta \text{ price}}{\Delta \text{ days}} = \dfrac{\$548.75 - \$544.40}{28} = \dfrac{\$4.35}{28} \approx \$0.16/\text{day}.$

**11.** $\dfrac{\Delta f}{\Delta x} = \dfrac{f(2.5) - f(1.8)}{2.5 - 1.8} = \dfrac{325 - 240}{2.5 - 1.8} = \dfrac{85}{0.7} \approx 121.4$; If spending on advertising is increased from 1.8 to 2.5 million

dollars, then the number of vehicles sold increases at an average rate of about 121,400 vehicles per million dollars spent.

**13.** The average rate of change is $\dfrac{\Delta \text{ balance}}{\Delta t} = \dfrac{9500(1.064)^{4.5} - 9500(1.064)^{2.5}}{4.5 - 2.5} \approx \dfrac{1465.44}{2} = 732.72$; From 2.5 to 4.5 years

after being opened, the balance of the account increased at an average rate of $732.72 per year.

**15.** We estimate that the points $(3, 70)$ and $(6, 90)$ lie on the graph, so the average rate of change for $3 \leq x \leq 6$ is

$\dfrac{\Delta y}{\Delta x} = \dfrac{90 - 70}{6 - 3} = \dfrac{20}{3} \approx 6.67$. (Note that this is the slope of the secant line through the two points.) This means that from 3

to 6 hours of charging time, the percentage of full charge increases at an average rate of about 6.67 percentage points per hour.

**17.** (a) The secant line passes through the points on the graph corresponding

to $x = 2$ and $x = 6$.

From the graph we estimate that $f(2) = 1.6$ and $f(6) = 1$, so the

slope of the secant line, and the average rate of change, is

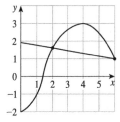

$\dfrac{\Delta y}{\Delta x} = \dfrac{f(6) - f(2)}{6 - 2} \approx \dfrac{1 - 1.6}{6 - 2} = \dfrac{-0.6}{4} = -0.15.$

(b) The secant line passing through the points on the graph corresponding to $x = 0$ and $x = 3$ has positive slope, so the average rate of change is positive.

(c) The secant line through the points $(1, f(1))$ and $(2, f(2))$ has a larger slope than the secant line through $(3, f(3))$ and $(4, f(4))$, so $[1, 2]$ gives a larger average rate of change.

19. (a) The average speed for $t_1 \leq t \leq t_2$ is the average rate of change of the height $h$: $\dfrac{\Delta h}{\Delta t} = \dfrac{h(t_2) - h(t_1)}{t_2 - t_1}$. (Note that $h$ is increasing for the given time intervals.)

    (i) $0 \leq t \leq 1$: $\quad \dfrac{\Delta h}{\Delta t} = \dfrac{h(1) - h(0)}{1 - 0} = \dfrac{20 - 0}{1 - 0} = 20$ ft/s

    (ii) $0.5 \leq t \leq 1$: $\quad \dfrac{\Delta h}{\Delta t} = \dfrac{h(1) - h(0.5)}{1 - 0.5} = \dfrac{20 - 14}{1 - 0.5} = \dfrac{6}{0.5} = 12$ ft/s

    (iii) $0.9 \leq t \leq 1$: $\quad \dfrac{\Delta h}{\Delta t} = \dfrac{h(1) - h(0.9)}{1 - 0.9} = \dfrac{20 - 19.44}{1 - 0.9} = \dfrac{0.56}{0.1} = 5.6$ ft/s

    (iv) $0.99 \leq t \leq 1$: $\quad \dfrac{\Delta h}{\Delta t} = \dfrac{h(1) - h(0.99)}{1 - 0.99} = \dfrac{20 - 19.9584}{1 - 0.99} = \dfrac{0.0416}{0.01} = 4.16$ ft/s

    (b) It appears that as we shorten the time period ending at 1 second, the average speed is getting closer to 4 ft/s, so we estimate that the speed of the ball after 1 second is 4 ft/s.

21. (a) The height $y = f(t) = 40t - 16t^2$ is decreasing for the given intervals, so the average rate of change is negative. In this case, the average speed is the absolute value of the average rate of change.

    (i) $2 \leq t \leq 2.5$: $\quad \left| \dfrac{\Delta y}{\Delta t} \right| = \left| \dfrac{f(2.5) - f(2)}{2.5 - 2} \right| = \left| \dfrac{0 - 16}{0.5} \right| = \left| \dfrac{-16}{0.5} \right| = 32$ ft/s

    (ii) $2 \leq t \leq 2.1$: $\quad \left| \dfrac{\Delta y}{\Delta t} \right| = \left| \dfrac{f(2.1) - f(2)}{2.1 - 2} \right| = \left| \dfrac{13.44 - 16}{0.1} \right| = \left| \dfrac{-2.56}{0.1} \right| = 25.6$ ft/s

    (iii) $2 \leq t \leq 2.05$: $\quad \left| \dfrac{\Delta y}{\Delta t} \right| = \left| \dfrac{f(2.05) - f(2)}{2.05 - 2} \right| = \left| \dfrac{14.76 - 16}{0.05} \right| = \left| \dfrac{-1.24}{0.05} \right| = 24.8$ ft/s

    (iv) $2 \leq t \leq 2.01$: $\quad \left| \dfrac{\Delta y}{\Delta t} \right| = \left| \dfrac{f(2.01) - f(2)}{2.01 - 2} \right| = \left| \dfrac{15.7584 - 16}{0.01} \right| = \left| \dfrac{-0.2416}{0.01} \right| = 24.16$ ft/s

    (b) It appears that as we shorten the time interval beginning at 2 seconds, the average speed is getting closer to 24 ft/s, so we estimate that the speed of the ball after 2 seconds is 24 ft/s.

## 2.2 Limits

**Prepare Yourself**

1. $f(0.01) = \dfrac{3^{0.01} - 2^{0.01}}{0.01} \approx 0.4091$

2. $g(0.1) = \dfrac{\sqrt{(0.1)^2 + 1} - 1}{(0.1)^2} = \dfrac{\sqrt{1.01} - 1}{0.01} \approx 0.4988$

**3.** (a) $x^2 - 5x - 24 = (x - 8)(x + 3)$  (b) $a^2 - 25 = (a + 5)(a - 5)$

  (c) $2w^2 - 7w - 15 = (2w + 3)(w - 5)$  (d) $b^3 + 1 = (b + 1)(b^2 - b + 1)$

**4.** (a) $\dfrac{x^2 - 2x - 3}{x^2 - 7x + 12} = \dfrac{(x - 3)(x + 1)}{(x - 3)(x - 4)} = \dfrac{x + 1}{x - 4}$

  (b) $\dfrac{(c + 2)^2 - 4}{c} = \dfrac{(c^2 + 4c + 4) - 4}{c} = \dfrac{c^2 + 4c}{c} = \dfrac{c(c + 4)}{c} = c + 4$

  (c) $\dfrac{\frac{1}{q} - \frac{1}{3}}{q - 3} = \dfrac{\frac{1}{q} - \frac{1}{3}}{q - 3} \cdot \dfrac{3q}{3q} = \dfrac{\frac{3q}{q} - \frac{3q}{3}}{3q(q - 3)} = \dfrac{3 - q}{3q(q - 3)} = \dfrac{-(q - 3)}{3q(q - 3)} = -\dfrac{1}{3q}$

  Or: $\dfrac{\frac{1}{q} - \frac{1}{3}}{q - 3} = \dfrac{\frac{3}{3q} - \frac{q}{3q}}{q - 3} = \dfrac{\frac{3 - q}{3q}}{q - 3} = \dfrac{3 - q}{3q(q - 3)} = \dfrac{-(q - 3)}{3q(q - 3)} = -\dfrac{1}{3q}$

**5.** $\dfrac{\sqrt{x + 1} - 2}{x - 3} \cdot \dfrac{\sqrt{x + 1} + 2}{\sqrt{x + 1} + 2} = \dfrac{\left(\sqrt{x + 1}\right)^2 + 2\sqrt{x + 1} - 2\sqrt{x + 1} - 4}{(x - 3)\left(\sqrt{x + 1} + 2\right)}$

$$= \dfrac{(x + 1) - 4}{(x - 3)\left(\sqrt{x + 1} + 2\right)} = \dfrac{x - 3}{(x - 3)\left(\sqrt{x + 1} + 2\right)} = \dfrac{1}{\sqrt{x + 1} + 2}$$

**6.** (a) $A(3) = 2^3 = 8$

  (b) $A(-2) = 1 - (-2)^2 = 1 - 4 = -3$

  (c) $A(1) = 2^1 = 2$

### Exercises

**1.** As $x$ gets closer and closer to 2, $f(x)$ gets closer and closer to 5. Yes, the graph could approach the point $(2, 5)$ but have a hole there, and instead have a point at $(2, 3)$. Then $f(2) = 3$.

**3.**

| $x$ | $\dfrac{x^2 - x - 2}{x^2 - 2x}$ |
|-----|-----|
| 2.1 | 1.476190 |
| 2.05 | 1.487805 |
| 2.01 | 1.497512 |
| 2.005 | 1.498753 |
| 2.001 | 1.499750 |

| $x$ | $\dfrac{x^2 - x - 2}{x^2 - 2x}$ |
|-----|-----|
| 1.9 | 1.526316 |
| 1.95 | 1.512821 |
| 1.99 | 1.502513 |
| 1.995 | 1.501253 |
| 1.999 | 1.500250 |

As $x$ gets closer and closer to 2, the function values approach 1.5, so it appears that

$$\lim_{x \to 2} \dfrac{x^2 - x - 2}{x^2 - 2x} = 1.5.$$

**5.** For $f(x) = \dfrac{\sqrt{x + 4} - 2}{x}$ :

| $x$ | $f(x)$ |
|-----|--------|
| 0.1 | 0.248457 |
| 0.05 | 0.249224 |
| 0.01 | 0.249844 |
| 0.001 | 0.249984 |

| $x$ | $f(x)$ |
|-----|--------|
| −0.1 | 0.251582 |
| −0.05 | 0.250786 |
| −0.01 | 0.250156 |
| −0.001 | 0.250016 |

It appears that $\lim\limits_{x \to 0} \dfrac{\sqrt{x + 4} - 2}{x} = 0.25.$

**7.** For $f(x) = \dfrac{x^6 - 1}{x^{10} - 1}$:

| $x$ | $f(x)$ |
|-----|--------|
| 0.9 | 0.719397 |
| 0.95 | 0.660186 |
| 0.99 | 0.612018 |
| 0.999 | 0.601200 |

| $x$ | $f(x)$ |
|-----|--------|
| 1.1 | 0.484119 |
| 1.05 | 0.540783 |
| 1.01 | 0.588022 |
| 1.001 | 0.598800 |

It appears that $\displaystyle\lim_{x \to 1} \frac{x^6 - 1}{x^{10} - 1} = 0.6$.

**9.** (a) For $h(x) = \dfrac{\sqrt{x^4 + 1} - 1}{x^4}$:

| $x$ | $h(x)$ |
|-----|--------|
| 1 | 0.414214 |
| 0.5 | 0.492423 |
| 0.2 | 0.499800 |
| 0.1 | 0.499988 |

(b) It appears that $\displaystyle\lim_{x \to 0} \frac{\sqrt{x^4 + 1} - 1}{x^4} = 0.5$.

(c)

| $x$ | $h(x)$ |
|-----|--------|
| 0.05 | 0.4999992 |
| 0.01 | 0.5 |
| 0.001 | 0 |
| 0.0001 | 0 |

The values of 0 do not seem correct. Here the values may vary from one calculator to another; however, every calculator will eventually give *false values*.

(d) According to a calculator, the numerator of $h(x)$ is 0 for $x = 0.001$ and $x = 0.0001$. This does not seem correct; it is likely that the 0 value is a result of the calculator rounding extremely small values.

(e)

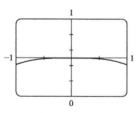

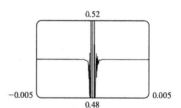

The first graph does corroborate our guess of 0.5 for the limit from part (b). The second graph shows incorrect output values that fluctuate.

**11.**
$$\lim_{x \to 2} (x^3 + 2x^2 + 1) = \lim_{x \to 2} (x^3) + \lim_{x \to 2} (2x^2) + \lim_{x \to 2} 1 \qquad \text{[Limit Law 1]}$$

$$= \lim_{x \to 2} (x^3) + 2 \lim_{x \to 2} (x^2) + \lim_{x \to 2} 1 \qquad \text{[Limit Law 3]}$$

$$= \left(\lim_{x \to 2} x\right)^3 + 2\left(\lim_{x \to 2} x\right)^2 + \lim_{x \to 2} 1 \qquad \text{[Limit Law 6]}$$

We have $\displaystyle\lim_{x \to 2} x = 2$, because the output is the same as the input, and $\displaystyle\lim_{x \to 2} 1 = 1$ because 1 is a constant. Thus

$$\lim_{x \to 2} (x^3 + 2x^2 + 1) = (2)^3 + 2(2)^2 + (1) = 8 + 8 + 1 = 17.$$

**13.** $\lim\limits_{v \to 1} \dfrac{v^2 - 5}{v} = \dfrac{\lim\limits_{v \to 1}(v^2 - 5)}{\lim\limits_{v \to 1} v} = \dfrac{\lim\limits_{v \to 1}(v^2) - \lim\limits_{v \to 1} 5}{\lim\limits_{v \to 1} v}$   [Limit Law 5, Limit Law 2]

$\qquad = \dfrac{\left(\lim\limits_{v \to 1} v\right)^2 - \lim\limits_{v \to 1} 5}{\lim\limits_{v \to 1} v}$   [Limit Law 6]

$\qquad = \dfrac{(1)^2 - 5}{1} = -4$

because $\lim\limits_{v \to 1} v = 1$ (the output is the same as the input) and $\lim\limits_{v \to 1} 5 = 5$ (since 5 is a constant).

**15.** The function $3e^t - 4$ is continuous for all real numbers, so we can evaluate the limit by direct substitution:

$\lim\limits_{t \to 1}(3e^t - 4) = 3e^1 - 4 = 3e - 4 \approx 4.155$.

**17.** The function $\dfrac{\ln m}{m + 2}$ is continuous on its domain $\{m \mid m > 0, m \neq -2\}$. Because $m = 2$ is in the domain, we can evaluate

the limit by direct substitution: $\lim\limits_{m \to 2}\left(\dfrac{\ln m}{m + 2}\right) = \dfrac{\ln 2}{2 + 2} = \dfrac{\ln 2}{4} \approx 0.173$.

**19.** (a) The domain of $f(x) = \dfrac{x^2 - 4}{x - 2}$ is $\{x \mid x \neq 2\}$.

(b) Because $f$ is a rational function, it is continuous on its domain. $x = 1$ is in the domain of $f$, so we evaluate the limit by

direct substitution: $\lim\limits_{x \to 1} f(x) = f(1) = \dfrac{1^2 - 4}{1 - 2} = \dfrac{-3}{-1} = 3$.

(c) $x = 2$ is not in the domain of $f$, so we can't evaluate the limit by direct substitution. Instead we first simplify the function:

$\dfrac{x^2 - 4}{x - 2} = \dfrac{(x + 2)(x - 2)}{x - 2} = x + 2 \ (x \neq 2)$. Because we do not consider $x = 2$ itself when evaluating the limit as

$x \to 2$, we have $\lim\limits_{x \to 2} f(x) = \lim\limits_{x \to 2} \dfrac{x^2 - 4}{x - 2} = \lim\limits_{x \to 2}(x + 2) = 2 + 2 = 4$. We are justified in using direct substitution in the

last limit because $x + 2$ is continuous everywhere.

**21.** (a) The domain of $A(z) = \dfrac{2z - 6}{z^2 - 5z + 6} = \dfrac{2(z - 3)}{(z - 3)(z - 2)}$ is $\{z \mid z \neq 2, z \neq 3\}$.

(b) $A$ is a rational function and hence continuous on its domain. Since $z = 0$ is in the domain, we use direct substitution to

evaluate the limit: $\lim\limits_{z \to 0} A(z) = A(0) = \dfrac{-6}{6} = -1$.

(c) $z = 3$ is not in the domain of $A$, so we can't evaluate the limit by direct substitution. But

$\lim\limits_{z \to 3} A(z) = \lim\limits_{z \to 3} \dfrac{2(z - 3)}{(z - 3)(z - 2)} = \lim\limits_{z \to 3} \dfrac{2}{z - 2}$   [since we assume $z \neq 3$]. Now $\dfrac{2}{z - 2}$ is a rational function and 3 is in

its domain, so we evaluate the last limit using direct substitution: $\lim\limits_{z \to 3} A(z) = \lim\limits_{z \to 3} \dfrac{2}{z - 2} = \dfrac{2}{3 - 2} = 2$.

**23.** $3t - 7$ is linear function and hence continuous everywhere, so direct substitution is appropriate and gives

$\lim\limits_{t \to 4}(3t - 7) = 3(4) - 7 = 5$.

**25.**  $\dfrac{x^2+5}{x+5}$ is a rational function and hence continuous on its domain $\{x \mid x \neq -5\}$. Because $x = 3$ is in the domain, we have

$$\lim_{x \to 3} \frac{x^2+5}{x+5} = \frac{3^2+5}{3+5} = \frac{14}{8} = \frac{7}{4}.$$

**27.**  2 is not in the domain of $\dfrac{x^2+x-6}{x-2}$, but we can evaluate the limit by direct substitution after simplifying:

$$\lim_{x \to 2} \frac{x^2+x-6}{x-2} = \lim_{x \to 2} \frac{(x+3)(x-2)}{x-2} = \lim_{x \to 2}(x+3) = 2+3 = 5.$$

**29.**  $\displaystyle\lim_{t \to -3} \frac{t^2-9}{2t^2+7t+3} = \lim_{t \to -3} \frac{(t+3)(t-3)}{(2t+1)(t+3)} = \lim_{t \to -3} \frac{t-3}{2t+1} = \frac{-3-3}{2(-3)+1} = \frac{-6}{-5} = \frac{6}{5}$

**31.**  $\displaystyle\lim_{h \to 0} \frac{(4+h)^2-16}{h} = \lim_{h \to 0} \frac{(16+8h+h^2)-16}{h} = \lim_{h \to 0} \frac{8h+h^2}{h} = \lim_{h \to 0} \frac{h(8+h)}{h} = \lim_{h \to 0}(8+h) = 8+0 = 8$

**33.**  We can use the formula $a^3+b^3 = (a+b)(a^2-ab+b^2)$ [see Appendix A] to factor the denominator:

$$\lim_{x \to -2} \frac{x+2}{x^3+8} = \lim_{x \to -2} \frac{x+2}{(x+2)(x^2-2x+4)} = \lim_{x \to -2} \frac{1}{x^2-2x+4} = \frac{1}{4+4+4} = \frac{1}{12}.$$

**35.**  $x = 7$ is not in the domain of $\dfrac{\sqrt{x+2}-3}{x-7}$, so we first rewrite the function by rationalizing the numerator (see Example 5):

$$\lim_{x \to 7} \frac{\sqrt{x+2}-3}{x-7} = \lim_{x \to 7} \frac{\sqrt{x+2}-3}{x-7} \cdot \frac{\sqrt{x+2}+3}{\sqrt{x+2}+3} = \lim_{x \to 7} \frac{(x+2)-9}{(x-7)(\sqrt{x+2}+3)}$$

$$= \lim_{x \to 7} \frac{x-7}{(x-7)(\sqrt{x+2}+3)} = \lim_{x \to 7} \frac{1}{\sqrt{x+2}+3} = \frac{1}{\sqrt{7+2}+3} = \frac{1}{6}$$

**37.**  $\displaystyle\lim_{x \to -4} \frac{\frac{1}{4}+\frac{1}{x}}{4+x} = \lim_{x \to -4} \frac{\frac{1}{4}+\frac{1}{x}}{4+x} \cdot \frac{4x}{4x} = \lim_{x \to -4} \frac{x+4}{4x(4+x)} = \lim_{x \to -4} \frac{1}{4x} = \frac{1}{4(-4)} = -\frac{1}{16}$

*Or:*  $\displaystyle\lim_{x \to -4} \frac{\frac{1}{4}+\frac{1}{x}}{4+x} = \lim_{x \to -4} \frac{\frac{1}{4}\cdot\frac{x}{x}+\frac{1}{x}\cdot\frac{4}{4}}{4+x} = \lim_{x \to -4} \frac{\frac{x+4}{4x}}{4+x} = \lim_{x \to -4} \frac{x+4}{4x(4+x)} = \lim_{x \to -4} \frac{1}{4x} = \frac{1}{4(-4)} = -\frac{1}{16}$

**39.**

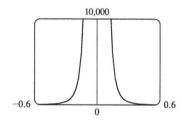

| $x$ | $\dfrac{3}{x^4}$ |
|---|---|
| $\pm 1$ | 3 |
| $\pm 0.5$ | 48 |
| $\pm 0.1$ | 30,000 |
| $\pm 0.01$ | 300,000,000 |

Notice from the table that as as $x$ becomes close to 0, the output values get very large. The graph also suggests that as $x$

approaches 0, the function values get larger and larger. Thus the function values do not approach any particular number, and

$\displaystyle\lim_{x \to 0}\left(3/x^4\right)$ does not exist.

**41.**

| $t$ | $\dfrac{e^t}{t}$ | $t$ | $\dfrac{e^t}{t}$ |
|---|---|---|---|
| 1 | $\approx 2.7$ | $-1$ | $\approx -0.4$ |
| 0.1 | $\approx 11$ | $-0.1$ | $\approx -9$ |
| 0.01 | $\approx 101$ | $-0.01$ | $\approx -99$ |
| 0.001 | $\approx 1001$ | $-0.001$ | $\approx -999$ |
| 0.0001 | $\approx 10{,}001$ | $-0.0001$ | $\approx -9999$ |

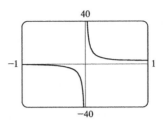

Notice from the tables that as as $t$ becomes close to 0, the output values get very large, positive or negative. The graph also suggests that as $t$ approaches 0, the function values get larger and larger (positive or negative). Thus the function values do not approach any particular number, and $\lim\limits_{t \to 0} \left(e^t/t\right)$ does not exist.

**43.** (a) $f(x)$ approaches 2 as $x$ approaches 1 from the left, so $\lim\limits_{x \to 1^-} f(x) = 2$.

(b) $f(x)$ approaches 3 as $x$ approaches 1 from the right, so $\lim\limits_{x \to 1^+} f(x) = 3$.

(c) $\lim\limits_{x \to 1} f(x)$ does not exist because the limits in part (a) and part (b) are not equal.

(d) $f(x)$ approaches 4 as $x$ approaches 5 from the left and from the right, so $\lim\limits_{x \to 5} f(x) = 4$.

(e) $f(5)$ is not defined, so it doesn't exist.

**45.** As $x$ approaches 1 from the left (so we use only values of $x$ smaller than 1), $f(x)$ approaches 3; and as $x$ approaches 1 from the right (we use values larger than 1), $f(x)$ approaches 7. No, the limit does not exist because the left- and right-hand limits are different.

**47.** (a) (i) If $x \to 1^+$, then $x > 1$ and $g(x) = x - 1$. Thus, $\lim\limits_{x \to 1^+} g(x) = \lim\limits_{x \to 1^+} (x - 1) = 1 - 1 = 0$.

(ii) We first determine the left-hand limit at 1: If $x \to 1^-$, then $x < 1$ and $g(x) = 1 - x^2$. Thus,

$\lim\limits_{x \to 1^-} g(x) = \lim\limits_{x \to 1^-} (1 - x^2) = 1 - 1^2 = 0$.

Since the left- and right-hand limits of $g$ at 1 are equal, $\lim\limits_{x \to 1} g(x) = 0$.

(iii) If $x \to 0$, then $-1 < x < 1$ and $g(x) = 1 - x^2$. Thus, $\lim\limits_{x \to 0} g(x) = \lim\limits_{x \to 0} (1 - x^2) = 1 - 0^2 = 1$.

(iv) If $x \to -1^-$, then $x < -1$ and $g(x) = -x$. Thus, $\lim\limits_{x \to -1^-} g(x) = \lim\limits_{x \to -1^-} (-x) = -(-1) = 1$.

(v) If $x \to -1^+$, then $-1 < x < 1$ and $g(x) = 1 - x^2$. Thus,

$$\lim\limits_{x \to -1^+} g(x) = \lim\limits_{x \to -1^+} (1 - x^2) = 1 - (-1)^2 = 1 - 1 = 0$$

(vi) $\lim\limits_{x \to -1} g(x)$ does not exist because the limits in part (iv) and part (v) are not equal.

(b)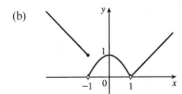

**49.** (a) If $x \to 0^+$, then $x > 0$ and $f(x) = |x|/x = x/x = 1$. Thus, $\lim\limits_{x \to 0^+} f(x) = \lim\limits_{x \to 0^+} 1 = 1$.

(b) If $x \to 0^-$, then $x < 0$ and $f(x) = |x|/x = -x/x = -1$. Thus, $\lim\limits_{x \to 0^-} f(x) = \lim\limits_{x \to 0^-} (-1) = -1$.

(c) Because the one-sided limits in parts (a) and (b) are not equal, $\lim\limits_{x \to 0} f(x)$ does not exist.

**51.** (a)

Cost (in dollars) / Time (in hours)

(b) There are discontinuities at times $t = 1, 2, 3$, and 4. A person

parking in the lot would want to keep in mind that the charge will

jump at the beginning of each hour.

**53.** (a)

1.5

$-1$ / 1

$-0.5$

$$\lim_{x \to 0} \frac{x}{\sqrt{1 + 3x} - 1} \approx \frac{2}{3}$$

(b)

| $x$ | $f(x)$ |
|---|---|
| $-0.001$ | 0.6661663 |
| $-0.0001$ | 0.6666167 |
| $-0.00001$ | 0.6666617 |
| $-0.000001$ | 0.6666662 |
| 0.000001 | 0.6666672 |
| 0.00001 | 0.6666717 |
| 0.0001 | 0.6667167 |
| 0.001 | 0.6671663 |

The limit appears to be $0.\overline{6}$ or $\dfrac{2}{3}$.

(c) $\lim\limits_{x \to 0} \left( \dfrac{x}{\sqrt{1 + 3x} - 1} \cdot \dfrac{\sqrt{1 + 3x} + 1}{\sqrt{1 + 3x} + 1} \right) = \lim\limits_{x \to 0} \dfrac{x\left(\sqrt{1 + 3x} + 1\right)}{(1 + 3x) - 1} = \lim\limits_{x \to 0} \dfrac{x\left(\sqrt{1 + 3x} + 1\right)}{3x}$

$= \lim\limits_{x \to 0} \left[ \tfrac{1}{3}\left(\sqrt{1 + 3x} + 1\right) \right] = \tfrac{1}{3}\left(\sqrt{1 + 3 \cdot 0} + 1\right) = \tfrac{1}{3}(1 + 1) = \tfrac{2}{3}$

since $\tfrac{1}{3}\left(\sqrt{1 + 3x} + 1\right)$ is continuous for $x \geq -\tfrac{1}{3}$.

**55.** (a) $\lim\limits_{x \to 0^-} f(x) = 1$

(b) $\lim\limits_{x \to 0^+} f(x) = 0$

(c) $\lim\limits_{x \to 0} f(x)$ does not exist because the limits in part (a)

and part (b) are not equal.

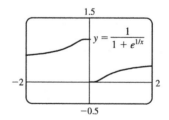

1.5

$y = \dfrac{1}{1 + e^{1/x}}$

$-2$ / 2

$-0.5$

**57.** $f(x) = \begin{cases} cx^2 + 2x & \text{if } x < 2 \\ x^3 - cx & \text{if } x \geq 2 \end{cases}$

$f$ is continuous on $(-\infty, 2)$ and $(2, \infty)$. Now $\lim\limits_{x \to 2^-} f(x) = \lim\limits_{x \to 2^-} \left(cx^2 + 2x\right) = c(2)^2 + 2(2) = 4c + 4$ and

$\lim\limits_{x \to 2^+} f(x) = \lim\limits_{x \to 2^+} \left(x^3 - cx\right) = (2)^3 - c(2) = 8 - 2c$. So $f$ is continuous $\Leftrightarrow 4c + 4 = 8 - 2c \Leftrightarrow 6c = 4 \Leftrightarrow$

$c = \tfrac{2}{3}$. Thus, for $f$ to be continuous on $(-\infty, \infty)$, $c = \tfrac{2}{3}$.

**59.** If we write $f(x) - 8$ as $\dfrac{f(x) - 8}{x - 1} \cdot (x - 1)$, then we have

$$\lim_{x \to 1} [f(x) - 8] = \lim_{x \to 1} \left[ \frac{f(x) - 8}{x - 1} \cdot (x - 1) \right] = \lim_{x \to 1} \frac{f(x) - 8}{x - 1} \cdot \lim_{x \to 1} (x - 1) = 10 \cdot (1 - 1) = 0. \text{ Thus}$$

$$\lim_{x \to 1} [f(x) - 8] = \lim_{x \to 1} f(x) - \lim_{x \to 1} 8 = \lim_{x \to 1} f(x) - 8 = 0 \quad \Rightarrow \quad \lim_{x \to 1} f(x) = 8.$$

*Note:* The value of $\lim\limits_{x \to 1} \dfrac{f(x) - 8}{x - 1}$ does not affect the answer since it's multiplied by 0. What's important is that

$\lim\limits_{x \to 1} \dfrac{f(x) - 8}{x - 1}$ exists.

## 2.3   Rates of Change and Derivatives

> **Prepare Yourself**

**1.** (a) $\dfrac{f(a) - f(2)}{a - 2} = \dfrac{(a^2 - 5a) - (-6)}{a - 2} = \dfrac{a^2 - 5a + 6}{a - 2} = \dfrac{(a - 3)(a - 2)}{a - 2} = a - 3$

(b) $\dfrac{f(3 + v) - f(3)}{v} = \dfrac{[(3 + v)^2 - 5(3 + v)] - (-6)}{v} = \dfrac{(9 + 6v + v^2) - 15 - 5v + 6}{v}$

$$= \dfrac{v^2 + v}{v} = \dfrac{v(v + 1)}{v} = v + 1$$

**2.** (a) $\dfrac{\dfrac{1}{5 + t} - \dfrac{1}{5}}{t} = \dfrac{\dfrac{5}{5(5 + t)} - \dfrac{(5 + t)}{5(5 + t)}}{t} = \dfrac{\dfrac{5 - (5 + t)}{5(5 + t)}}{t} = \dfrac{5 - 5 - t}{5(5 + t)(t)} = \dfrac{-t}{5t(5 + t)} = -\dfrac{1}{5(5 + t)}$

(b) $\dfrac{\dfrac{1}{r} - \dfrac{1}{4}}{r - 4} = \dfrac{\dfrac{1}{r} - \dfrac{1}{4}}{r - 4} \cdot \dfrac{4r}{4r} = \dfrac{\dfrac{4r}{r} - \dfrac{4r}{4}}{4r(r - 4)} = \dfrac{4 - r}{4r(r - 4)} = \dfrac{-(r - 4)}{4r(r - 4)} = -\dfrac{1}{4r}$

**3.** (a) $\lim\limits_{x \to 4} \dfrac{x^2 - 16}{x - 4} = \lim\limits_{x \to 4} \dfrac{(x + 4)(x - 4)}{x - 4} = \lim\limits_{x \to 4} (x + 4) = 4 + 4 = 8$

(b) $\lim\limits_{a \to 4} \dfrac{a^2 + 2a - 24}{a - 4} = \lim\limits_{a \to 4} \dfrac{(a + 6)(a - 4)}{a - 4} = \lim\limits_{a \to 4} (a + 6) = 4 + 6 = 10$

(c) First we have $(-2 + b)^3 = (-2 + b)(-2 + b)^2 = (-2 + b)(4 - 4b + b^2) = b^3 - 6b^2 + 12b - 8$, so

$$\lim_{b \to 0} \frac{(-2 + b)^3 + 8}{b} = \lim_{b \to 0} \frac{(b^3 - 6b^2 + 12b - 8) + 8}{b} = \lim_{b \to 0} \frac{b^3 - 6b^2 + 12b}{b}$$

$$= \lim_{b \to 0} \frac{b(b^2 - 6b + 12)}{b} = \lim_{b \to 0} (b^2 - 6b + 12) = 0^2 - 6(0) + 12 = 12$$

(d) Rewrite the function in the limit by rationalizing the numerator:

$$\lim_{t \to 9} \frac{\sqrt{t} - 3}{t - 9} = \lim_{t \to 9} \left( \frac{\sqrt{t} - 3}{t - 9} \cdot \frac{\sqrt{t} + 3}{\sqrt{t} + 3} \right) = \lim_{t \to 9} \left( \frac{t - 9}{(t - 9)(\sqrt{t} + 3)} \right) = \lim_{t \to 9} \frac{1}{\sqrt{t} + 3} = \frac{1}{\sqrt{9} + 3} = \frac{1}{6}.$$

**4.** Using the point-slope form of the equation of a line $y - y_1 = m(x - x_1)$, we have

$$y - (-5) = \tfrac{3}{4}(x - 2) \quad \text{or} \quad y + 5 = \tfrac{3}{4}x - \tfrac{3}{2} \quad \text{or} \quad y = \tfrac{3}{4}x - \tfrac{13}{2}.$$

**5.** The average rate of change is $\dfrac{\Delta f}{\Delta x} = \dfrac{f(4) - f(-1)}{4 - (-1)} = \dfrac{35 - 5}{4 - (-1)} = \dfrac{30}{5} = 6.$

**1.** Using Definition 1, the instantaneous rate of change is

$$\lim_{t_2 \to t_1} \frac{f(t_2) - f(t_1)}{t_2 - t_1} = \lim_{t \to 5} \frac{f(t) - f(5)}{t - 5} = \lim_{t \to 5} \frac{t^2 - 25}{t - 5} = \lim_{t \to 5} \frac{(t+5)(t-5)}{t - 5} = \lim_{t \to 5}(t+5) = 5 + 5 = 10, \text{ where we have}$$

used $t$ in place of $t_2$ for convenience. Using Definition 2, the instantaneous rate of change is

$$\lim_{h \to 0} \frac{f(5+h) - f(5)}{h} = \lim_{h \to 0} \frac{(5+h)^2 - 25}{h} = \lim_{h \to 0} \frac{(25 + 10h + h^2) - 25}{h} = \lim_{h \to 0} \frac{10h + h^2}{h}$$

$$= \lim_{h \to 0} \frac{h(10 + h)}{h} = \lim_{h \to 0}(10 + h) = 10 + 0 = 10$$

**3.** Here Definition 1 gives a more straightforward computation for the instantaneous rate of change:

$$\lim_{x_2 \to x_1} \frac{N(x_2) - N(x_1)}{x_2 - x_1} = \lim_{x \to 3} \frac{N(x) - N(3)}{x - 3} = \lim_{x \to 3} \frac{\frac{1}{x} - \frac{1}{3}}{x - 3} = \lim_{x \to 3} \frac{\frac{3}{3x} - \frac{x}{3x}}{x - 3} = \lim_{x \to 3} \frac{\frac{3-x}{3x}}{x - 3}$$

$$= \lim_{x \to 3} \frac{-(x - 3)}{3x(x - 3)} = \lim_{x \to 3} \frac{-1}{3x} = \frac{-1}{3(3)} = -\frac{1}{9}$$

**5.** The instantaneous speed is the instantaneous rate of change of $h$ at $t = 30$. By Definition 2, this is

$$\lim_{k \to 0} \frac{h(30 + k) - h(30)}{k} = \lim_{k \to 0} \frac{\left[1400(30 + k) - 16(30 + k)^2\right] - 27{,}600}{k}$$

$$= \lim_{k \to 0} \frac{\left[42{,}000 + 1400k - 16(900 + 60k + k^2)\right] - 27{,}600}{k}$$

$$= \lim_{k \to 0} \frac{42{,}000 + 1400k - 14{,}400 - 960k - 16k^2 - 27{,}600}{k}$$

$$= \lim_{k \to 0} \frac{440k - 16k^2}{k} = \lim_{k \to 0} \frac{k(440 - 16k)}{k} = \lim_{k \to 0}(440 - 16k) = 440 - 16(0) = 440$$

(Here we have used $k$ in place of $h$ in Definition 2 to avoid confusion with the function $h$.) Thus the instantaneous speed of the bullet 30 seconds after firing is 440 ft/s.

**7.** (a) The slope of the secant line is also the average rate of change: $m = \dfrac{\Delta y}{\Delta x} = \dfrac{f(x_2) - f(x_1)}{x_2 - x_1} = \dfrac{f(x) - f(3)}{x - 3}$.

(b) By Definition 3, the slope of the tangent line is $m = \lim_{x_2 \to x_1} \dfrac{f(x_2) - f(x_1)}{x_2 - x_1} = \lim_{x \to 3} \dfrac{f(x) - f(3)}{x - 3}$. This is the limit of the

slope of the secant line from part (a) as the point $(x, f(x))$ approaches the point $(3, f(3))$. Equivalently, by Definition 4,

the slope is $m = \lim_{h \to 0} \dfrac{f(3 + h) - f(3)}{h}$.

**9.** Using Definition 5 with $h(x) = 2x^2 + 1$ and $a = 3$, and using $k$ in place of $h$ to avoid confusion with the function $h$, we have

$$h'(3) = \lim_{k \to 0} \frac{h(3 + k) - h(3)}{k} = \lim_{k \to 0} \frac{\left[2(3 + k)^2 + 1\right] - 19}{k} = \lim_{k \to 0} \frac{\left[2(9 + 6k + k^2) + 1\right] - 19}{k}$$

$$= \lim_{k \to 0} \frac{18 + 12k + 2k^2 + 1 - 19}{k} = \lim_{k \to 0} \frac{12k + 2k^2}{k} = \lim_{k \to 0} \frac{k(12 + 2k)}{k} = \lim_{k \to 0}(12 + 2k) = 12 + 2(0) = 12$$

**11.** $g'(4) = \lim\limits_{h \to 0} \dfrac{g(4+h) - g(4)}{h} = \lim\limits_{h \to 0} \dfrac{[(4+h)^2 + 5(4+h) - 2] - 34}{h} = \lim\limits_{h \to 0} \dfrac{16 + 8h + h^2 + 20 + 5h - 2 - 34}{h}$

$= \lim\limits_{h \to 0} \dfrac{13h + h^2}{h} = \lim\limits_{h \to 0} \dfrac{h(13+h)}{h} = \lim\limits_{h \to 0}(13+h) = 13 + 0 = 13$

**13.** Using Definition 6 with $f(x) = 4x^2 - x$ and $a = 3$, we have

$f'(3) = \lim\limits_{x \to 3} \dfrac{f(x) - f(3)}{x - 3} = \lim\limits_{x \to 3} \dfrac{4x^2 - x - 33}{x - 3} = \lim\limits_{x \to 3} \dfrac{(4x + 11)(x - 3)}{x - 3} = \lim\limits_{x \to 3}(4x + 11) = 4(3) + 11 = 23.$

**15.** $g'(1) = \lim\limits_{t \to 1} \dfrac{g(t) - g(1)}{t - 1} = \lim\limits_{t \to 1} \dfrac{\dfrac{t}{t+1} - \dfrac{1}{2}}{t - 1} = \lim\limits_{t \to 1} \dfrac{\dfrac{2t}{2(t+1)} - \dfrac{t+1}{2(t+1)}}{t - 1} = \lim\limits_{t \to 1} \dfrac{\dfrac{2t - (t+1)}{2(t+1)}}{t - 1}$

$= \lim\limits_{t \to 1} \dfrac{2t - t - 1}{2(t+1)(t-1)} = \lim\limits_{t \to 1} \dfrac{t - 1}{2(t+1)(t-1)} = \lim\limits_{t \to 1} \dfrac{1}{2(t+1)} = \dfrac{1}{2(1+1)} = \dfrac{1}{4}$

**17.** (a) Using Definition 5,

$g'(3) = \lim\limits_{h \to 0} \dfrac{g(3+h) - g(3)}{h} = \lim\limits_{h \to 0} \dfrac{[2(3+h)^2 + 6] - 24}{h} = \lim\limits_{h \to 0} \dfrac{[2(9 + 6h + h^2) + 6] - 24}{h}$

$= \lim\limits_{h \to 0} \dfrac{18 + 12h + 2h^2 + 6 - 24}{h} = \lim\limits_{h \to 0} \dfrac{h(12 + 2h)}{h} = \lim\limits_{h \to 0}(12 + 2h) = 12 + 2(0) = 12$

(b) $g'(a) = \lim\limits_{h \to 0} \dfrac{g(a+h) - g(a)}{h} = \lim\limits_{h \to 0} \dfrac{[2(a+h)^2 + 6] - (2a^2 + 6)}{h} = \lim\limits_{h \to 0} \dfrac{[2(a^2 + 2ah + h^2) + 6] - (2a^2 + 6)}{h}$

$= \lim\limits_{h \to 0} \dfrac{2a^2 + 4ah + 2h^2 + 6 - 2a^2 - 6}{h} = \lim\limits_{h \to 0} \dfrac{4ah + 2h^2}{h} = \lim\limits_{h \to 0} \dfrac{h(4a + 2h)}{h}$

$= \lim\limits_{h \to 0}(4a + 2h) = 4a + 2(0) = 4a$

**19.** (a) Using Definition 5 with $f(x) = 3 - 2x + 4x^2$ gives

$f'(a) = \lim\limits_{h \to 0} \dfrac{f(a+h) - f(a)}{h} = \lim\limits_{h \to 0} \dfrac{[3 - 2(a+h) + 4(a+h)^2] - (3 - 2a + 4a^2)}{h}$

$= \lim\limits_{h \to 0} \dfrac{[3 - 2a - 2h + 4(a^2 + 2ah + h^2)] - (3 - 2a + 4a^2)}{h}$

$= \lim\limits_{h \to 0} \dfrac{3 - 2a - 2h + 4a^2 + 8ah + 4h^2 - 3 + 2a - 4a^2}{h} = \lim\limits_{h \to 0} \dfrac{-2h + 8ah + 4h^2}{h}$

$= \lim\limits_{h \to 0} \dfrac{h(-2 + 8a + 4h)}{h} = \lim\limits_{h \to 0}(-2 + 8a + 4h) = -2 + 8a + 4(0) = -2 + 8a$

(b) The instantaneous rate of change of $f$ when $x = 5$ is $f'(5)$, and from part (a) we have $f'(a) = -2 + 8a$, so

$f'(5) = -2 + 8(5) = 38.$

(c) The slope of the tangent line to the graph of $f$ at the point $(-1, 9)$ is $f'(-1) = -2 + 8(-1) = -10.$

**21.**

The line from $P(2, f(2))$ to $Q(2 + h, f(2 + h))$ is the line

that has slope $\dfrac{f(2 + h) - f(2)}{h}.$

**23.** $g'(0)$ is the only negative value. The slope at $x = 4$ is smaller than the slope at $x = 2$, and both are smaller than the slope at $x = -2$. Thus, $g'(0) < 0 < g'(4) < g'(2) < g'(-2)$.

**25.** The sketch shows the graph for a room temperature of $72°$ and a refrigerator temperature of $38°$. The initial rate of change is greater in magnitude (the curve is steeper) than the rate of change after an hour.

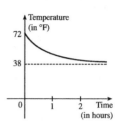

**27.** Using Definition 5 with $f(x) = 3x^2 - 5x$, we have

$$f'(2) = \lim_{h \to 0} \frac{f(2+h) - f(2)}{h} = \lim_{h \to 0} \frac{[3(2+h)^2 - 5(2+h)] - 2}{h} = \lim_{h \to 0} \frac{[3(4 + 4h + h^2) - 5(2+h)] - 2}{h}$$

$$= \lim_{h \to 0} \frac{12 + 12h + 3h^2 - 10 - 5h - 2}{h} = \lim_{h \to 0} \frac{3h^2 + 7h}{h} = \lim_{h \to 0} \frac{h(3h + 7)}{h} = \lim_{h \to 0} (3h + 7) = 3(0) + 7 = 7$$

The slope of the tangent line at $(2, 2)$ is $f'(2) = 7$, so an equation of the line is $y - 2 = 7(x - 2)$ $\Leftrightarrow$ $y - 2 = 7x - 14$ $\Leftrightarrow$ $y = 7x - 12$.

**29.** (a) $F'(2) = \lim_{h \to 0} \frac{F(2+h) - F(2)}{h} = \lim_{h \to 0} \frac{[(2+h)^2 - 2(2+h)] - 0}{h} = \lim_{h \to 0} \frac{4 + 4h + h^2 - 4 - 2h}{h}$

$$= \lim_{h \to 0} \frac{2h + h^2}{h} = \lim_{h \to 0} \frac{h(2 + h)}{h} = \lim_{h \to 0} (2 + h) = 2 + 0 = 2$$

The slope of the tangent line at $(2, 0)$ is $F'(2) = 2$, and an equation is $y - 0 = 2(x - 2)$ or $y = 2x - 4$.

(b)

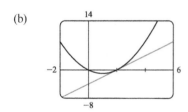

**31.** The slope of the tangent line to the curve at $(2, 4.4)$ is the derivative at $x = 2$. For $f(x) = 4.2x - x^2$,

$$f'(2) = \lim_{h \to 0} \frac{f(2+h) - f(2)}{h} = \lim_{h \to 0} \frac{[4.2(2+h) - (2+h)^2] - 4.4}{h} = \lim_{h \to 0} \frac{[8.4 + 4.2h - (4 + 4h + h^2)] - 4.4}{h}$$

$$= \lim_{h \to 0} \frac{8.4 + 4.2h - 4 - 4h - h^2 - 4.4}{h} = \lim_{h \to 0} \frac{0.2h - h^2}{h} = \lim_{h \to 0} \frac{h(0.2 - h)}{h} = \lim_{h \to 0} (0.2 - h) = 0.2 - 0 = 0.2$$

Thus an equation of the tangent line is $y - 4.4 = 0.2(x - 2)$ or $y = 0.2x + 4$.

**33.** Using Definition 6 with $f(x) = \dfrac{x - 1}{x - 2}$, we have

$$f'(3) = \lim_{x \to 3} \frac{f(x) - f(3)}{x - 3} = \lim_{x \to 3} \frac{\dfrac{x - 1}{x - 2} - 2}{x - 3} = \lim_{x \to 3} \left( \frac{\dfrac{x - 1}{x - 2} - 2}{x - 3} \cdot \frac{x - 2}{x - 2} \right) = \lim_{x \to 3} \frac{x - 1 - 2(x - 2)}{(x - 3)(x - 2)}$$

$$= \lim_{x \to 3} \frac{x - 1 - 2x + 4}{(x - 3)(x - 2)} = \lim_{x \to 3} \frac{-x + 3}{(x - 3)(x - 2)} = \lim_{x \to 3} \frac{-(x - 3)}{(x - 3)(x - 2)} = \lim_{x \to 3} \frac{-1}{x - 2} = \frac{-1}{3 - 2} = -1.$$

[continued]

The slope of the tangent line to the curve at $(3, 2)$ is $f'(3) = -1$, so an equation of the line is $y - 2 = -1(x - 3)$   or $y = -x + 5$.

35. The velocity of the particle at $t = 5$ is $f'(5)$. Using Definition 5, this is

$$f'(5) = \lim_{h \to 0} \frac{f(5 + h) - f(5)}{h} = \lim_{h \to 0} \frac{[40(5 + h) - 6(5 + h)^2] - 50}{h} = \lim_{h \to 0} \frac{[200 + 40h - 6(25 + 10h + h^2)] - 50}{h}$$

$$= \lim_{h \to 0} \frac{200 + 40h - 150 - 60h - 6h^2 - 50}{h} = \lim_{h \to 0} \frac{-20h - 6h^2}{h} = \lim_{h \to 0} \frac{h(-20 - 6h)}{h}$$

$$= \lim_{h \to 0} (-20 - 6h) = -20 - 6(0) = -20$$

Thus the velocity is $-20$ m/s, and the speed is $|-20| = 20$ m/s.

37. (a) (i) $\dfrac{\Delta C}{\Delta x} = \dfrac{C(105) - C(100)}{105 - 100} = \dfrac{6601.25 - 6500}{5} = \$20.25/\text{unit}$

(ii) $\dfrac{\Delta C}{\Delta x} = \dfrac{C(101) - C(100)}{101 - 100} = \dfrac{6520.05 - 6500}{1} = \$20.05/\text{unit}$

(b) By Definition 2, the instantaneous rate of change is

$$\lim_{h \to 0} \frac{C(100 + h) - C(100)}{h} = \lim_{h \to 0} \frac{\left[5000 + 10(100 + h) + 0.05(100 + h)^2\right] - 6500}{h}$$

$$= \lim_{h \to 0} \frac{5000 + 1000 + 10h + 0.05(10{,}000 + 200h + h^2) - 6500}{h}$$

$$= \lim_{h \to 0} \frac{6000 + 10h + 500 + 10h + 0.05h^2 - 6500}{h} = \lim_{h \to 0} \frac{20h + 0.05h^2}{h}$$

$$= \lim_{h \to 0} \frac{h(20 + 0.05h)}{h} = \lim_{h \to 0} (20 + 0.05h) = 20 + 0.05(0) = \$20/\text{unit}$$

39. (a) $f'(x)$ is the rate of change of the production cost with respect to the number of ounces of gold produced. Its units are dollars per ounce.

(b) After 800 ounces of gold have been produced, the rate at which the production cost is increasing is $\$17$/ounce. So the cost of producing the 800th (or 801st) ounce is about $\$17$.

(c) In the short term, the values of $f'(x)$ will decrease because more efficient use is made of start-up costs as $x$ increases. But eventually $f'(x)$ might increase due to large-scale operations.

41. (a) $f'(v)$ is the rate at which the fuel consumption is changing with respect to the speed. Its units are $(\text{gal}/\text{h})/(\text{mi}/\text{h})$.

(b) The fuel consumption is decreasing by $0.05\ (\text{gal}/\text{h})/(\text{mi}/\text{h})$ as the car's speed reaches 20 mi/h. So if you increase your speed to 21 mi/h, you could expect to decrease your fuel consumption by about $0.05\ (\text{gal}/\text{h})/(\text{mi}/\text{h})$.

43. (a) $H'(8)$ is the rate at which the daily heating cost changes with respect to temperature when the outside temperature is $58\,°\text{F}$. The units are dollars$/\,°\text{F}$.

(b) If the outside temperature increases, the building should require less heating, so we would expect $H'(58)$ to be negative.

**45.**  (a) $P(80) = -125$ means that the shop loses $125 in profit if it sells only 80 mugs in a week.

(b) $P'(80)$ is the rate of change of profit with respect to the number of mugs sold when 80 mugs are sold in a week, so

$P'(80) = 1.5$ means that the weekly profit is increasing at a rate of $1.50/mug if the number sold increases from 80.

**47.**  (a) By Definition 5, $f'(1) = \lim\limits_{h \to 0} \dfrac{f(1+h) - f(1)}{h} = \lim\limits_{h \to 0} \dfrac{3^{1+h} - 3}{h}$. We calculate $\dfrac{3^{1+h} - 3}{h}$ for increasingly smaller

values of $h$, both positive and negative:

| $h$ | $\dfrac{3^{1+h} - 3}{h}$ | $h$ | $\dfrac{3^{1+h} - 3}{h}$ |
|---|---|---|---|
| 0.1 | 3.484 | −0.1 | 3.121 |
| 0.01 | 3.314 | −0.01 | 3.278 |
| 0.001 | 3.298 | −0.001 | 3.294 |
| 0.0001 | 3.296 | −0.0001 | 3.296 |

We estimate that $f'(1) \approx 3.296$.

(b) We zoom in until the curve looks virtually like a straight line, namely

the tangent line to the curve.

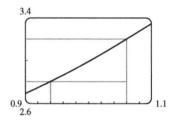

From the graph, it appears that the line passes through the points $(0.94, 2.8)$ and $(1.06, 3.2)$, so we estimate that the slope

of the tangent is about $\dfrac{3.2 - 2.8}{1.06 - 0.94} = \dfrac{0.4}{0.12} \approx 3.3$. Thus $f'(1) \approx 3.3$.

**49.**  $T'(8)$ is the rate at which the temperature is changing at 8:00 AM. We can estimate the value of $T'(8)$ by finding the average

rates of change for $6 \le t \le 8$ and $8 \le t \le 10$, and averaging the results.

$6 \le t \le 8$:  $\dfrac{T(8) - T(6)}{8 - 6} = \dfrac{84 - 75}{2} = 4.5$  $\qquad$  $8 \le t \le 10$:  $\dfrac{T(10) - T(8)}{10 - 8} = \dfrac{90 - 84}{2} = 3$

Then $T'(8) \approx \dfrac{4.5 + 3}{2} = 3.75°\text{F}/\text{h}$.

**51.**  (a) The average speed is the average rate of change of $s$ (with respect to $t$).

(i) On the interval $[1, 3]$, average speed $= \dfrac{s(3) - s(1)}{3 - 1} = \dfrac{10.7 - 1.4}{2} = \dfrac{9.3}{2} = 4.65 \text{ m/s}$.

(ii) On the interval $[2, 3]$, average speed $= \dfrac{s(3) - s(2)}{3 - 2} = \dfrac{10.7 - 5.1}{1} = 5.6 \text{ m/s}$.

(iii) On the interval $[3, 5]$, average speed $= \dfrac{s(5) - s(3)}{5 - 3} = \dfrac{25.8 - 10.7}{2} = \dfrac{15.1}{2} = 7.55 \text{ m/s}$.

(iv) On the interval $[3, 4]$, average speed $= \dfrac{s(4) - s(3)}{4 - 3} = \dfrac{17.7 - 10.7}{1} = 7 \text{ m/s}$.

(b)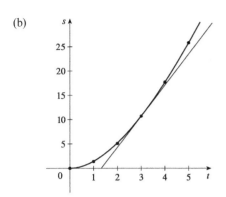

We sketch an approximate tangent line to the curve at $t = 3$. It appears that the tangent line passes through the points $(2, 4)$ and $(5, 23)$, so its slope is approximately $\dfrac{23 - 4}{5 - 2} \approx 6.3$. Thus we estimate the (instantaneous) speed when $t = 3$ to be 6.3 m/s.

**53.** (a)  (i) 2003 to 2007: $\dfrac{\Delta P}{\Delta t} = \dfrac{P(2007) - P(2003)}{2007 - 2003} = \dfrac{63.1 - 25.6}{2007 - 2003} = \dfrac{37.5}{4} = 9.375$ percentage points/year

(ii) 2003 to 2005: $\dfrac{\Delta P}{\Delta t} = \dfrac{P(2005) - P(2003)}{2005 - 2003} = \dfrac{46.3 - 25.6}{2005 - 2003} = \dfrac{20.7}{2} = 10.35$ percentage points/year

(iii) 2001 to 2003: $\dfrac{\Delta P}{\Delta t} = \dfrac{P(2003) - P(2001)}{2003 - 2001} = \dfrac{25.6 - 16.3}{2003 - 2001} = \dfrac{9.3}{2} = 4.65$ percentage points/year

(b) We estimate the instantaneous rate of growth in 2003 by averaging the average rates of growth for 2001 to 2003 and 2003 to 2005. Thus, we estimate the rate to be $\frac{4.65 + 10.35}{2} = 7.5$ percentage points/year.

(c) We draw a scatter plot and sketch a smooth curve through the points. Then we sketch an approximate tangent line to the curve at 2003.

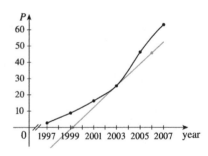

It appears that this line has $x$-intercept 1999.2 and passes through the point $(2006, 46)$, so we estimate its slope to be $m = \frac{46 - 0}{2006 - 1999.2} = \frac{46}{6.8} \approx 6.8$. Thus the instantaneous rate of growth in 2003 was approximately 6.8 percentage points/year.

**55.** (a) $S'(T)$ is the rate at which the oxygen solubility changes with respect to the water temperature. Its units are $(\text{mg/L})/^\circ\text{C}$.

(b) We sketch an approximate tangent line to the curve at $T = 16$. It appears that this line goes through the points $(0, 14)$ and $(32, 6)$, so its slope is approximately $\dfrac{6 - 14}{32 - 0} = \dfrac{-8}{32} = -0.25$. Thus $S'(16) \approx -0.25 \; (\text{mg/L})/^\circ\text{C}$. This means that as the temperature increases past $16^\circ\text{C}$, the oxygen solubility is decreasing at a rate of $0.25 \; (\text{mg/L})/^\circ\text{C}$.

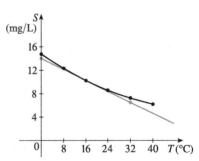

**57.** (a) A graphing calculator gives a quadratic model as approximately $V(t) = 1.11t^2 - 66.6t + 999$, where $V$ is the volume of water (in gallons) remaining in the tank, and $t$ is the number of minutes.

(b) By Definition 5,

$$V'(15) = \lim_{h \to 0} \frac{V(15+h) - V(15)}{h} = \lim_{h \to 0} \frac{[1.11(15+h)^2 - 66.6(15+h) + 999] - 249.75}{h}$$

$$= \lim_{h \to 0} \frac{[1.11(225 + 30h + h^2) - 66.6(15+h) + 999] - 249.75}{h}$$

$$= \lim_{h \to 0} \frac{249.75 + 33.3h + 1.11h^2 - 999 - 66.6h + 999 - 249.75}{h}$$

$$= \lim_{h \to 0} \frac{-33.3h + 1.11h^2}{h} = \lim_{h \to 0} \frac{h(-33.3 + 1.11h)}{h} = \lim_{h \to 0} (-33.3 + 1.11h) = -33.3$$

This means that the water volume is decreasing at a rate of 33.3 gallons per minute after 15 minutes.

**59.** The graph must have positive slope at $x = 1$ and $x = 3$, but the slope of the secant line through the corresponding points must be negative.

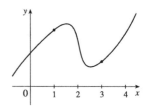

**61.** (a) The average rate of change on the interval $[1995, 2005]$ is

$$\frac{\Delta D}{\Delta t} = \frac{D(2005) - D(1995)}{2005 - 1995} = \frac{7932.7 - 4974.0}{2005 - 1995} = \frac{2958.7}{10} = 295.87 \approx 296 \text{ billion dollars per year. This is the same}$$

answer we found in Example 10.

(b) The symmetric difference quotient is the average rate of change of $f$ on the interval $[a - d, a + d]$:

$$\frac{\Delta f}{\Delta x} = \frac{f(a+d) - f(a-d)}{(a+d) - (a-d)} = \frac{f(a+d) - f(a-d)}{a + d - a + d} = \frac{f(a+d) - f(a-d)}{2d}.$$

The average rate of change on $[a - d, a]$ is $\dfrac{f(a) - f(a-d)}{a - (a-d)} = \dfrac{f(a) - f(a-d)}{d}$ and

on $[a, a + d]$ it's $\dfrac{f(a+d) - f(a)}{(a+d) - a} = \dfrac{f(a+d) - f(a)}{d}$. Averaging these results gives

$$\frac{1}{2}\left(\frac{f(a) - f(a-d)}{d} + \frac{f(a+d) - f(a)}{d}\right) = \frac{1}{2}\left(\frac{f(a) - f(a-d) + f(a+d) - f(a)}{d}\right) = \frac{f(a+d) - f(a-d)}{2d},$$

which agrees with the symmetric difference quotient.

(c) Using the symmetric difference quotient for $f(x) = x^3 - 2x^2 + 2$ with $a = 1$ and $d = 0.4$, we estimate

$$f'(1) \approx \frac{f(1+0.4) - f(1-0.4)}{2(0.4)} = \frac{f(1.4) - f(0.6)}{0.8} = \frac{0.824 - 1.496}{0.8} = \frac{-0.672}{0.8} = -0.84. \text{ This is the slope of the}$$

secant line through the points $(0.6, f(0.6))$ and $(1.4, f(1.4))$. Looking at the graph, we see that the direction of the tangent line more closely matches the direction of this secant line than the directions of the secant lines corresponding to the intervals $[0.6, 1]$ or $[1, 1.4]$. Thus the symmetric difference quotient gives a better estimate than the slopes of either of the other secant lines for the derivative value.

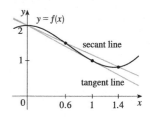

## 2.4   The Derivative as a Function

**1.**
$$\frac{g(t+h) - g(t)}{h} = \frac{[6(t+h) - 3(t+h)^2] - (6t - 3t^2)}{h} = \frac{[6(t+h) - 3(t^2 + 2th + h^2)] - (6t - 3t^2)}{h}$$

$$= \frac{6t + 6h - 3t^2 - 6th - 3h^2 - 6t + 3t^2}{h} = \frac{6h - 6th - 3h^2}{h} = \frac{h(6 - 6t - 3h)}{h} = 6 - 6t - 3h$$

**2.**
$$\frac{f(x+c) - f(x)}{c} = \frac{\dfrac{1}{x+c} - \dfrac{1}{x}}{c} = \frac{\dfrac{x}{x(x+c)} - \dfrac{1(x+c)}{x(x+c)}}{c} = \frac{\dfrac{x - (x+c)}{x(x+c)}}{c} = \frac{\dfrac{-c}{x(x+c)}}{c}$$

$$= \frac{-c}{cx(x+c)} = -\frac{1}{x(x+c)}$$

**3.**
$$\lim_{h \to 0} \frac{h^2 + 2xh}{h} = \lim_{h \to 0} \frac{h(h + 2x)}{h} = \lim_{h \to 0} (h + 2x) = 0 + 2x = 2x$$

**4.**
$$\lim_{h \to 0} \frac{(x+h)^3 + 2(x+h) - x^3 - 2x}{h} = \lim_{h \to 0} \frac{(x^3 + 3x^2h + 3xh^2 + h^3) + 2x + 2h - x^3 - 2x}{h}$$

$$= \lim_{h \to 0} \frac{3x^2h + 3xh^2 + h^3 + 2h}{h} = \lim_{h \to 0} \frac{h(3x^2 + 3xh + h^2 + 2)}{h}$$

$$= \lim_{h \to 0} (3x^2 + 3xh + h^2 + 2) = 3x^2 + 3x(0) + (0)^2 + 2 = 3x^2 + 2$$

**5.**
$$\frac{\dfrac{x+a}{x+a+2} - \dfrac{x}{x+2}}{a} = \frac{\dfrac{x+a}{x+a+2} - \dfrac{x}{x+2}}{a} \cdot \frac{(x+a+2)(x+2)}{(x+a+2)(x+2)}$$

$$= \frac{(x+a)(x+2) - x(x+a+2)}{a(x+a+2)(x+2)} = \frac{x^2 + 2x + ax + 2a - x^2 - ax - 2x}{a(x+a+2)(x+2)}$$

$$= \frac{2a}{a(x+a+2)(x+2)} = \frac{2}{(x+a+2)(x+2)}$$

*Or:*

$$\frac{\dfrac{x+a}{x+a+2} - \dfrac{x}{x+2}}{a} = \frac{\dfrac{(x+a)(x+2)}{(x+a+2)(x+2)} - \dfrac{x(x+a+2)}{(x+2)(x+a+2)}}{a}$$

$$= \frac{\dfrac{(x+a)(x+2) - x(x+a+2)}{(x+a+2)(x+2)}}{a} = \frac{x^2 + 2x + ax + 2a - x^2 - ax - 2x}{a(x+a+2)(x+2)}$$

$$= \frac{2a}{a(x+a+2)(x+2)} = \frac{2}{(x+a+2)(x+2)}$$

**6.** Using Definition 5 in Section 2.3, we have

$$f'(4) = \lim_{h \to 0} \frac{f(4+h) - f(4)}{h} = \lim_{h \to 0} \frac{\sqrt{4+h} - 2}{h} = \lim_{h \to 0} \left( \frac{\sqrt{4+h} - 2}{h} \cdot \frac{\sqrt{4+h} + 2}{\sqrt{4+h} + 2} \right)$$

$$= \lim_{h \to 0} \frac{4+h-4}{h \left( \sqrt{4+h} + 2 \right)} = \lim_{h \to 0} \frac{h}{h \left( \sqrt{4+h} + 2 \right)} = \lim_{h \to 0} \frac{1}{\sqrt{4+h} + 2} = \frac{1}{\sqrt{4+0} + 2} = \frac{1}{4}.$$

**Exercises**

**1.** To find the value of $f'(a)$, sketch a tangent line to the curve at $x = a$ and estimate the slope of the line. The value of the slope becomes the $y$-value on the graph of $f'$.

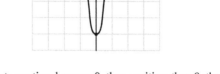

(a) $f'(-3) \approx 1.5$      (b) $f'(-2) \approx 1$

(c) $f'(-1) \approx 0$      (d) $f'(0) \approx -4$

(e) $f'(1) \approx 0$      (f) $f'(2) \approx 1$

(g) $f'(3) \approx 1.5$

**3.** (a)$' =$ II, since from left to right, the slopes of the tangents to graph (a) start out negative, become 0, then positive, then 0, then negative again. The actual function values in graph II follow the same pattern.

(b)$' =$ IV, since from left to right, the slopes of the tangents to graph (b) start out at a fixed positive value, then suddenly become negative, then positive again. The jumps in graph IV indicate sudden changes in the slopes of the tangents.

(c)$' =$ I, since the slopes of the tangents to graph (c) are negative for $x < 0$ and positive for $x > 0$, as are the function values of graph I.

(d)$' =$ III, since from left to right, the slopes of the tangents to graph (d) are positive, then 0, then negative, then 0, then positive, then 0, then negative again, and the function values in graph III follow the same pattern.

Hints for Exercises 4–7: If $f$ has a horizontal tangent line at a number, then $f'(x) = 0$ there, so the graph of $f'$ will have an $x$-intercept at the same number. On any interval where $f$ has tangent lines with positive slope, the graph of $f'$ will be positive (so above the $x$-axis). If $f$ has negative slope, the graph of $f'$ will be below the $x$-axis. The steeper the slope, positive or negative, the larger the value of $f'$ (positive or negative).

**5.**

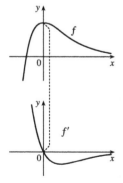

**7.**

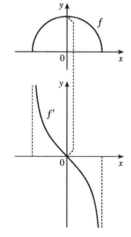

**9.**

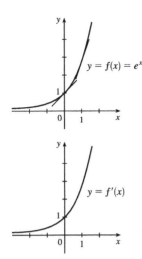

The slope at 0 appears to be about 1, and the slope at 1 appears to be between 2.5 and 3. As $x$ increases the slope gets larger, and as $x$ decreases the slope gets closer to 0. Since the graphs are so similar, we might guess that $f'(x) = e^x$.

**11.** It appears that there are horizontal tangents on the graph of $M$ for $t = 1963$ and $t = 1971$. Thus, there are $t$-intercepts for those values of $t$ on the graph of $M'$. The derivative is negative for the years 1963 to 1971, and positive elsewhere. Thus the graph of $M'$ is below the $t$-axis for those years and above it elsewhere.

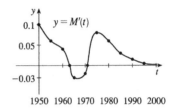

**13.** (a) $U'(t)$ is the rate at which the unemployment rate is changing with respect to time. Its units are percent unemployed per year.

(b) As in Example 2, $U'(t) = \lim\limits_{h \to 0} \dfrac{U(t+h) - U(t)}{h} \approx \dfrac{U(t+h) - U(t)}{h}$ for small values of $h$.

**For 1999:** Using $h = 1$, we get $U'(1999) \approx \dfrac{U(2000) - U(1999)}{2000 - 1999} = \dfrac{4.0 - 4.2}{1} = -0.2.$

**For 2000:** We estimate $U'(2000)$ by using $h = -1$ and $h = 1$, and then average the two results to obtain a final estimate.

$h = -1 \quad \Rightarrow \quad U'(2000) \approx \dfrac{U(1999) - U(2000)}{1999 - 2000} = \dfrac{4.2 - 4.0}{-1} = -0.2;$

$h = 1 \quad \Rightarrow \quad U'(2000) \approx \dfrac{U(2001) - U(2000)}{2001 - 2000} = \dfrac{4.7 - 4.0}{1} = 0.7.$

So we estimate that $U'(2000) \approx \frac{1}{2}[(-0.2) + 0.7] = 0.25$. Making similar calculations for the remaining years, we get the estimates in the following table.

| $t$ | 1999 | 2000 | 2001 | 2002 | 2003 | 2004 | 2005 | 2006 | 2007 | 2008 |
|---|---|---|---|---|---|---|---|---|---|---|
| $U'(t)$ | −0.2 | 0.25 | 0.9 | 0.65 | −0.15 | −0.45 | −0.45 | −0.25 | 0.6 | 1.2 |

**15.** (a) By zooming in, we estimate that $f'(0) = 0$, $f'\left(\frac{1}{2}\right) = 1$, $f'(1) = 2$, and $f'(2) = 4$.

(b) By symmetry, the value of $f'(-x)$ is the same as $f'(x)$ but opposite in sign: $f'(-x) = -f'(x)$. So $f'\left(-\frac{1}{2}\right) = -1$, $f'(-1) = -2$, and $f'(-2) = -4$.

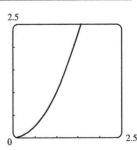

(c) It appears that $f'(x)$ is twice the value of $x$, so we guess that $f'(x) = 2x$.

(d) $f'(x) = \lim\limits_{h \to 0} \dfrac{f(x+h) - f(x)}{h} = \lim\limits_{h \to 0} \dfrac{(x+h)^2 - x^2}{h}$

$\quad = \lim\limits_{h \to 0} \dfrac{(x^2 + 2hx + h^2) - x^2}{h} = \lim\limits_{h \to 0} \dfrac{2hx + h^2}{h} = \lim\limits_{h \to 0} \dfrac{h(2x + h)}{h} = \lim\limits_{h \to 0} (2x + h) = 2x$

**17.** By Definition 2,

$f'(x) = \lim\limits_{h \to 0} \dfrac{f(x+h) - f(x)}{h} = \lim\limits_{h \to 0} \dfrac{\left[\frac{1}{2}(x+h) - \frac{1}{3}\right] - \left(\frac{1}{2}x - \frac{1}{3}\right)}{h} = \lim\limits_{h \to 0} \dfrac{\frac{1}{2}x + \frac{1}{2}h - \frac{1}{3} - \frac{1}{2}x + \frac{1}{3}}{h}$

$\quad = \lim\limits_{h \to 0} \dfrac{\frac{1}{2}h}{h} = \lim\limits_{h \to 0} \frac{1}{2} = \frac{1}{2}.$

**19.** Using Definition 2 with $k$ in place of $h$ (to avoid confusion with the function $h$) gives

$h'(v) = \lim\limits_{k \to 0} \dfrac{h(v+k) - h(v)}{k} = \lim\limits_{k \to 0} \dfrac{\left[4(v+k)^2 - 2\right] - \left(4v^2 - 2\right)}{k}$

$\quad = \lim\limits_{k \to 0} \dfrac{\left[4(v^2 + 2vk + k^2) - 2\right] - \left(4v^2 - 2\right)}{k} = \lim\limits_{k \to 0} \dfrac{4v^2 + 8vk + 4k^2 - 2 - 4v^2 + 2}{k}$

$\quad = \lim\limits_{k \to 0} \dfrac{8vk + 4k^2}{k} = \lim\limits_{k \to 0} \dfrac{k(8v + 4k)}{k} = \lim\limits_{k \to 0} (8v + 4k) = 8v.$

**21.** $f'(x) = \lim\limits_{h \to 0} \dfrac{f(x+h) - f(x)}{h} = \lim\limits_{h \to 0} \dfrac{\left[(x+h)^3 - 3(x+h) + 5\right] - (x^3 - 3x + 5)}{h}$

$\quad = \lim\limits_{h \to 0} \dfrac{(x^3 + 3x^2 h + 3xh^2 + h^3 - 3x - 3h + 5) - x^3 + 3x - 5}{h} = \lim\limits_{h \to 0} \dfrac{3x^2 h + 3xh^2 + h^3 - 3h}{h}$

$\quad = \lim\limits_{h \to 0} \dfrac{h\left(3x^2 + 3xh + h^2 - 3\right)}{h} = \lim\limits_{h \to 0} \left(3x^2 + 3xh + h^2 - 3\right) = 3x^2 - 3$

**23.** $B'(p) = \lim\limits_{h \to 0} \dfrac{B(p+h) - B(p)}{h} = \lim\limits_{h \to 0} \dfrac{\dfrac{3}{p+h} - \dfrac{3}{p}}{h} = \lim\limits_{h \to 0} \dfrac{\dfrac{3p}{(p+h)p} - \dfrac{3(p+h)}{p(p+h)}}{h}$

$\quad = \lim\limits_{h \to 0} \dfrac{\dfrac{3p - 3(p+h)}{p(p+h)}}{h} = \lim\limits_{h \to 0} \dfrac{-3h}{ph(p+h)} = \lim\limits_{h \to 0} \dfrac{-3}{p(p+h)} = \dfrac{-3}{p(p+0)} = -\dfrac{3}{p^2}$

**25.** For $f(x) = \sqrt{x}$,

$f'(x) = \lim\limits_{h \to 0} \dfrac{f(x+h) - f(x)}{h} = \lim\limits_{h \to 0} \dfrac{\sqrt{x+h} - \sqrt{x}}{h} = \lim\limits_{h \to 0} \left(\dfrac{\sqrt{x+h} - \sqrt{x}}{h} \cdot \dfrac{\sqrt{x+h} + \sqrt{x}}{\sqrt{x+h} + \sqrt{x}}\right)$

$\quad = \lim\limits_{h \to 0} \dfrac{x+h - x}{h\left(\sqrt{x+h} + \sqrt{x}\right)} = \lim\limits_{h \to 0} \dfrac{h}{h\left(\sqrt{x+h} + \sqrt{x}\right)} = \lim\limits_{h \to 0} \dfrac{1}{\sqrt{x+h} + \sqrt{x}} = \dfrac{1}{\sqrt{x+0} + \sqrt{x}} = \dfrac{1}{2\sqrt{x}}.$

Thus $\dfrac{dy}{dx} = \dfrac{1}{2\sqrt{x}}.$

**27.** $G'(t) = \lim\limits_{h \to 0} \dfrac{G(t+h) - G(t)}{h} = \lim\limits_{h \to 0} \dfrac{\dfrac{4(t+h)}{(t+h)+1} - \dfrac{4t}{t+1}}{h}$

$= \lim\limits_{h \to 0} \dfrac{\dfrac{4(t+h)(t+1)}{(t+h+1)(t+1)} - \dfrac{4t(t+h+1)}{(t+1)(t+h+1)}}{h} = \lim\limits_{h \to 0} \dfrac{\dfrac{4(t+h)(t+1) - 4t(t+h+1)}{(t+h+1)(t+1)}}{h}$

$= \lim\limits_{h \to 0} \dfrac{4(t^2 + t + ht + h) - 4t(t+h+1)}{h(t+h+1)(t+1)}$

$= \lim\limits_{h \to 0} \dfrac{4t^2 + 4t + 4ht + 4h - 4t^2 - 4ht - 4t}{h(t+h+1)(t+1)} = \lim\limits_{h \to 0} \dfrac{4h}{h(t+h+1)(t+1)}$

$= \lim\limits_{h \to 0} \dfrac{4}{(t+h+1)(t+1)} = \dfrac{4}{(t+0+1)(t+1)} = \dfrac{4}{(t+1)^2}$

For both $G$ and $G'$, we require that $t + 1 \neq 0 \iff t \neq -1$, so the domain of $G$ and the domain of $G'$ is $(-\infty, -1) \cup (-1, \infty)$.

**29.** $g'(x) = \lim\limits_{h \to 0} \dfrac{g(x+h) - g(x)}{h} = \lim\limits_{h \to 0} \dfrac{\sqrt{1 + 2(x+h)} - \sqrt{1 + 2x}}{h}$

$= \lim\limits_{h \to 0} \left( \dfrac{\sqrt{1 + 2x + 2h} - \sqrt{1 + 2x}}{h} \cdot \dfrac{\sqrt{1 + 2x + 2h} + \sqrt{1 + 2x}}{\sqrt{1 + 2x + 2h} + \sqrt{1 + 2x}} \right)$

$= \lim\limits_{h \to 0} \dfrac{(1 + 2x + 2h) - (1 + 2x)}{h\left(\sqrt{1 + 2x + 2h} + \sqrt{1 + 2x}\right)}$

$= \lim\limits_{h \to 0} \dfrac{2h}{h\left(\sqrt{1 + 2x + 2h} + \sqrt{1 + 2x}\right)} = \lim\limits_{h \to 0} \dfrac{2}{\sqrt{1 + 2x + 2h} + \sqrt{1 + 2x}}$

$= \dfrac{2}{\sqrt{1 + 2x + 0} + \sqrt{1 + 2x}} = \dfrac{2}{2\sqrt{1 + 2x}} = \dfrac{1}{\sqrt{1 + 2x}}$

$g$ is defined only when $1 + 2x \geq 0 \iff x \geq -\frac{1}{2}$, so its domain is $\left[-\frac{1}{2}, \infty\right)$. $g'$ is defined only when $1 + 2x > 0$, so its domain is $\left(-\frac{1}{2}, \infty\right)$.

**31.**

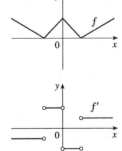

**33.** (a) $dP/dt$ is the rate at which the percentage of the city's power produced by solar panels changes with respect to time $t$, measured in percentage points per year.

(b) 2 years after January 1, 2000 (January 1, 2002), the percentage of power produced by solar panels was increasing at a rate of 3.5 percentage points per year.

**35.** $f'(x) = \lim\limits_{h\to 0} \dfrac{f(x+h) - f(x)}{h} = \lim\limits_{h\to 0} \dfrac{\left[1 + 4(x+h) - (x+h)^2\right] - (1 + 4x - x^2)}{h}$

$= \lim\limits_{h\to 0} \dfrac{\left[1 + 4(x+h) - (x^2 + 2xh + h^2)\right] - (1 + 4x - x^2)}{h}$

$= \lim\limits_{h\to 0} \dfrac{1 + 4x + 4h - x^2 - 2xh - h^2 - 1 - 4x + x^2}{h} = \lim\limits_{h\to 0} \dfrac{4h - 2xh - h^2}{h}$

$= \lim\limits_{h\to 0} \dfrac{h(4 - 2x - h)}{h} = \lim\limits_{h\to 0} (4 - 2x - h) = 4 - 2x.$

$f''$ is the derivative of $f'$, so

$f''(x) = \lim\limits_{h\to 0} \dfrac{f'(x+h) - f'(x)}{h} = \lim\limits_{h\to 0} \dfrac{[4 - 2(x+h)] - (4 - 2x)}{h} = \lim\limits_{h\to 0} \dfrac{4 - 2x - 2h - 4 + 2x}{h}$

$= \lim\limits_{h\to 0} \dfrac{-2h}{h} = \lim\limits_{h\to 0} (-2) = -2.$

Alternatively, we can note that $f'$ is a linear function with constant slope $-2$, so its derivative is $-2$ for every $x$-value. Thus $f''(x) = -2$.

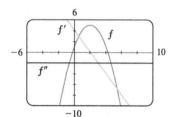

We see from the graph that our answers are reasonable because the graph of $f'$ is positive where $f$ is increasing and negative where $f$ is decreasing. The graph of $f'$ is linear with constant negative slope, so it makes sense that the graph of $f''$ is a horizontal line.

**37.** (a) Since $f'(x) > 0$ on $(1, 5)$, $f$ is increasing on this interval. Since $f'(x) < 0$ on $(0, 1)$ and $(5, 6)$, $f$ is decreasing on these intervals.

(b) Since $f'(x) = 0$ at $x = 1$ and $f'$ changes from negative to positive there, $f$ changes from decreasing to increasing and has a local minimum at $x = 1$. Since $f'(x) = 0$ at $x = 5$ and $f'$ changes from positive to negative there, $f$ changes from increasing to decreasing and has a local maximum at $x = 5$.

(c) Since $f(0) = 0$, start at the origin. Draw a curve that decreases on $(0, 1)$ with a local minimum at $x = 1$, where the curve has a horizontal tangent line. The curve increases on $(1, 5)$, and the steepest slope should occur at $x = 3$, since that's where the largest value of $f'$ occurs. At $x = 5$ the curve has a local maximum and a horizontal tangent, and it decreases on $(5, 6)$.

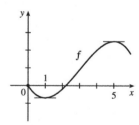

**39.** The derivative $f'$ is increasing when the slopes of the tangent lines of $f$ are becoming larger as $x$ increases. This appears to be the case on the interval $(2, 5)$. (Also, the graph is concave upward there.) The derivative is decreasing when the slopes of the tangent lines of $f$ are becoming smaller as $x$ increases, and this appears to be the case on $(-\infty, 2)$ and $(5, \infty)$. (The graph of $f$ is concave downward.) So $f'$ is increasing on $(2, 5)$ and decreasing on $(-\infty, 2)$ and $(5, \infty)$.

**41.** The function must be always decreasing (since the first derivative is always

negative) and concave downward (since the second derivative is always

negative).

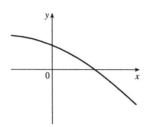

**43.** If $D(t)$ is the size of the deficit as a function of time, then at the time of the speech $D'(t) > 0$ because the deficit is increasing.

Because $D''(t)$ is the derivative of $D'(t)$, it measures the rate of change of $D'(t)$, which is decreasing, so $D''(t) < 0$.

**45.**  (a) It is helpful to first plot the points. Looking at the graph, we see that

the rate of increase of the population is initially very small, then

increases rapidly until about 1932, when it starts decreasing. The

rate becomes negative by 1936, peaks in magnitude in 1937, and

approaches 0 in 1940.

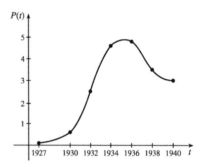

   (b) Inflection points appear to be at about $(1932, 2.5)$ and $(1937, 4.3)$. The rate of change of population density starts to

   decrease in 1932 and starts to increase in 1937. The rates of population increase and decrease have their maximum values

   at those points.

**47.** Velocity is the derivative of position: $v(t) = s'(t)$. Thus

$$v(t) = \lim_{h \to 0} \frac{s(t+h) - s(t)}{h} = \lim_{h \to 0} \frac{\left[1.7(t+h)^2 - 3.1(t+h) + 6.8\right] - (1.7t^2 - 3.1t + 6.8)}{h}$$

$$= \lim_{h \to 0} \frac{\left[1.7(t^2 + 2th + h^2) - 3.1(t+h) + 6.8\right] - (1.7t^2 - 3.1t + 6.8)}{h}$$

$$= \lim_{h \to 0} \frac{1.7t^2 + 3.4th + 1.7h^2 - 3.1t - 3.1h + 6.8 - 1.7t^2 + 3.1t - 6.8}{h}$$

$$= \lim_{h \to 0} \frac{3.4th + 1.7h^2 - 3.1h}{h} = \lim_{h \to 0} \frac{h(3.4t + 1.7h - 3.1)}{h} = \lim_{h \to 0} (3.4t + 1.7h - 3.1) = 3.4t - 3.1$$

Acceleration is the derivative of velocity; thus $a(t) = v'(t) = 3.4$ because $v$ is a linear function with constant slope 3.4. (We

can also find $v'$ using Definition 2.)

**49.** Most students learn more in the third hour of studying than in the eighth hour, so $K(3) - K(2)$ is larger than $K(8) - K(7)$.

In other words, as you begin studying for a test, the rate of knowledge gain is large and then starts to taper off, so $K'(t)$

decreases and the graph of $K$ is concave downward.

**51.** $f'(x) = \lim\limits_{h \to 0} \dfrac{f(x+h) - f(x)}{h} = \lim\limits_{h \to 0} \dfrac{[2(x+h)^2 - (x+h)^3] - (2x^2 - x^3)}{h}$

$\qquad = \lim\limits_{h \to 0} \dfrac{[2(x^2 + 2xh + h^2) - (x^3 + 3x^2h + 3xh^2 + h^3)] - (2x^2 - x^3)}{h}$

$\qquad = \lim\limits_{h \to 0} \dfrac{2x^2 + 4xh + 2h^2 - x^3 - 3x^2h - 3xh^2 - h^3 - 2x^2 + x^3}{h} = \lim\limits_{h \to 0} \dfrac{4xh + 2h^2 - 3x^2h - 3xh^2 - h^3}{h}$

$\qquad = \lim\limits_{h \to 0} \dfrac{h(4x + 2h - 3x^2 - 3xh - h^2)}{h} = \lim\limits_{h \to 0} (4x + 2h - 3x^2 - 3xh - h^2) = 4x - 3x^2$

$f''(x) = \lim\limits_{h \to 0} \dfrac{f'(x+h) - f'(x)}{h} = \lim\limits_{h \to 0} \dfrac{[4(x+h) - 3(x+h)^2] - (4x - 3x^2)}{h}$

$\qquad = \lim\limits_{h \to 0} \dfrac{[4x + 4h - 3(x^2 + 2xh + h^2)] - 4x + 3x^2}{h} = \lim\limits_{h \to 0} \dfrac{4x + 4h - 3x^2 - 6xh - 3h^2 - 4x + 3x^2}{h}$

$\qquad = \lim\limits_{h \to 0} \dfrac{4h - 6xh - 3h^2}{h} = \lim\limits_{h \to 0} \dfrac{h(4 - 6x - 3h)}{h} = \lim\limits_{h \to 0} (4 - 6x - 3h) = 4 - 6x$

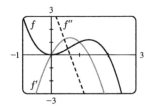

The graphs are consistent with the geometric interpretations of the derivatives because $f'$ has $x$-intercepts where $f$ has a horizontal tangent, and $f'$ is positive where $f$ is increasing and is negative where $f$ is decreasing. $f''$ has an $x$-intercept where $f'$ has a horizontal tangent, and $f''$ is positive where $f'$ is increasing and negative where $f'$ is decreasing.

**53.** As we zoom in toward $(-1, 0)$, the curve appears more and more like a straight line, so $f(x) = x + \sqrt{|x|}$ is differentiable at $x = -1$. But no matter how much we zoom in toward the origin, the curve doesn't straighten out—we can't eliminate the sharp point (a cusp). So no tangent line exists there, and $f$ is not differentiable at $x = 0$.

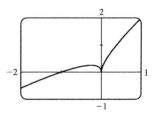

**55. (a)**

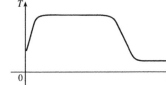

**(b)** The initial temperature of the water is close to room temperature because of the water that was in the pipes. When the water from the hot water tank starts coming out, $dT/dt$ is large and positive as $T$ increases to the temperature of the water in the tank. In the next phase, $dT/dt = 0$ as the water comes out at a constant, high temperature. After some time, $dT/dt$ becomes negative as the contents of the hot water tank are exhausted. Finally, when the hot water has run out, $dT/dt$ is once again 0 as the water maintains its (cold) temperature.

**(c)**

**57.** Notice that at the points where $c$ has a horizontal tangent, neither $a$ nor $b$ is equal to 0. Thus the derivative of curve $c$ does not appear here, so $c$ must be the graph of $f''$. Next, $c$ is 0 where $b$ has a horizontal tangent, so $c$ must be the derivative of $b$, and hence $b$ is the graph of $f'$. This leaves $a$ as the graph of $f$. This is confirmed by the fact that $b$ is 0 where $a$ has a horizontal tangent.

## 2  Review

**1.** $\dfrac{\Delta h}{\Delta x} = \dfrac{h(4) - h(1)}{4 - 1} = \dfrac{\sqrt{17} - 2\sqrt{2}}{3} \approx 0.432$

**3.** $\dfrac{\Delta R}{\Delta t} = \dfrac{R(9) - R(6)}{9 - 6} = \dfrac{179.7 - 154.2}{9 - 6} = \dfrac{25.5}{3} = 8.5$; From the end of the 6th month to the end of the 9th month after opening, the restaurant's revenue increased at an average rate of \$8500 per month.

**5.** For $f(t) = \dfrac{2^t - 8}{t - 3}$:

| $t$ | $f(t)$ | $t$ | $f(t)$ |
|------|--------|--------|--------|
| 2.9 | 5.357 | 3.1 | 5.742 |
| 2.95 | 5.450 | 3.05 | 5.642 |
| 2.99 | 5.526 | 3.01 | 5.564 |
| 2.999 | 5.543 | 3.001 | 5.547 |
| 2.9999 | 5.545 | 3.0001 | 5.545 |

It appears that $\lim\limits_{t \to 3} \dfrac{2^t - 8}{t - 3} \approx 5.545$.

**7.** $5x^2 - 4x + 5$ is a polynomial and hence continuous everywhere, so $\lim\limits_{x \to 1}(5x^2 - 4x + 5) = 5(1)^2 - 4(1) + 5 = 6$.

**9.** $-3$ is not in the domain of $\dfrac{x^2 - 9}{x^2 + 2x - 3}$, but we can evaluate the limit by direct substitution after simplifying:

$\lim\limits_{x \to -3} \dfrac{x^2 - 9}{x^2 + 2x - 3} = \lim\limits_{x \to -3} \dfrac{(x + 3)(x - 3)}{(x + 3)(x - 1)} = \lim\limits_{x \to -3} \dfrac{x - 3}{x - 1} = \dfrac{-3 - 3}{-3 - 1} = \dfrac{-6}{-4} = \dfrac{3}{2}$.

**11.** The function $4e^{-2t}$ is continuous for all real numbers, so we can evaluate the limit by direct substitution:

$\lim\limits_{t \to 0} 4e^{-2t} = 4e^{-2(0)} = 4e^0 = 4$.

**13.** $\lim\limits_{h \to 0} \dfrac{(h - 1)^3 + 1}{h} = \lim\limits_{h \to 0} \dfrac{(h^3 - 3h^2 + 3h - 1) + 1}{h} = \lim\limits_{h \to 0} \dfrac{h^3 - 3h^2 + 3h}{h}$

$= \lim\limits_{h \to 0} \dfrac{h(h^2 - 3h + 3)}{h} = \lim\limits_{h \to 0}(h^2 - 3h + 3) = (0)^2 - 3(0) + 3 = 3$

**15.** (a) (i) $\lim\limits_{x \to 2^+} f(x) = 3$

    (ii) $\lim\limits_{x \to -3^+} f(x) = 0$

(iii) $\lim\limits_{x \to -3} f(x)$ does not exist since the left and right limits are not equal. (The left limit is $-2$.)

(iv) $\lim\limits_{x \to 4} f(x) = 2$

(v) $\lim\limits_{x \to 0} f(x)$ does not exist since the function values do not approach any particular number.

(b) $f$ is not continuous at $x = -3, 0, 2$, and $4$, where we observe breaks in the graph. Note that $\lim\limits_{x \to 4} f(x)$ exists, but the function value does not equal the limit there, so we have a hole in the graph.

**17.** (a) The position function is $s(t) = 1 + 2t + t^2/4$. Note that here $s$ is an increasing function for $t \geq 0$, so the average speed is the average rate of change of position.

(i) $[1, 3]$: $\quad \dfrac{\Delta s}{\Delta t} = \dfrac{s(3) - s(1)}{3 - 1} = \dfrac{9.25 - 3.25}{3 - 1} = \dfrac{6}{2} = 3 \text{ m/s}$

(ii) $[1, 2]$: $\quad \dfrac{\Delta s}{\Delta t} = \dfrac{s(2) - s(1)}{2 - 1} = \dfrac{6 - 3.25}{2 - 1} = \dfrac{2.75}{1} = 2.75 \text{ m/s}$

(iii) $[1, 1.5]$: $\quad \dfrac{\Delta s}{\Delta t} = \dfrac{s(1.5) - s(1)}{1.5 - 1} = \dfrac{4.5625 - 3.25}{1.5 - 1} = \dfrac{1.3125}{0.5} = 2.625 \text{ m/s}$

(iv) $[1, 1.1]$: $\quad \dfrac{\Delta s}{\Delta t} = \dfrac{s(1.1) - s(1)}{1.1 - 1} \approx \dfrac{3.5025 - 3.25}{1.1 - 1} = \dfrac{0.2525}{0.1} = 2.525 \text{ m/s}$

(b) When $t = 1$, the instantaneous speed is

$$\lim_{h \to 0} \frac{s(1 + h) - s(1)}{h} = \lim_{h \to 0} \frac{\left[1 + 2(1 + h) + \frac{1}{4}(1 + h)^2\right] - 3.25}{h} = \lim_{h \to 0} \frac{\left[1 + 2 + 2h + 0.25(1 + 2h + h^2)\right] - 3.25}{h}$$

$$= \lim_{h \to 0} \frac{3 + 2h + 0.25 + 0.5h + 0.25h^2 - 3.25}{h} = \lim_{h \to 0} \frac{2.5h + 0.25h^2}{h}$$

$$= \lim_{h \to 0} \frac{h(2.5 + 0.25h)}{h} = \lim_{h \to 0} (2.5 + 0.25h) = 2.5 \text{ m/s}$$

**19.** By Definition 5 in Section 2.3,

$$g'(3) = \lim_{h \to 0} \frac{g(3 + h) - g(3)}{h} = \lim_{h \to 0} \frac{[0.5(3 + h)^2 + 4] - 8.5}{h} = \lim_{h \to 0} \frac{[0.5(9 + 6h + h^2) + 4] - 8.5}{h}$$

$$= \lim_{h \to 0} \frac{4.5 + 3h + 0.5h^2 + 4 - 8.5}{h} = \lim_{h \to 0} \frac{3h + 0.5h^2}{h} = \lim_{h \to 0} \frac{h(3 + 0.5h)}{h} = \lim_{h \to 0} (3 + 0.5h) = 3$$

**21.** Estimating the slopes of the tangent lines at $x = 2, 3$, and $5$, we obtain approximate values $0.4$, $2$, and $0.1$. Arranging the numbers in increasing order, we have: $0 < f'(5) < f'(2) < 1 < f'(3)$.

**23.** (a) $f'(r)$ is the rate at which the total cost changes with respect to the interest rate. Its units are dollars/(percent per year).

(b) The total cost of paying off the loan is increasing by $\$1200$/(percent per year) as the interest rate reaches $10\%$. So if the interest rate goes up from $10\%$ to $11\%$, the cost goes up approximately $\$1200$.

(c) As $r$ increases, $C$ increases. So $f'(r)$ will always be positive.

**25.** (a) (i) By Definition 5 in Section 2.3, $f'(1) = \lim\limits_{h \to 0} \dfrac{f(1+h) - f(1)}{h} = \lim\limits_{h \to 0} \dfrac{e^{0.5(1+h)} - e^{0.5}}{h}$. We calculate

$\dfrac{e^{0.5(1+h)} - e^{0.5}}{h}$ for increasingly smaller values of $h$, both positive and negative:

| $h$ | $\dfrac{e^{0.5(1+h)} - e^{0.5}}{h}$ | $h$ | $\dfrac{e^{0.5(1+h)} - e^{0.5}}{h}$ |
|---|---|---|---|
| 0.1 | 0.8453 | $-0.1$ | 0.8041 |
| 0.01 | 0.8264 | $-0.01$ | 0.8223 |
| 0.001 | 0.8246 | $-0.001$ | 0.8242 |
| 0.0001 | 0.8244 | $-0.0001$ | 0.8243 |

We estimate that $f'(1) \approx 0.824$.

(ii) We zoom in until the curve looks virtually like a straight line, namely the tangent line to the curve.

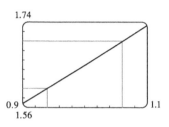

From the graph, it appears that the line passes through the points $(0.94, 1.6)$ and $(1.06, 1.7)$, so we estimate that the

slope of the tangent is about $\dfrac{1.7 - 1.6}{1.06 - 0.94} = \dfrac{0.1}{0.12} \approx 0.8$. Thus $f'(1) \approx 0.8$.

(b) From part (a)(i) the slope of the tangent line is $f'(1) \approx 0.824$, and $f(1) = e^{0.5} = \sqrt{e} \approx 1.649$, so an equation of the

tangent line is approximately $y - 1.649 = 0.824(x - 1) \quad \Leftrightarrow \quad y = 0.824x + 0.825$.

**27.** $C'(2000) = \lim\limits_{h \to 0} \dfrac{C(2000 + h) - C(2000)}{h} \approx \dfrac{C(2000 + h) - C(2000)}{h}$ for small values of $h$. For $h = 5$, we get

$C'(2000) \approx \dfrac{C(2005) - C(2000)}{5} = \dfrac{758.8 - 568.6}{5} = \dfrac{190.2}{5} = 38.04$ and for $h = -5$, we have

$C'(2000) \approx \dfrac{C(1995) - C(2000)}{-5} = \dfrac{409.3 - 568.6}{-5} = \dfrac{-159.3}{-5} = 31.86$. (These are average rates of change on the

intervals $[2000, 2005]$ and $[1995, 2000]$.) Averaging these two results, we estimate $C'(2000) \approx \frac{1}{2}(38.04 + 31.86) = 34.95$.

$C'(t)$ is the rate at which the total value of US currency in circulation is changing, in billions of dollars per year, so

$C'(2000) \approx 34.95$ means that in 2000 the total value of US currency in circulation was increasing at a rate of

$\$34.95$ billion/year.

**29.** (a) We estimate that the points $(1997, 19.8)$ and $(2001, 16.2)$ lie on the graph, so the average rate of change for

$1997 \le t \le 2001$ is $\dfrac{\Delta P}{\Delta t} = \dfrac{16.2 - 19.8}{2001 - 1997} = \dfrac{-3.6}{4} = -0.9$ percentage points per year. (Note that this is the slope of the

secant line through the two points.)

(b) $P'(2002)$ is the slope of the tangent line to the graph at $t = 2002$. We draw an approximate tangent line that appears to pass through the points $(2002, 16.7)$ and $(2004, 18.4)$, so its slope is $\frac{18.4 - 16.7}{2004 - 2002} = 0.85$. Thus we estimate that $P'(2002) \approx 0.85$, which means that in 2002 the percentage of children living below the poverty level was increasing at a rate of approximately 0.85 percentage points per year.

**31.** $f$ is not differentiable: at $x = -4$ because $f$ is not continuous, at $x = -1$ because $f$ has a corner, at $x = 2$ because $f$ is not continuous, and at $x = 5$ because $f$ has a vertical tangent.

**33.**

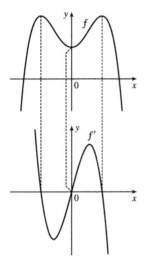

**35.** By Definition 2 in Section 2.4,

$$f'(x) = \lim_{h \to 0} \frac{f(x+h) - f(x)}{h} = \lim_{h \to 0} \frac{[m(x+h) + b] - (mx + b)}{h} = \lim_{h \to 0} \frac{mx + mh + b - mx - b}{h}$$

$$= \lim_{h \to 0} \frac{mh}{h} = \lim_{h \to 0} m = m.$$

**37.** $g'(x) = \lim_{h \to 0} \dfrac{g(x+h) - g(x)}{h} = \lim_{h \to 0} \dfrac{\left[2(x+h)^2 - 3(x+h) + 1\right] - \left(2x^2 - 3x + 1\right)}{h}$

$= \lim_{h \to 0} \dfrac{\left[2(x^2 + 2xh + h^2) - 3(x+h) + 1\right] - \left(2x^2 - 3x + 1\right)}{h}$

$= \lim_{h \to 0} \dfrac{2x^2 + 4xh + 2h^2 - 3x - 3h + 1 - 2x^2 + 3x - 1}{h} = \lim_{h \to 0} \dfrac{4xh + 2h^2 - 3h}{h}$

$= \lim_{h \to 0} \dfrac{h(4x + 2h - 3)}{h} = \lim_{h \to 0} (4x + 2h - 3) = 4x - 3$

**39.** $A'(w) = \lim\limits_{h \to 0} \dfrac{A(w+h) - A(w)}{h} = \lim\limits_{h \to 0} \dfrac{\dfrac{(w+h)+1}{2(w+h)-1} - \dfrac{w+1}{2w-1}}{h} = \lim\limits_{h \to 0} \dfrac{\dfrac{w+h+1}{2w+2h-1} - \dfrac{w+1}{2w-1}}{h}$

$= \lim\limits_{h \to 0} \left[ \dfrac{\dfrac{w+h+1}{2w+2h-1} - \dfrac{w+1}{2w-1}}{h} \cdot \dfrac{(2w+2h-1)(2w-1)}{(2w+2h-1)(2w-1)} \right]$

$= \lim\limits_{h \to 0} \dfrac{(w+h+1)(2w-1) - (w+1)(2w+2h-1)}{h(2w+2h-1)(2w-1)}$

$= \lim\limits_{h \to 0} \dfrac{\left(2w^2 - w + 2hw - h + 2w - 1\right) - \left(2w^2 + 2hw - w + 2w + 2h - 1\right)}{h(2w+2h-1)(2w-1)} = \lim\limits_{h \to 0} \dfrac{-3h}{h(2w+2h-1)(2w-1)}$

$= \lim\limits_{h \to 0} \dfrac{-3}{(2w+2h-1)(2w-1)} = \dfrac{-3}{(2w+0-1)(2w-1)} = -\dfrac{3}{(2w-1)^2}$

**41.** From Exercise 37, $g'(x) = 4x - 3$. $g''$ is the derivative of $g'$, and since $g'$ is a linear function with constant slope 4, $g''(x) = 4$. (We can also establish this result using the definition of derivative.)

**43.** (a) Using the data closest to $t = 6$, we compute average speeds over the intervals $[4, 6]$ and $[6, 8]$.

$[4, 6]: \quad \dfrac{s(6) - s(4)}{6 - 4} = \dfrac{95 - 40}{2} = 27.5 \qquad [6, 8] : \quad \dfrac{s(8) - s(6)}{8 - 6} = \dfrac{180 - 95}{2} = 42.5$

Averaging these two values gives us $\dfrac{27.5 + 42.5}{2} = 35$ ft/s as an estimate for the speed of the car after 6 seconds.

(b) We draw a scatter plot of the data and sketch a smooth curve through the points. It appears that the inflection point is at approximately $(8, 180)$.

(c) The velocity of the car is at a maximum at the inflection point. It is where the graph changes from concave upward (velocity is increasing) to concave downward (velocity is decreasing).

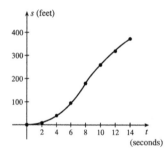

**45.** (a) $f'(x) > 0$ on $(-2, 0)$ and $(2, \infty)$, so $f$ is increasing on those intervals. $f'(x) < 0$ on $(-\infty, -2)$ and $(0, 2)$, so $f$ is decreasing on those intervals.

(b) $f'(x) = 0$ at $x = -2$ and $x = 2$, and $f'$ changes from negative to positive there, so $f$ has a local minimum at each of these values. At $x = 0$, $f'(x) = 0$ and $f'$ changes from positive to negative, so $f$ has a local maximum there.

(c) $f'$ is increasing on $(-\infty, -1)$ and $(1, \infty)$, so $f'' > 0$ and $f$ is concave upward on those intervals.

$f'$ is decreasing on $(-1, 1)$, so $f'' < 0$ and $f$ is concave downward on this interval.

# 3 TECHNIQUES OF DIFFERENTIATION

## 3.1 Shortcuts to Finding Derivatives

---

**Prepare Yourself**

1. (a) $f(x) = \sqrt[3]{x} = x^{1/3}$ 　　　　　　　　(b) $g(w) = 1/w = w^{-1}$

　　(c) $A(t) = 4/\sqrt{t} = 4/t^{1/2} = 4t^{-1/2}$ 　　(d) $B(v) = 8/v^3 = 8v^{-3}$

　　(e) $y = \sqrt[4]{x^3} = x^{3/4}$

2. The tangent line has slope $-2$ and passes through the point $(3, 8)$, so an equation is $y - 8 = -2(x - 3)$ $\Leftrightarrow$ $y - 8 = -2x + 6$ $\Leftrightarrow$ $y = -2x + 14$.

**Exercises**

1. $f(x) = 186.5$ is a constant function, so its derivative is 0, that is, $f'(x) = 0$.

3. $y = x^9$ $\Rightarrow$ $y' = 9x^8$ by the Power Rule.

5. $g(t) = t^{7/2}$ $\Rightarrow$ $g'(t) = \frac{7}{2}t^{(7/2)-1} = \frac{7}{2}t^{5/2}$ by the Power Rule.

7. $y = x^{-2/5}$ $\Rightarrow$ $dy/dx = -\frac{2}{5}x^{(-2/5)-1} = -\frac{2}{5}x^{-7/5}$

9. First write $L(t) = \sqrt[4]{t}$ as a power function: $L(t) = t^{1/4}$. Then $L'(t) = \frac{1}{4}t^{(1/4)-1} = \frac{1}{4}t^{-3/4}$.

11. $y = 8x^6$, and using the Constant Multiple Rule and the Power Rule, we have $\dfrac{dy}{dx} = 8\dfrac{d}{dx}(x^6) = 8(6x^5) = 48x^5$.

13. First write $f(x) = 7/x^2$ as a power function: $f(x) = 7x^{-2}$. Then, using the Constant Multiple Rule and the Power Rule,
$f'(x) = 7(-2x^{-2-1}) = -14x^{-3} = -14/x^3$.

15. $y = 5x - 3$, and using the Difference Rule, $\dfrac{dy}{dx} = \dfrac{d}{dx}(5x) - \dfrac{d}{dx}(3) = 5\dfrac{d}{dx}(x) - \dfrac{d}{dx}(3) = 5(1) - 0 = 5$.

17. $f(x) = x^3 - 4x + 6$ $\Rightarrow$
$$f'(x) = \frac{d}{dx}(x^3) - \frac{d}{dx}(4x) + \frac{d}{dx}(6) = \frac{d}{dx}(x^3) - 4\frac{d}{dx}(x) + \frac{d}{dx}(6)$$
$$= 3x^2 - 4(1) + 0 = 3x^2 - 4$$

19. $y = 0.7x^4 - 1.8x^3 + 5.1x$ $\Rightarrow$
$$\frac{dy}{dx} = \frac{d}{dx}(0.7x^4) - \frac{d}{dx}(1.8x^3) + \frac{d}{dx}(5.1x) = 0.7\frac{d}{dx}(x^4) - 1.8\frac{d}{dx}(x^3) + 5.1\frac{d}{dx}(x)$$
$$= 0.7(4x^3) - 1.8(3x^2) + 5.1(1) = 2.8x^3 - 5.4x^2 + 5.1$$

21. $q = e^r + 3.4$ $\Rightarrow$ $\dfrac{dq}{dr} = \dfrac{d}{dr}(e^r) + \dfrac{d}{dr}(3.4) = e^r + 0 = e^r$

**23.** $G(x) = x^3 - 4e^x \Rightarrow G'(x) = \frac{d}{dx}(x^3) - \frac{d}{dx}(4e^x) = \frac{d}{dx}(x^3) - 4\frac{d}{dx}(e^x) = 3x^2 - 4e^x$

**25.** $f(t) = \frac{1}{4}(t^4 + 8) \Rightarrow$

$$f'(t) = \frac{1}{4}\frac{d}{dt}(t^4 + 8) = \frac{1}{4}\left[\frac{d}{dt}(t^4) + \frac{d}{dt}(8)\right] = \frac{1}{4}(4t^3 + 0) = t^3$$

Or: $f(t) = \frac{1}{4}(t^4 + 8) = \frac{1}{4}t^4 + 2 \Rightarrow f'(t) = \frac{1}{4}(4t^3) + 0 = t^3$

**27.** $f(q) = \dfrac{6}{q} - \dfrac{3}{q^2} = 6q^{-1} - 3q^{-2} \Rightarrow f'(q) = 6(-1q^{-1-1}) - 3(-2q^{-2-1}) = -6q^{-2} + 6q^{-3} = -\dfrac{6}{q^2} + \dfrac{6}{q^3}$

**29.** $y = x\left(\sqrt{x} + 1/\sqrt{x}\right) = x\sqrt{x} + x/\sqrt{x} = x \cdot x^{1/2} + x/x^{1/2} = x^{3/2} + x^{1/2} \Rightarrow$

$$\frac{dy}{dx} = \frac{3}{2}x^{(3/2)-1} + \frac{1}{2}x^{(1/2)-1} = \frac{3}{2}x^{1/2} + \frac{1}{2}x^{-1/2} = \frac{3}{2}\sqrt{x} + \frac{1}{2\sqrt{x}}$$

**31.** $F(x) = \left(\frac{1}{2}x\right)^5 = \left(\frac{1}{2}\right)^5 x^5 = \frac{1}{32}x^5 \Rightarrow F'(x) = \frac{1}{32}(5x^4) = \frac{5}{32}x^4$

**33.** $y = \dfrac{7x^2 - 3x + 5}{x} = \dfrac{7x^2}{x} - \dfrac{3x}{x} + \dfrac{5}{x} = 7x - 3 + 5x^{-1} \Rightarrow$

$$\frac{dy}{dx} = 7(1) - 0 + 5(-1x^{-1-1}) = 7 - 5x^{-2} = 7 - \frac{5}{x^2}$$

**35.** $f(y) = \dfrac{A}{y^{10}} + Be^y = Ay^{-10} + Be^y \Rightarrow$

$$f'(y) = A(-10y^{-10-1}) + B(e^y) = -10Ay^{-11} + Be^y = -\frac{10A}{y^{11}} + Be^y$$

**37.** $v = t^2 - \dfrac{1}{\sqrt[4]{t^3}} = t^2 - t^{-3/4} \Rightarrow \dfrac{dv}{dt} = 2t - \left[-\frac{3}{4}t^{(-3/4-1)}\right] = 2t + \frac{3}{4}t^{-7/4}$

**39.** $y = 2x^2 + 3/x = 2x^2 + 3x^{-1} \Rightarrow dy/dx = 2(2x) + 3(-1x^{-2}) = 4x - 3/x^2$, and the slope of the curve at $(3, 19)$ is

$$\left.\frac{dy}{dx}\right|_{x=3} = 4(3) - \frac{3}{3^2} = 12 - \frac{1}{3} = \frac{35}{3}$$

**41.** $y = x^4 + 2e^x \Rightarrow dy/dx = 4x^3 + 2e^x$. At $(0, 2)$, $dy/dx = 4(0)^3 + 2e^0 = 2$ and an equation of the tangent line is

$y - 2 = 2(x - 0)$ or $y = 2x + 2$.

**43.** $y = x + \sqrt{x} = x + x^{1/2} \Rightarrow dy/dx = 1 + \frac{1}{2}x^{-1/2} = 1 + 1/(2\sqrt{x})$.

Then the slope of the tangent line at $(1, 2)$ is

$dy/dx = 1 + 1/(2\sqrt{1}) = 1 + \frac{1}{2} = \frac{3}{2}$, and its equation is

$y - 2 = \frac{3}{2}(x - 1)$ or $y = \frac{3}{2}x + \frac{1}{2}$.

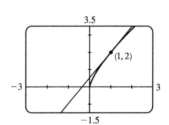

**45.** $p(t) = 0.12t^2 - 1.6t + 59.5 \Rightarrow p'(t) = 0.12(2t) - 1.6(1) + 0 = 0.24t - 1.6$, and the instantaneous rate of change after

the 14th week is $p'(14) = 0.24(14) - 1.6 = 1.76$ (dollars per barrel)/week.

**47.** (a) A graphing calculator gives a cubic model as approximately $s(t) = -2.597t^3 + 69.88t^2 + 0.7817t + 1.778$, where $t$ is the time in seconds and $s(t)$ is the distance in feet.

(b) The velocity of the car is $v(t) = s'(t) = -2.597(3t^2) + 69.88(2t) + 0.7817(1) + 0 = -7.791t^2 + 139.76t + 0.7817$ ft/s. After 4 seconds, $v(4) = -7.791(4)^2 + 139.76(4) + 0.7817 \approx 435.2$, so the speed was about $435.2$ ft/s.

**49.** We zoom in on the graph of $f$ at $x = 1$ until the curve looks like a straight line. The graph below uses the viewing rectangle $[0.9, 1.1]$ by $[f(0.9) = 1.701, f(1.1) = 2.299]$. (If the graph still looked curved, we could try graphing for $0.99 \le x \le 1.01$, for instance.) Using the points $(0.9, f(0.9))$ and $(1.1, f(1.1))$ on the curve, we estimate that

$$f'(1) \approx \frac{f(1.1) - f(0.9)}{1.1 - 0.9} = \frac{2.299 - 1.701}{1.1 - 0.9} = \frac{0.589}{0.2} = 2.99.$$

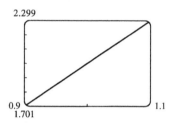

Now $f(x) = 3x^2 - x^3 \ \Rightarrow \ f'(x) = 3(2x) - (3x^2) = 6x - 3x^2$, so the exact value is $f'(1) = 6(1) - 3(1)^2 = 3$.

**51.** $f(x) = e^x - 5x \ \Rightarrow \ f'(x) = e^x - 5$.

Notice that $f'(x) = 0$ when $f$ has a horizontal tangent, $f'$ is positive when $f$ is increasing, and $f'$ is negative when $f$ is decreasing.

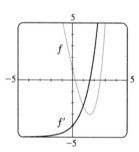

**53.** $f(x) = x - 3x^{1/3} \ \Rightarrow$

$f'(x) = 1 - 3\left[\frac{1}{3}x^{(1/3)-1}\right] = 1 - x^{-2/3} = 1 - 1/x^{2/3}$.

Note that $f'(x) = 0$ when $f$ has a horizontal tangent, $f'$ is positive when $f$ is increasing, and $f'$ is negative when $f$ is decreasing.

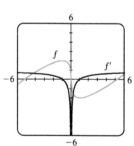

**55.** (a)

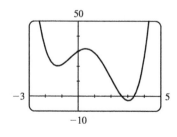

(b) From the graph in part (a), it appears that $f'$ is zero at $x \approx -1.25$, $x \approx 0.5$,

and $x \approx 3$. The slopes are negative (so $f'$ is negative) on approximately

$(-\infty, -1.25)$ and $(0.5, 3)$. The slopes are positive (so $f'$ is positive) on

$(-1.25, 0.5)$ and $(3, \infty)$.

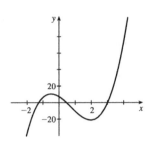

(c) $f(x) = x^4 - 3x^3 - 6x^2 + 7x + 30 \implies$

$f'(x) = 4x^3 - 3(3x^2) - 6(2x) + 7(1) + 0$

$\quad = 4x^3 - 9x^2 - 12x + 7$

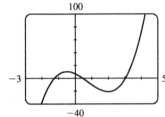

**57.** $f(x) = x^4 - 3x^3 + 16x \implies f'(x) = 4x^3 - 3(3x^2) + 16(1) = 4x^3 - 9x^2 + 16$,

$f''(x) = 4(3x^2) - 9(2x) + 0 = 12x^2 - 18x$

**59.** $f(x) = 0.2x^4 - 1.1x^3 + 0.6x^2 + 2.2x \implies$

$f'(x) = 0.2(4x^3) - 1.1(3x^2) + 0.6(2x) + 2.2(1)$

$\quad = 0.8x^3 - 3.3x^2 + 1.2x + 2.2$,

$f''(x) = 0.8(3x^2) - 3.3(2x) + 1.2(1) + 0 = 2.4x^2 - 6.6x + 1.2$

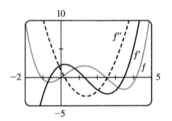

Note that $f'(x) = 0$ when $f$ has a horizontal tangent, and $f''(x) = 0$ when $f'$ has a horizontal tangent.

**61.** (a) The velocity is $v = ds/dt = 4.2(3t^2) + 3.4(2t) - 6(1) + 0 = 12.6t^2 + 6.8t - 6$, and after 4 seconds,

$v = 12.6(4)^2 + 6.8(4) - 6 = 222.8$ ft/s.

(b) The acceleration is given by $a = dv/dt = 12.6(2t) + 6.8(1) - 0 = 25.2t + 6.8$, and after 6 seconds,

$a = 25.2(6) + 6.8 = 158$ ft/s$^2$.

**63.** (a) $s = t^3 - 3t \implies v(t) = ds/dt = 3t^2 - 3$, $a(t) = v'(t) = 3(2t) - 0 = 6t$

(b) $a(2) = 6(2) = 12$ m/s$^2$

(c) If we consider only $t \geq 0$, then the velocity is 0 when $v(t) = 3t^2 - 3 = 0 \implies t^2 = 1 \implies t = 1$, and

$a(1) = 6(1) = 6$ m/s$^2$.

**65.** $y = 2x^3 + 3x^2 - 12x + 1 \implies y' = 2(3x^2) + 3(2x) - 12(1) + 0 = 6x^2 + 6x - 12$, and the tangent line is horizontal

when $y' = 0 \iff 6x^2 + 6x - 12 = 0 \iff 6(x^2 + x - 2) = 0 \iff 6(x + 2)(x - 1) = 0 \iff x = -2$ or $x = 1$.

The points on the curve are $(-2, 21)$ and $(1, -6)$.

**67.** $y = 6x^3 + 5x - 3 \implies y' = 6(3x^2) + 5(1) - 0 = 18x^2 + 5$. Note that $x^2 \geq 0$ for all values of $x$, so $18x^2 \geq 0$ and

$18x^2 + 5 \geq 5$. Thus $y' \geq 5$ for all values of $x$, and since $y'$ is the slope of the tangent line, we can't have a tangent line with

slope 4.

**69.** $y = ax^2$, so the slope of the tangent line is $y' = a(2x) = 2ax$, and when $x = 2$, the slope is $y' = 4a$. The given line

$2x + y = b \iff y = -2x + b$ has slope $-2$, so we must have $4a = -2 \Rightarrow a = -\frac{1}{2}$. Then the parabola is $y = -\frac{1}{2}x^2$,

and when $x = 2$, $y = -\frac{1}{2} \cdot 2^2 = -2$, so the given line $2x + y = b$ must pass through the point $(2, -2)$:

$2(2) + (-2) = b \Rightarrow b = 2$. Thus $a = -\frac{1}{2}$ and $b = 2$.

## 3.2 Introduction to Marginal Analysis

---

**Prepare Yourself**

**1.** (a) $f(x) = 120 + 2.6x + 0.02x^2 \Rightarrow f'(x) = 0 + 2.6(1) + 0.02(2x) = 2.6 + 0.04x$

(b) $f'(50) = 2.6 + 0.04(50) = 4.6$

**2.** $A(q) = 0.001q^3 + 0.05q^2 + 20q + 350 \Rightarrow A'(q) = 0.001(3q^2) + 0.05(2q) + 20(1) + 0 = 0.003q^2 + 0.1q + 20$, and

$A'(400) = 0.003(400)^2 + 0.1(400) + 20 = 540$.

**3.** Draw a tangent line to the graph at $x = 60$.

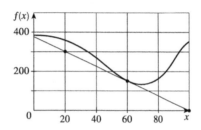

It appears that the tangent line passes through the points $(20, 300)$ and $(100, 0)$, so its slope is $m = \dfrac{0 - 300}{100 - 20} = -3.75$.

Thus we estimate that $f'(60) \approx -3.75$.

**4.** To solve $\dfrac{250 + 5x + 0.002x^2}{x} = 0.004x + 5$, first multiply both sides of the equation by $x$, giving

$250 + 5x + 0.002x^2 = 0.004x^2 + 5x$. Then $250 = 0.002x^2 \Rightarrow x^2 = \dfrac{250}{0.002} = 125{,}000 \Rightarrow$

$x = \sqrt{125{,}000} \approx 353.55$. ($x \approx -353.55$ is also a solution to the equation, but here $x > 0$.)

**5.** (a) $18e^{-0.5x} = 8 \Rightarrow e^{-0.5x} = \frac{8}{18} = \frac{4}{9} \Rightarrow \ln\frac{4}{9} = -0.5x \Rightarrow x = \frac{1}{-0.5}\ln\frac{4}{9} = -2\ln\frac{4}{9} \approx 1.6$

(b) $200x^{-1.4} = 75 \Rightarrow x^{-1.4} = \frac{75}{200} = \frac{3}{8} \Rightarrow (x^{-1.4})^{1/(-1.4)} = (\frac{3}{8})^{1/(-1.4)} \Rightarrow x = (\frac{3}{8})^{-1/1.4} \approx 2.0$

---

**Exercises**

**1.** If $q$ is the number of units, then the cost is $C(q) = 2000 + 15q$ dollars.

**3.** $C(q) = 1800 + 0.12q + 0.003q^2$

(a) The average cost is

$$\frac{C(q)}{q} = \frac{1800 + 0.12q + 0.003q^2}{q} = \frac{1800}{q} + \frac{0.12q}{q} + \frac{0.003q^2}{q} = \frac{1800}{q} + 0.12 + 0.003q \text{ dollars/unit}$$

(b) The marginal cost is $C'(q) = 0 + 0.12(1) + 0.003(2q) = 0.12 + 0.006q$ dollars/unit.

(c) When 500 bars have been produced, the average cost is $\dfrac{C(500)}{500} = \dfrac{1800}{500} + 0.12 + 0.003(500) = \$5.22/\text{bar}$ and the

marginal cost is $C'(500) = 0.12 + 0.006(500) = \$3.12/\text{bar}$. The actual cost of the 501st bar is

$C(501) - C(500) = 2613.123 - 2610 = \$3.123$, very close to the marginal cost at this production level.

**5.** (a) $C(x) = 2000 + 3x + 0.01x^2 + 0.0002x^3 \;\Rightarrow\; C'(x) = 0 + 3(1) + 0.01(2x) + 0.0002(3x^2) = 3 + 0.02x + 0.0006x^2$

(b) $C'(100) = 3 + 0.02(100) + 0.0006(100)^2 = 3 + 2 + 6 = \$11/\text{pair}$. $C'(100)$ is the rate at which the cost is increasing as

the 100th pair of jeans is produced. It predicts the (approximate) cost of the 101st pair.

(c) The cost of manufacturing the 101st pair of jeans is

$C(101) - C(100) = 2611.0702 - 2600 = 11.0702 \approx \$11.07$. This is close to the marginal cost from part (b).

**7.** (a) $\dfrac{C(60)}{60} = \dfrac{8.8}{60} \approx 0.14667$ million dollars per thousand units, or $146.67 per unit.

(b) $C'(40) \approx \dfrac{C(40 + h) - C(40)}{h}$ for small values of $h$. For $h = 10$, we get

$C'(40) \approx \dfrac{C(50) - C(40)}{10} = \dfrac{7.3 - 6.3}{10} = \dfrac{1}{10} = 0.1$, and for $h = -10$, we have

$C'(40) \approx \dfrac{C(30) - C(40)}{-10} = \dfrac{5.6 - 6.3}{-10} = \dfrac{-0.7}{-10} = 0.07$. Averaging these two results, we estimate that

$C'(40) \approx \frac{1}{2}(0.1 + 0.07) = 0.085$. This means that after 40,000 units have been produced, the cost is increasing at a rate

of about 0.085 million dollars per thousand units, or $85/unit.

(c) The cost of producing the 40,001st unit is approximately the marginal cost after producing 40,000 units. From part (b), this

is $85.

**9.** (a) Since we know the values of $f(40)$ and $f'(40)$, we use Equation 1 with $a = 40$ and $\Delta x = 2$:

$f(40 + 2) \approx f(40) + f'(40)(2) \;\Rightarrow\; f(42) \approx 378 + 6(2) = 390.$

(b) Using Equation 1 with $a = 40$ and $\Delta x = -1.5$, we have

$f(38.5) = f(40 + (-1.5)) \approx f(40) + f'(40)(-1.5) = 378 + 6(-1.5) = 369.$

**11.** After 10 tons of paper have been produced, the cost is increasing at the rate $C'(10) = 350$ thousand dollars per ton.

There are 2000 pounds in a ton, so an additional 500 lb of paper is 0.25 ton and will cost approximately

$(\$350,000/\text{ton}) \times 0.25\text{ ton} = \$87,500.$

**13.** (a) The marginal cost $C'$ is the slope of the graph of $C$. The curve is steeper at $q = 100$ than at $q = 200$, so the marginal cost

is higher when 100 units are produced.

(b) We draw a tangent line to the graph at $q = 600$.

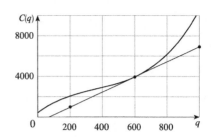

It appears that the tangent line passes through the points $(200, 1000)$ and $(1000, 7000)$, so its slope is

$$m = \frac{7000 - 1000}{1000 - 200} = \frac{6000}{800} = 7.5. \text{ Thus we estimate that } C'(600) \approx 7.5.$$

(c) Marginal cost is minimized where the slope of the graph of $C$ is smallest (which occurs at an inflection point). Looking at the graph, the curve is the least steep at about $q = 400$.

**15.** The average cost is $\dfrac{C(q)}{q} = \dfrac{0.01q^2 + 2q + 250}{q}$ and the marginal cost is $C'(q) = 0.01(2q) + 2 + 0 = 0.02q + 2$. Average

cost is minimized when the average cost and marginal costs are equal:

$$\frac{0.01q^2 + 2q + 250}{q} = 0.02q + 2$$

$$0.01q^2 + 2q + 250 = 0.02q^2 + 2q$$

$$250 = 0.01q^2$$

$$25,000 = q^2$$

$$q = \sqrt{25,000} \approx 158$$

Thus 158 loaves should be baked in order to minimize average cost.

**17.** We graph the average cost function $\dfrac{C(q)}{q} = \dfrac{6200 + 23q - 0.02q^2 + 0.0001q^3}{q}$ and the marginal cost function

$C'(q) = 23 - 0.04q + 0.0003q^2$.

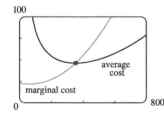

Average cost is minimized where the curves intersect, at $q \approx 351$, so we estimate that 351 pairs of shoes should be produced each month.

**19.** (a) The profit function is

$$P(q) = R(q) - C(q) = \left(16q - 0.002q^2\right) - \left(2500 + 4q + 0.005q^2\right)$$

$$= 16q - 0.002q^2 - 2500 - 4q - 0.005q^2 = -0.007q^2 + 12q - 2500$$

(b) Marginal cost: $C'(q) = 4 + 0.01q$; Marginal revenue: $R'(q) = 16 - 0.004q$

(c) Note that initially $R'(q) > C'(q)$, and profit is maximized when marginal revenue and marginal cost are equal:

$R'(q) = C'(q) \Rightarrow 16 - 0.004q = 4 + 0.01q \Rightarrow 12 = 0.014q \Rightarrow q = \frac{12}{0.014} \approx 857$. Thus to maximize profit,

the manufacturer should produce 857 power supplies.

**21.** (a) 4500 units corresponds to $q = 4.5$, and according to the demand function, the price the company should charge for each

unit is $p = 32e^{-0.6(4.5)} = 32e^{-2.7} \approx \$2.15$.

(b) Substitute $p = 7.50$ into the demand equation and solve for $q$: $7.50 = 32e^{-0.6q} \Rightarrow \frac{7.5}{32} = e^{-0.6q} \Rightarrow$

$-0.6q = \ln \frac{7.5}{32} \Rightarrow q = \frac{1}{-0.6} \ln \frac{7.5}{32} \approx 2.418$. Thus 2418 units should sell ($q$ is measured in thousands).

**23.** $C(q) = 680 + 4q + 0.01q^2$ and the revenue function is $R(q) = q \cdot p = q(12 - q/500) = 12q - \frac{1}{500}q^2 = 12q - 0.002q^2$.

Profit is maximized when $R'(q) = C'(q) \Rightarrow 12 - 0.004q = 4 + 0.02q \Rightarrow 8 = 0.024q \Rightarrow q = \frac{8}{0.024} \approx 333$ units.

[Note that initially $R'(q) > C'(q)$.]

**25.** (a) Revenue = number of bracelets × price per bracelet, and profit = revenue − cost. We compute these amounts in the table

below.

| Number of bracelets | Price per bracelet | Revenue | Cost | Profit |
|---|---|---|---|---|
| 100 | $8000 | $800,000 | $215,000 | $585,000 |
| 200 | $7500 | $1,500,000 | $420,000 | $1,080,000 |
| 300 | $6000 | $1,800,000 | $625,000 | $1,175,000 |
| 400 | $5000 | $2,000,000 | $820,000 | $1,180,000 |
| 500 | $4200 | $2,100,000 | $1,015,000 | $1,085,000 |
| 600 | $3600 | $2,160,000 | $1,205,000 | $955,000 |

Of the production levels listed in the table, making 400 bracelets generates the highest profit.

(b) The marginal cost is the instantaneous rate of change of cost $C$ when the quantity $q$ produced is 400, and we can estimate

its value by averaging the average rates of change for the intervals $[300, 400]$ and $[400, 500]$.

$$[300, 400]: \quad \frac{C(400) - C(300)}{400 - 300} = \frac{820,000 - 625,000}{100} = 1950$$

$$[400, 500]: \quad \frac{C(500) - C(400)}{500 - 400} = \frac{1,015,000 - 820,000}{100} = 1950$$

The values are identical and so their average is also 1950. Thus we estimate the marginal cost to be $1950/bracelet.

We estimate the marginal revenue similarly.

$$[300, 400]: \quad \frac{R(400) - R(300)}{400 - 300} = \frac{2,000,000 - 1,800,000}{100} = 2000$$

$$[400, 500]: \quad \frac{R(500) - R(400)}{500 - 400} = \frac{2,100,000 - 2,000,000}{100} = 1000$$

Thus we estimate that the marginal revenue is approximately $\frac{1}{2}[2000 + 1000] = 1500$ dollars/bracelet.

(c) No. Because marginal revenue < marginal cost, increasing production would decrease profit.

**27.** (a) From the graph, we estimate that $R(800) = 14$ and $C(800) = 11.2$, so we estimate the profit to be

$R(800) - C(800) = 14 - 11.2 = 2.8$ thousand dollars or \$2800.

(b) We draw tangent lines to the revenue and cost curves at $q = 1400$.

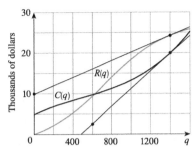

The tangent line to the graph of $R$ appears to pass through the points $(0, 10)$ and $(1400, 24.5)$ with slope $\frac{24.5-10}{1400-0} \approx 0.01$,

and the tangent line to the graph of $C$ appears to pass through $(600, 5)$ and $(1400, 20)$ with slope $\frac{20-5}{1400-600} \approx 0.02$, so we

estimate the marginal cost to be $C'(1400) \approx 0.02$ thousand dollars/unit and the marginal revenue to be $R'(1400) \approx 0.01$

thousand dollars/unit. Since $R'(1400) < C'(1400)$, production should not be increased, as this would decrease profit.

(c) Profit is maximized when $R'(q) = C'(q)$ [with $R'(q) > C'(q)$ for smaller values of $q$].

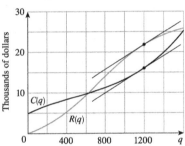

Looking at the graph, tangent lines to the curves are parallel (and

hence have the same slope) at approximately $q = 1200$ units.

(Note also that the vertical distance between the curves, representing

the difference between revenue and profit, is greatest at this value.)

**29.** The marginal revenue is $R'(q) = 65$. Then $R'(q) = C'(q) \iff 65 = 0.16q + 37 \iff q = \frac{28}{0.16} = 175$. Initially we have

$R'(q) > C'(q)$, so the maximum profit occurs when 175 cars are waxed.

## 3.3   The Product and Quotient Rules

**Prepare Yourself**

**1.** (a) $f(x) = 5x^3 + 3x \Rightarrow f'(x) = 5(3x^2) + 3(1) = 15x^2 + 3$

(b) $g(x) = 1/x^2 = x^{-2} \Rightarrow g'(x) = -2x^{-2-1} = -2x^{-3} = -2/x^3$

(c) $r(x) = \sqrt{x} = x^{1/2} \Rightarrow r'(x) = \frac{1}{2}x^{(1/2)-1} = \frac{1}{2}x^{-1/2} = \frac{1}{2\sqrt{x}}$

(d) $U(t) = e^t \Rightarrow U'(t) = e^t$

**2.** (a) $5t^{-2} = \dfrac{5}{t^2}$ \qquad (b) $\frac{1}{2}x^{-1/2} = \dfrac{1}{2x^{1/2}} = \dfrac{1}{2\sqrt{x}}$ \qquad (c) $4x^{1/3} = 4\sqrt[3]{x}$

**3.** (a) $s(t) = 0.2t^3 + 14t + 3$, so the velocity function is $v(t) = s'(t) = 0.6t^2 + 14$. Then $v(2) = 0.6(2)^2 + 14 = 16.4$, so the speed after 2 minutes is 16.4 ft/min.

(b) The acceleration function is $a(t) = v'(t) = 1.2t$, and the acceleration after 2 minutes is $a(2) = 1.2(2) = 2.4$ ft/min$^2$.

---

**Exercises**

**1.** Product Rule: $y = (x^2 + 1)(x^3 + 1) \Rightarrow$

$$\frac{dy}{dx} = (x^2 + 1)\frac{d}{dx}(x^3 + 1) + (x^3 + 1)\frac{d}{dx}(x^2 + 1)$$

$$= (x^2 + 1)(3x^2) + (x^3 + 1)(2x) = 3x^4 + 3x^2 + 2x^4 + 2x = 5x^4 + 3x^2 + 2x$$

Multiplying first: $y = (x^2 + 1)(x^3 + 1) = x^5 + x^3 + x^2 + 1 \Rightarrow dy/dx = 5x^4 + 3x^2 + 2x$ (equivalent).

For this problem, multiplying first seems to be the more convenient method.

**3.** By the Product Rule, $f(x) = x^2 e^x \Rightarrow f'(x) = x^2 \frac{d}{dx}(e^x) + e^x \frac{d}{dx}(x^2) = x^2 e^x + e^x(2x) = (x^2 + 2x)e^x$.

**5.** By the Quotient Rule, $y = \dfrac{e^x}{x^2} \Rightarrow$

$$\frac{dy}{dx} = \frac{x^2 \dfrac{d}{dx}(e^x) - e^x \dfrac{d}{dx}(x^2)}{(x^2)^2} = \frac{x^2(e^x) - e^x(2x)}{x^4} = \frac{(x^2 - 2x)e^x}{x^4} = \frac{x(x - 2)e^x}{x^4} = \frac{(x - 2)e^x}{x^3}.$$

The notations $\overset{PR}{\Rightarrow}$ and $\overset{QR}{\Rightarrow}$ indicate the use of the Product and Quotient Rules, respectively.

**7.** $g(x) = \dfrac{3x - 1}{2x + 1} \overset{QR}{\Rightarrow}$

$$g'(x) = \frac{(2x + 1)\dfrac{d}{dx}(3x - 1) - (3x - 1)\dfrac{d}{dx}(2x + 1)}{(2x + 1)^2} = \frac{(2x + 1)(3) - (3x - 1)(2)}{(2x + 1)^2}$$

$$= \frac{6x + 3 - 6x + 2}{(2x + 1)^2} = \frac{5}{(2x + 1)^2}$$

**9.** $F(y) = \left(\dfrac{1}{y^2} - \dfrac{3}{y^4}\right)(y + 5y^3) = (y^{-2} - 3y^{-4})(y + 5y^3) \overset{PR}{\Rightarrow}$

$$F'(y) = (y^{-2} - 3y^{-4})\frac{d}{dx}(y + 5y^3) + (y + 5y^3)\frac{d}{dx}(y^{-2} - 3y^{-4})$$

$$= (y^{-2} - 3y^{-4})(1 + 15y^2) + (y + 5y^3)(-2y^{-3} + 12y^{-5})$$

$$= (y^{-2} + 15 - 3y^{-4} - 45y^{-2}) + (-2y^{-2} + 12y^{-4} - 10 + 60y^{-2})$$

$$= 5 + 14y^{-2} + 9y^{-4} \text{ or } 5 + 14/y^2 + 9/y^4$$

Alternatively, we can multiply and simplify before taking the derivative:

$$F(y) = (y^{-2} - 3y^{-4})(y + 5y^3) = y^{-1} + 5y - 3y^{-3} - 15y^{-1} = 5y - 14y^{-1} - 3y^{-3} \Rightarrow F'(y) = 5 + 14y^{-2} + 9y^{-4}.$$

**11.** $y = \dfrac{t^2}{3t^2 - 2t + 1}$ $\overset{\text{QR}}{\Rightarrow}$

$$\frac{dy}{dt} = \frac{(3t^2 - 2t + 1)(2t) - t^2(6t - 2)}{(3t^2 - 2t + 1)^2} = \frac{6t^3 - 4t^2 + 2t - 6t^3 + 2t^2}{(3t^2 - 2t + 1)^2}$$

$$= \frac{2t - 2t^2}{(3t^2 - 2t + 1)^2} = \frac{2t(1 - t)}{(3t^2 - 2t + 1)^2}$$

**13.** $y = (r^2 - 2r)e^r$ $\overset{\text{PR}}{\Rightarrow}$ $y' = (r^2 - 2r)(e^r) + e^r(2r - 2) = e^r(r^2 - 2r + 2r - 2) = (r^2 - 2)e^r$

**15.** $P = \dfrac{5e^t}{2 + 3t^2}$ $\overset{\text{QR}}{\Rightarrow}$ $\dfrac{dP}{dt} = \dfrac{(2 + 3t^2)(5e^t) - 5e^t(6t)}{(2 + 3t^2)^2} = \dfrac{5e^t(2 + 3t^2 - 6t)}{(2 + 3t^2)^2} = \dfrac{5(3t^2 - 6t + 2)e^t}{(2 + 3t^2)^2}$

**17.** Rather than using the Quotient Rule, we first simplify the function:

$$y = \frac{v^3 - 2v\sqrt{v}}{v} = \frac{v^3}{v} - \frac{2v\sqrt{v}}{v} = v^2 - 2\sqrt{v} = v^2 - 2v^{1/2}. \text{ Then } y' = 2v - 2\left(\tfrac{1}{2}v^{-1/2}\right) = 2v - v^{-1/2} = 2v - \frac{1}{\sqrt{v}}.$$

**19.** $f(x) = \dfrac{A}{B + Ce^x}$ $\overset{\text{QR}}{\Rightarrow}$ $f'(x) = \dfrac{(B + Ce^x) \cdot 0 - A(Ce^x)}{(B + Ce^x)^2} = -\dfrac{ACe^x}{(B + Ce^x)^2}$

**21.** $y = \dfrac{x}{x^2 - 1}$ $\overset{\text{QR}}{\Rightarrow}$ $\dfrac{dy}{dx} = \dfrac{(x^2 - 1)(1) - x(2x)}{(x^2 - 1)^2} = \dfrac{-x^2 - 1}{(x^2 - 1)^2}.$ The slope of the curve at $x = 2$ is

$$\left.\frac{dy}{dx}\right|_{x=2} = \frac{-2^2 - 1}{(2^2 - 1)^2} = \frac{-4 - 1}{3^2} = -\frac{5}{9} \approx -0.556.$$

**23.** $y = 2xe^x$ $\overset{\text{PR}}{\Rightarrow}$ $y' = 2x(e^x) + e^x(2) = 2e^x(x + 1).$ At $(0, 0)$, $y' = 2e^0(0 + 1) = 2 \cdot 1 \cdot 1 = 2$, and an equation of the tangent line is $y - 0 = 2(x - 0)$, or $y = 2x$.

**25. (a)** $A(x) = \dfrac{p(x)}{x}$ $\Rightarrow$ $A'(x) = \dfrac{x \cdot p'(x) - p(x) \cdot 1}{x^2} = \dfrac{xp'(x) - p(x)}{x^2}.$

If $A'(x) > 0$, then $A(x)$ is increasing; that is, the average productivity per worker increases as the size of the workforce increases.

**(b)** If $p'(x)$ is greater than the average productivity, then $p'(x) > A(x)$ $\Rightarrow$ $p'(x) > \dfrac{p(x)}{x}$ $\Rightarrow$ $xp'(x) > p(x)$

(since $x > 0$) $\Rightarrow$ $xp'(x) - p(x) > 0$ $\Rightarrow$ $\dfrac{xp'(x) - p(x)}{x^2} > 0$ $\Rightarrow$ $A'(x) > 0$.

**27.** If the position of the object after $t$ seconds is $s(t) = \dfrac{3t}{2 + t}$ ft, then the velocity is

$$v(t) = s'(t) = \frac{(2 + t)(3) - 3t(1)}{(2 + t)^2} = \frac{6}{(2 + t)^2} = \frac{6}{t^2 + 4t + 4} \text{ ft/s and the acceleration is}$$

$$a(t) = v'(t) = \frac{(t^2 + 4t + 4)(0) - 6(2t + 4)}{(t^2 + 4t + 4)^2} = \frac{-12t - 24}{(t^2 + 4t + 4)^2} \text{ ft/s}^2. \text{ After 4 seconds, the acceleration is}$$

$$a(4) = \frac{-12(4) - 24}{[(4)^2 + 4(4) + 4]^2} = \frac{-72}{1296} = -\frac{1}{18} \approx -0.0556 \text{ ft/s}^2.$$

**29.** (a) $y = f(x) = \dfrac{1}{1+x^2}$ $\Rightarrow$ $f'(x) = \dfrac{(1+x^2)(0) - 1(2x)}{(1+x^2)^2} = \dfrac{-2x}{(1+x^2)^2}$.

(b)

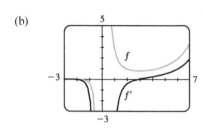

So the slope of the tangent line at the point $\left(-1, \frac{1}{2}\right)$ is $f'(-1) = \dfrac{2}{2^2} = \dfrac{1}{2}$

and its equation is $y - \frac{1}{2} = \frac{1}{2}(x+1)$ or $y = \frac{1}{2}x + 1$.

**31.** (a) $f(x) = \dfrac{e^x}{x^3}$ $\Rightarrow$ $f'(x) = \dfrac{x^3(e^x) - e^x(3x^2)}{(x^3)^2} = \dfrac{x^2e^x(x-3)}{x^6} = \dfrac{e^x(x-3)}{x^4}$

(b)

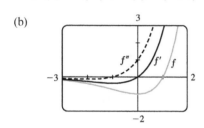

$f' = 0$ when $f$ has a horizontal tangent line, $f'$ is negative when $f$ is decreasing, and $f'$ is positive when $f$ is increasing.

**33.** (a) $f(x) = (x-1)e^x$ $\Rightarrow$ $f'(x) = (x-1)e^x + e^x(1) = e^x(x-1+1) = xe^x$,

$f''(x) = x(e^x) + e^x(1) = (x+1)e^x$.

(b)

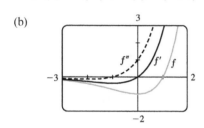

$f' = 0$ when $f$ has a horizontal tangent and $f'' = 0$ when $f'$ has a horizontal tangent. $f'$ is negative when $f$ is decreasing and positive when $f$ is increasing. $f''$ is negative when $f'$ is decreasing and positive when $f'$ is increasing. $f''$ is negative when $f$ is concave down and positive when $f$ is concave up.

**35.** $f(x) = \dfrac{x^2}{1+x}$ $\Rightarrow$ $f'(x) = \dfrac{(1+x)(2x) - x^2(1)}{(1+x)^2} = \dfrac{2x + 2x^2 - x^2}{(1+x)^2} = \dfrac{x^2 + 2x}{x^2 + 2x + 1}$ $\Rightarrow$

$f''(x) = \dfrac{(x^2 + 2x + 1)(2x + 2) - (x^2 + 2x)(2x + 2)}{(x^2 + 2x + 1)^2} = \dfrac{(2x+2)(x^2 + 2x + 1 - x^2 - 2x)}{[(x+1)^2]^2}$

$= \dfrac{2(x+1)(1)}{(x+1)^4} = \dfrac{2}{(x+1)^3}$,

so $f''(1) = \dfrac{2}{(1+1)^3} = \dfrac{2}{8} = \dfrac{1}{4}$.

**37.** We are given that $f(5) = 1$, $f'(5) = 6$, $g(5) = -3$, and $g'(5) = 2$.

(a) $A(x) = f(x)\,g(x)$ $\Rightarrow$ $A'(x) = f(x)\,g'(x) + g(x)\,f'(x)$, so

$A'(5) = f(5)\,g'(5) + g(5)\,f'(5) = (1)(2) + (-3)(6) = 2 - 18 = -16$.

(b) $B(x) = \dfrac{f(x)}{g(x)}$ $\Rightarrow$ $B'(x) = \dfrac{g(x)\,f'(x) - f(x)\,g'(x)}{[g(x)]^2}$, so

$B'(5) = \dfrac{g(5)\,f'(5) - f(5)\,g'(5)}{[g(5)]^2} = \dfrac{(-3)(6) - (1)(2)}{(-3)^2} = -\dfrac{20}{9}$.

(c) $C(x) = \dfrac{g(x)}{f(x)}$ $\Rightarrow$ $C'(x) = \dfrac{f(x)\,g'(x) - g(x)\,f'(x)}{[f(x)]^2}$, so

$$C'(5) = \frac{f(5)\,g'(5) - g(5)\,f'(5)}{[f(5)]^2} = \frac{(1)(2) - (-3)(6)}{(1)^2} = 20.$$

**39.** $f(x) = e^x g(x)$ $\Rightarrow$ $f'(x) = e^x g'(x) + g(x)\,e^x = e^x[g'(x) + g(x)]$, so $f'(0) = e^0[g'(0) + g(0)] = 1(5 + 2) = 7$.

**41.** (a) From the graphs of $f$ and $g$, we obtain the following values: $f(1) = 2$ since the point $(1, 2)$ is on the graph of $f$;

$g(1) = 1$ since the point $(1, 1)$ is on the graph of $g$; $f'(1) = 2$ since the slope of the line segment between $(0, 0)$ and

$(2, 4)$ is $\dfrac{4 - 0}{2 - 0} = 2$; $g'(1) = -1$ since the slope of the line segment between $(-2, 4)$ and $(2, 0)$ is $\dfrac{0 - 4}{2 - (-2)} = -1$.

Now $u(x) = f(x)\,g(x)$, so $u'(1) = f(1)\,g'(1) + g(1)\,f'(1) = 2 \cdot (-1) + 1 \cdot 2 = 0$.

(b) From the graphs of $f$ and $g$, we have $f(5) = 3$, $g(5) = 2$, $f'(5) = -\frac{1}{3}$, and $g'(5) = \frac{2}{3}$.

$$v(x) = \frac{f(x)}{g(x)}, \text{ so } v'(5) = \frac{g(5)\,f'(5) - f(5)\,g'(5)}{[g(5)]^2} = \frac{2\left(-\frac{1}{3}\right) - 3\left(\frac{2}{3}\right)}{2^2} = \frac{-\frac{8}{3}}{4} = -\frac{2}{3}.$$

**43.** If $P(t)$ denotes the population at time $t$ and $A(t)$ the average annual income, then $T(t) = P(t)A(t)$ is the total personal

income. The rate at which $T(t)$ is rising is given by $T'(t) = P(t)A'(t) + A(t)P'(t)$ $\Rightarrow$

$$T'(1999) = P(1999)A'(1999) + A(1999)P'(1999) = (961{,}400)(\$1400/\text{yr}) + (\$30{,}593)(9200/\text{yr})$$

$$= \$1{,}345{,}960{,}000/\text{yr} + \$281{,}455{,}600/\text{yr} = \$1{,}627{,}415{,}600/\text{yr}$$

So the total personal income was rising by about $1.627 billion per year in 1999.

The term $P(t)A'(t) \approx \$1.346$ billion represents the portion of the rate of change of total income due to the existing

population's increasing income. The term $A(t)P'(t) \approx \$281$ million represents the portion of the rate of change of total

income due to increasing population.

**45.** (a) $y = xg(x)$ $\Rightarrow$ $y' = xg'(x) + g(x) \cdot 1 = xg'(x) + g(x)$

(b) $y = \dfrac{x}{g(x)}$ $\Rightarrow$ $y' = \dfrac{g(x) \cdot 1 - x \cdot g'(x)}{[g(x)]^2} = \dfrac{g(x) - xg'(x)}{[g(x)]^2}$

(c) $y = \dfrac{g(x)}{x}$ $\Rightarrow$ $y' = \dfrac{x \cdot g'(x) - g(x) \cdot 1}{(x)^2} = \dfrac{xg'(x) - g(x)}{x^2}$

**47.** $g(x) = xe^x$ $\Rightarrow$ $g'(x) = x(e^x) + e^x(1) = (x + 1)e^x$, $g''(x) = (x + 1)e^x + e^x(1) = (x + 2)e^x$, and the third derivative

is $g'''(x) = (x + 2)e^x + e^x(1) = (x + 3)e^x$. The pattern suggests that the $n$th derivative is $g^{(n)}(x) = (x + n)e^x$.

## 3.4   The Chain Rule

**Prepare Yourself**

**1.** $f(g(x)) = \sqrt{5 + 4x} = \sqrt{g(x)}$, so we must have $f(x) = \sqrt{x}$.

**2.** $f(g(x)) = e^{-3x} = e^{g(x)}$, so we must have $f(x) = e^x$.

**3.** (a) $\dfrac{d}{dx}(4\sqrt{x}) = \dfrac{d}{dx}(4x^{1/2}) = 4 \cdot \frac{1}{2}x^{-1/2} = 2/\sqrt{x}$

(b) $\dfrac{d}{dx}(5x^6) = 5 \cdot 6x^5 = 30x^5$

(c) $\dfrac{d}{dt}(2/\sqrt[3]{t}) = \dfrac{d}{dt}(2t^{-1/3}) = 2\left(-\frac{1}{3}t^{(-1/3)-1}\right) = -\frac{2}{3}t^{-4/3}$ or $-\dfrac{2}{3t^{4/3}}$

(d) $\dfrac{d}{dt}(-8e^t) = -8e^t$

(e) $\dfrac{d}{dx}(x^2e^x) = x^2 \cdot e^x + e^x \cdot 2x = (x^2 + 2x)e^x$ by the Product Rule.

(f) $\dfrac{d}{dw}\left(\dfrac{w}{w^2+1}\right) = \dfrac{(w^2+1)\cdot 1 - w \cdot 2w}{(w^2+1)^2} = \dfrac{1-w^2}{(w^2+1)^2}$ by the Quotient Rule.

**4.** $y = 6\sqrt{x} + x^2 = 6x^{1/2} + x^2 \;\Rightarrow\; y' = 6\left(\frac{1}{2}x^{-1/2}\right) + 2x = \dfrac{3}{\sqrt{x}} + 2x$, and at the point $(4, 28)$, the slope of the graph is

$y' = \dfrac{3}{\sqrt{4}} + 2(4) = \dfrac{19}{2}$.

---

**Exercises**

**1.** Let $u = g(x) = x^2 + 4$ and $y = f(u) = \sqrt{u} = u^{1/2}$. Then $f'(u) = \frac{1}{2}u^{-1/2} = \dfrac{1}{2\sqrt{u}}$, and by Equation 1,

$\dfrac{dy}{dx} = f'(g(x)) \cdot g'(x) = \dfrac{1}{2\sqrt{g(x)}} \cdot 2x = \dfrac{1}{2\sqrt{x^2+4}} \cdot 2x = \dfrac{x}{\sqrt{x^2+4}}$.

Or, using Equation 2, we have $\dfrac{dy}{dx} = \dfrac{dy}{du}\dfrac{du}{dx} = \dfrac{1}{2\sqrt{u}} \cdot 2x = \dfrac{1}{2\sqrt{x^2+4}} \cdot 2x = \dfrac{x}{\sqrt{x^2+4}}$.

**3.** Let $u = g(x) = 1 - x^2$ and $y = f(u) = u^{10}$. Then $\dfrac{dy}{dx} = \dfrac{dy}{du}\dfrac{du}{dx} = (10u^9)(-2x) = 10(1-x^2)^9(-2x) = -20x(1-x^2)^9$.

**5.** Let $u = g(x) = \sqrt{x} = x^{1/2}$ and $y = f(u) = e^u$. Then $\dfrac{dy}{dx} = \dfrac{dy}{du}\dfrac{du}{dx} = (e^u)\left(\frac{1}{2}x^{-1/2}\right) = e^{\sqrt{x}} \cdot \dfrac{1}{2\sqrt{x}} = \dfrac{e^{\sqrt{x}}}{2\sqrt{x}}$.

**7.** $f(x) = \sqrt{9 - x^2} = (9 - x^2)^{1/2} \;\Rightarrow$

$f'(x) = \frac{1}{2}(9 - x^2)^{-1/2} \cdot \dfrac{d}{dx}(9 - x^2) = \dfrac{1}{2(9-x^2)^{1/2}} \cdot (0 - 2x) = -\dfrac{x}{\sqrt{9-x^2}}$

**9.** $F(x) = \sqrt[4]{1 + 2x + x^3} = (1 + 2x + x^3)^{1/4} \;\Rightarrow$

$F'(x) = \frac{1}{4}(1 + 2x + x^3)^{-3/4} \cdot \dfrac{d}{dx}(1 + 2x + x^3) = \dfrac{1}{4(1+2x+x^3)^{3/4}} \cdot (2 + 3x^2) = \dfrac{2 + 3x^2}{4(1+2x+x^3)^{3/4}}$

**11.** $y = (2x^4 - 8x^2)^7 \;\Rightarrow\; \dfrac{dy}{dx} = 7(2x^4 - 8x^2)^6 \cdot \dfrac{d}{dx}(2x^4 - 8x^2) = 7(2x^4 - 8x^2)^6(8x^3 - 16x)$

**13.** $g(t) = \dfrac{1}{(t^4+1)^3} = (t^4 + 1)^{-3} \;\Rightarrow\; g'(t) = -3(t^4 + 1)^{-4}(4t^3) = -12t^3(t^4 + 1)^{-4} = -\dfrac{12t^3}{(t^4+1)^4}$

**15.** $A(x) = 5.3e^{0.8x} \;\Rightarrow\; A'(x) = 5.3e^{0.8x} \cdot \dfrac{d}{dx}(0.8x) = 5.3e^{0.8x}(0.8) = 4.24e^{0.8x}$   [See Equation 6]

**17.** $y = xe^{-x^2}$ and, first using the Product Rule, we have

$$y' = x \cdot \frac{d}{dx}(e^{-x^2}) + e^{-x^2} \cdot \frac{d}{dx}(x) = x\left[e^{-x^2}(-2x)\right] + e^{-x^2}(1) = -2x^2 e^{-x^2} + e^{-x^2} = (1 - 2x^2)e^{-x^2}.$$

**19.** $P(t) = 6^t + 8 \;\Rightarrow\; P'(t) = 6^t \ln 6 + 0 = 6^t \ln 6$

**21.** $A = 4500(1.124^t) \;\Rightarrow\; A' = 4500\left(1.124^t \ln 1.124\right) = 4500(\ln 1.124)1.124^t \approx 526.02(1.124^t)$

**23.** $g(x) = (1 + 4x)^5(3 + x - x^2)^8 \;\Rightarrow\;$

$$\begin{aligned}
g'(x) &= (1 + 4x)^5 \cdot \frac{d}{dx}(3 + x - x^2)^8 + (3 + x - x^2)^8 \cdot \frac{d}{dx}(1 + 4x)^5 \qquad \text{[Product Rule]}\\
&= (1 + 4x)^5 \cdot 8(3 + x - x^2)^7(1 - 2x) + (3 + x - x^2)^8 \cdot 5(1 + 4x)^4 \cdot 4\\
&= 4(1 + 4x)^4(3 + x - x^2)^7[2(1 + 4x)(1 - 2x) + 5(3 + x - x^2)]\\
&= 4(1 + 4x)^4(3 + x - x^2)^7[2(1 + 2x - 8x^2) + 5(3 + x - x^2)]\\
&= 4(1 + 4x)^4(3 + x - x^2)^7[2 + 4x - 16x^2 + 15 + 5x - 5x^2] = 4(1 + 4x)^4(3 + x - x^2)^7(17 + 9x - 21x^2)
\end{aligned}$$

**25.** Using Formula 7 and the Chain Rule, $P(t) = 4^{2+t/3} \;\Rightarrow\;$

$$P'(t) = 4^{2+t/3}(\ln 4) \cdot \frac{d}{dt}(2 + t/3) = 4^{2+t/3}(\ln 4)(0 + \tfrac{1}{3}) = \tfrac{1}{3}(\ln 4)\left(4^{2+t/3}\right) \text{ or equivalently}$$

$$\tfrac{1}{3}(\ln 4)(4^2 4^{t/3}) = \tfrac{16}{3}(\ln 4)(4^{t/3}).$$

**27.** $L(t) = e^{3 \cdot 2^t} \;\Rightarrow\; L'(t) = e^{3 \cdot 2^t} \cdot \dfrac{d}{dt}(3 \cdot 2^t) = e^{3 \cdot 2^t}(3 \cdot 2^t \ln 2) = 3(\ln 2)(2^t)e^{3 \cdot 2^t}$

**29.** $F(z) = \sqrt{\dfrac{z-1}{z+1}} = \left(\dfrac{z-1}{z+1}\right)^{1/2} \;\Rightarrow\;$

$$F'(z) = \frac{1}{2}\left(\frac{z-1}{z+1}\right)^{-1/2} \cdot \frac{d}{dz}\left(\frac{z-1}{z+1}\right) = \frac{1}{2}\left(\frac{z+1}{z-1}\right)^{1/2} \cdot \frac{(z+1)(1) - (z-1)(1)}{(z+1)^2}$$

$$= \frac{1}{2}\frac{(z+1)^{1/2}}{(z-1)^{1/2}} \cdot \frac{z+1-z+1}{(z+1)^2} = \frac{1}{2}\frac{(z+1)^{1/2}}{(z-1)^{1/2}} \cdot \frac{2}{(z+1)^2} = \frac{1}{(z-1)^{1/2}(z+1)^{3/2}}$$

**31.** $y = \dfrac{r}{\sqrt{r^2+1}} = \dfrac{r}{(r^2+1)^{1/2}} \;\Rightarrow\;$

$$y' = \frac{(r^2+1)^{1/2} \cdot 1 - r \cdot \frac{1}{2}(r^2+1)^{-1/2}(2r)}{[(r^2+1)^{1/2}]^2} = \frac{(r^2+1)^{1/2} - \dfrac{r^2}{(r^2+1)^{1/2}}}{r^2+1} = \frac{(r^2+1)^{1/2} \cdot \dfrac{(r^2+1)^{1/2}}{(r^2+1)^{1/2}} - \dfrac{r^2}{(r^2+1)^{1/2}}}{r^2+1}$$

$$= \frac{(r^2+1) - r^2}{(r^2+1)^{1/2}(r^2+1)} = \frac{1}{(r^2+1)^{3/2}} \text{ or } (r^2+1)^{-3/2}$$

*Another solution:* Write the function as a product and use the Product Rule: $y = r(r^2+1)^{-1/2} \;\Rightarrow\;$

$y' = r \cdot \left(-\frac{1}{2}\right)(r^2+1)^{-3/2}(2r) + (r^2+1)^{-1/2} \cdot 1 = -r^2(r^2+1)^{-3/2} + (r^2+1)^{-1/2}$. To simplify the expression, we can

factor out $(r^2+1)^{-3/2}$, since both terms have a power of $(r^2+1)$ and $-\frac{3}{2}$ is the *smallest* exponent that appears on the factor.

Thus $y' = (r^2+1)^{-3/2}[-r^2 + (r^2+1)^1] = (r^2+1)^{-3/2}(1) = (r^2+1)^{-3/2}$.

**33.** $y = \dfrac{10}{1 + 2e^{-0.3t}} \;\Rightarrow\; y' = \dfrac{(1 + 2e^{-0.3t})(0) - 10(2e^{-0.3t}(-0.3))}{(1 + 2e^{-0.3t})^2} = \dfrac{6e^{-0.3t}}{(1 + 2e^{-0.3t})^2}$

**35.** $Q(x) = \sqrt{e^{3x} + x} = (e^{3x} + x)^{1/2} \;\Rightarrow\; Q'(x) = \frac{1}{2}(e^{3x} + x)^{-1/2}(e^{3x}(3) + 1) = \dfrac{3e^{3x} + 1}{2\sqrt{e^{3x} + x}}$

**37.** $y = e^{\sqrt[3]{x^2+2}} = e^{(x^2+2)^{1/3}} \;\Rightarrow\; y' = e^{(x^2+2)^{1/3}} \cdot \frac{1}{3}(x^2 + 2)^{-2/3}(2x) = \frac{2}{3}x(x^2 + 2)^{-2/3}e^{\sqrt[3]{x^2+2}}$

**39.** (a) $y = 4^x \;\Rightarrow\; y' = 4^x \ln 4$. It appears that the graph of $y'$ is a vertical

stretch of the graph of $y$. In fact, $y' = 4^x \ln 4 \approx 1.386(4^x) = 1.386y$.

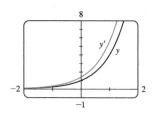

(b) $y = \left(\frac{1}{4}\right)^x \;\Rightarrow\; y' = \left(\frac{1}{4}\right)^x \ln\frac{1}{4}$. The relationship between the graphs is

not the same as in part (a), as the graph of $y'$ is both a vertical stretch of the

graph of $y$ and a reflection about the $x$-axis. In fact,

$y' = \left(\frac{1}{4}\right)^x \ln\frac{1}{4} \approx -1.386 \left(\frac{1}{4}\right)^x = -1.386y.$

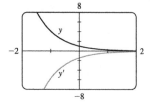

**41.** $y = e^{-0.5x} \;\Rightarrow\; y' = e^{-0.5x}(-0.5) = -0.5e^{-0.5x} \;\Rightarrow\; y'' = -0.5 \left(e^{-0.5x}(-0.5)\right) = 0.25e^{-0.5x}$

**43.** Chain Rule: $f(x) = (3x^5 + 1)^2 \;\Rightarrow\; f'(x) = 2(3x^5 + 1)^1(15x^4) = 30x^4(3x^5 + 1)$.

Expanding first: $f(x) = (3x^5 + 1)^2 = (3x^5 + 1)(3x^5 + 1) = 9x^{10} + 6x^5 + 1 \;\Rightarrow\; f'(x) = 90x^9 + 30x^4$.

If we expand the result after using the Chain Rule, we get $30x^4(3x^5 + 1) = 90x^9 + 30x^4$, so the answers are equivalent.

**45.** $f(x) = 3x - 3^x \;\Rightarrow\; f'(x) = 3 - 3^x \ln 3$, so the slope of the graph at $x = 2$ is $f'(2) = 3 - 3^2 \ln 3 = 3 - 9 \ln 3 \approx -6.888$.

**47.** $y = (1 + 2x)^{10} \;\Rightarrow\; y' = 10(1 + 2x)^9 \cdot 2 = 20(1 + 2x)^9$. At $(0, 1)$, the slope is $y' = 20(1 + 0)^9 = 20$, and an equation

of the tangent line is $y - 1 = 20(x - 0)$, or $y = 20x + 1$.

**49.** (a) $y = \dfrac{2}{1 + e^{-x}} \;\Rightarrow\; y' = \dfrac{(1 + e^{-x}) \cdot 0 - 2 \cdot e^{-x}(-1)}{(1 + e^{-x})^2} = \dfrac{2e^{-x}}{(1 + e^{-x})^2}$. At $(0, 1)$,

$y' = \dfrac{2e^0}{(1 + e^0)^2} = \dfrac{2(1)}{(1 + 1)^2} = \dfrac{2}{2^2} = \dfrac{1}{2}$. So an equation of the tangent line is $y - 1 = \frac{1}{2}(x - 0)$    or    $y = \frac{1}{2}x + 1$.

(b)

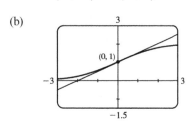

**51.** (a) $f(x) = x\sqrt{2 - x^2} = x(2 - x^2)^{1/2} \;\Rightarrow\;$

$f'(x) = x \cdot \frac{1}{2}(2 - x^2)^{-1/2}(-2x) + (2 - x^2)^{1/2} \cdot 1 = \dfrac{-x^2}{\sqrt{2 - x^2}} + \sqrt{2 - x^2}$

$= \dfrac{-x^2}{\sqrt{2 - x^2}} + \sqrt{2 - x^2} \cdot \dfrac{\sqrt{2 - x^2}}{\sqrt{2 - x^2}} = \dfrac{-x^2 + (2 - x^2)}{\sqrt{2 - x^2}} = \dfrac{2 - 2x^2}{\sqrt{2 - x^2}}$

(b)

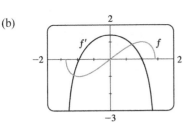

$f' = 0$ when $f$ has a horizontal tangent line, $f'$ is negative when $f$ is decreasing, and $f'$ is positive when $f$ is increasing.

**53.** $C(t) = 0.6t(0.98^t)$ $\Rightarrow$ $C'(t) = 0.6\left[t \cdot (0.98^t)\ln 0.98 + (0.98^t) \cdot 1\right] = 0.6(0.98^t)(1 + t\ln 0.98)$. Two hours is equivalent to 120 minutes, and $C'(120) = 0.6(0.98^{120})(1 + 120\ln 9.8) \approx -0.0757$. Thus two hours after the dose was taken, the drug concentration was decreasing at a rate of about 0.0757 $(\mu g/mL)/min$.

**55.** $A(t) = 26{,}800e^{0.07t}$ $\Rightarrow$ $A'(t) = 26{,}800e^{0.07t}(0.07) = 1876e^{0.07t}$ and
$A'(3.5) = 1876e^{0.07(3.5)} = 1876e^{0.245} \approx 2396.82$. This means that after 3.5 years, the value of the account is increasing at a rate of $2396.82 per year.

**57.** $F(t) = 75 + 105(0.62^t)$ $\Rightarrow$ $F'(t) = 105(0.62^t)\ln 0.62$. The instantaneous rate of change of the coffee's temperature after an hour and a half is $F'(1.5) = 105(0.62^{1.5})\ln 0.62 \approx -24.5\,°F/h$.

**59.** (a) A graphing calculator gives an exponential model as approximately $Q = f(t) = 100.01(0.00004515)^t$.

(b) $f'(t) = 100.01(0.00004515)^t \ln 0.00004515$, so $f'(0.04) = 100.01(0.00004515)^{0.04}\ln 0.00004515 \approx -670.6\ \mu A$.

**61.** (a) $F(x) = f(e^x)$ $\Rightarrow$ $F'(x) = f'(e^x)\dfrac{d}{dx}\left(e^x\right) = e^x f'(e^x)$

(b) $G(x) = e^{f(x)}$ $\Rightarrow$ $G'(x) = e^{f(x)}\dfrac{d}{dx}f(x) = e^{f(x)}f'(x)$

**63.** (a) $u(x) = f(g(x))$ $\Rightarrow$ $u'(x) = f'(g(x))g'(x)$. So $u'(1) = f'(g(1))g'(1) = f'(3)g'(1)$. To find $f'(3)$, note that $f$ is linear from $(2,4)$ to $(6,3)$, so its slope is $\dfrac{3-4}{6-2} = -\dfrac{1}{4}$. To find $g'(1)$, note that $g$ is linear from $(0,6)$ to $(2,0)$, so its slope is $\dfrac{0-6}{2-0} = -3$. Thus, $f'(3)g'(1) = \left(-\tfrac{1}{4}\right)(-3) = \tfrac{3}{4}$.

(b) $v(x) = g(f(x))$ $\Rightarrow$ $v'(x) = g'(f(x))f'(x)$. So $v'(1) = g'(f(1))f'(1) = g'(2)f'(1)$, which does not exist since $g'(2)$ does not exist. (The graph has a corner there.)

(c) $w(x) = g(g(x))$ $\Rightarrow$ $w'(x) = g'(g(x))g'(x)$. So $w'(1) = g'(g(1))g'(1) = g'(3)g'(1)$. To find $g'(3)$, note that $g$ is linear from $(2,0)$ to $(5,2)$, so its slope is $\dfrac{2-0}{5-2} = \dfrac{2}{3}$. Thus, $g'(3)g'(1) = \left(\tfrac{2}{3}\right)(-3) = -2$.

**65.** $y = e^{2x}$ $\Rightarrow$ $\dfrac{dy}{dx} = e^{2x}(2) = 2e^{2x}$, $\dfrac{d^2y}{dx^2} = 2 \cdot e^{2x}(2) = 4e^{2x}$ or $2^2e^{2x}$, $\dfrac{d^3y}{dx^3} = 2^2 \cdot e^{2x}(2) = 2^3e^{2x}$, $\dfrac{d^4y}{dx^4} = 2^4e^{2x}$, $\ldots$,

$\dfrac{d^{30}y}{dx^{30}} = 2^{30}e^{2x} = 1{,}073{,}741{,}824e^{2x}$.

## 3.5 Implicit Differentiation and Logarithms

---

**Prepare Yourself**

**1.** (a) $\dfrac{d}{dx}\,[3x + f(x)] = \dfrac{d}{dx}(3x) + \dfrac{d}{dx}f(x) = 3 + f'(x)$

(b) $\dfrac{d}{dx}\,[x\,f(x)] = x \cdot \dfrac{d}{dx}f(x) + f(x) \cdot \dfrac{d}{dx}(x) = x\,f'(x) + f(x) \cdot 1 = x\,f'(x) + f(x)$

(c) $\dfrac{d}{dx}[f(x)]^3 = 3[f(x)]^2 \cdot \dfrac{d}{dx}f(x) = 3[f(x)]^2 \cdot f'(x)$

(d) $\dfrac{d}{dx}e^{f(x)} = e^{f(x)} \cdot \dfrac{d}{dx}f(x) = e^{f(x)} \cdot f'(x)$

**2.** (a) $\dfrac{d}{dx}(x + y^4) = \dfrac{d}{dx}(x) + \dfrac{d}{dx}(y^4) = 1 + 4y^3\dfrac{dy}{dx}$

(b) $\dfrac{d}{dx}(\sqrt{x} + \sqrt{y}) = \dfrac{d}{dx}(x^{1/2}) + \dfrac{d}{dx}(y^{1/2}) = \tfrac{1}{2}x^{-1/2} + \tfrac{1}{2}y^{-1/2}\dfrac{dy}{dx} = \dfrac{1}{2\sqrt{x}} + \dfrac{1}{2\sqrt{y}}\dfrac{dy}{dx}$

(c) $\dfrac{d}{dx}(x^2y^2 + y) = \dfrac{d}{dx}(x^2y^2) + \dfrac{d}{dx}(y) = \left[x^2 \cdot \dfrac{d}{dx}(y^2) + y^2 \cdot \dfrac{d}{dx}(x^2)\right] + \dfrac{dy}{dx}$

$$= x^2 \cdot 2y\dfrac{dy}{dx} + y^2 \cdot 2x + \dfrac{dy}{dx} = 2x^2y\dfrac{dy}{dx} + 2xy^2 + \dfrac{dy}{dx}$$

(d) $\dfrac{d}{dx}(e^x - e^y) = \dfrac{d}{dx}(e^x) - \dfrac{d}{dx}(e^y) = e^x - e^y\dfrac{dy}{dx}$

**3.** $N = 5a - 2b + a^2\sqrt{b}$

(a) $\dfrac{dN}{da} = \dfrac{d}{da}(5a) - \dfrac{d}{da}(2b) + \dfrac{d}{da}(a^2\sqrt{b}) = 5 - 0 + 2a\sqrt{b} = 5 + 2a\sqrt{b}$

(b) $\dfrac{dN}{db} = \dfrac{d}{db}(5a) - \dfrac{d}{db}(2b) + \dfrac{d}{db}(a^2b^{1/2}) = 0 - 2 + a^2 \cdot \tfrac{1}{2}b^{-1/2} = -2 + \dfrac{a^2}{2\sqrt{b}}$

---

**Exercises**

**1.** (a) Differentiate both sides of the equation $xy + 2x + 3x^2 = 4$ with respect to $x$, treating $y$ as a function of $x$:

$$\dfrac{d}{dx}(xy + 2x + 3x^2) = \dfrac{d}{dx}(4)$$

$$\dfrac{d}{dx}(xy) + \dfrac{d}{dx}(2x) + \dfrac{d}{dx}(3x^2) = 0$$

$$\left(x \cdot \dfrac{dy}{dx} + y \cdot 1\right) + 2 + 6x = 0$$

$$x\dfrac{dy}{dx} = -(y + 2 + 6x)$$

$$\dfrac{dy}{dx} = -\dfrac{y + 2 + 6x}{x}$$

(b) $xy + 2x + 3x^2 = 4 \ \Rightarrow \ xy = 4 - 2x - 3x^2 \ \Rightarrow \ y = \dfrac{4 - 2x - 3x^2}{x} = \dfrac{4}{x} - \dfrac{2x}{x} - \dfrac{3x^2}{x} = \dfrac{4}{x} - 2 - 3x$ or

$4x^{-1} - 2 - 3x$. Thus $\dfrac{dy}{dx} = -4x^{-2} - 0 - 3 = -\dfrac{4}{x^2} - 3$.

(c) From part (a), $\dfrac{dy}{dx} = -\dfrac{y + 2 + 6x}{x} = -\dfrac{[(4/x) - 2 - 3x] + 2 + 6x}{x} = -\dfrac{(4/x) + 3x}{x} = -\dfrac{4}{x^2} - 3$.

**3.** Differentiate both sides of $x^2 + y^2 = 1$ with respect to $x$, treating $y$ as a function of $x$: $\dfrac{d}{dx}(x^2 + y^2) = \dfrac{d}{dx}(1) \ \Rightarrow$

$\dfrac{d}{dx}(x^2) + \dfrac{d}{dx}(y^2) = 0 \ \Rightarrow \ 2x + 2y\dfrac{dy}{dx} = 0 \ \Rightarrow \ 2y\dfrac{dy}{dx} = -2x \ \Rightarrow \ \dfrac{dy}{dx} = -\dfrac{2x}{2y} = -\dfrac{x}{y}$

**5.** $\dfrac{d}{dx}(x^3 + x^2y + 4y^2) = \dfrac{d}{dx}(6) \ \Rightarrow \ 3x^2 + \left(x^2 \cdot \dfrac{dy}{dx} + y \cdot 2x\right) + 8y\dfrac{dy}{dx} = 0 \ \Rightarrow \ x^2\dfrac{dy}{dx} + 8y\dfrac{dy}{dx} = -3x^2 - 2xy.$

To solve for $\dfrac{dy}{dx}$, first factor out $\dfrac{dy}{dx}$ in the expression on the left: $(x^2 + 8y)\dfrac{dy}{dx} = -3x^2 - 2xy.$ Then

$\dfrac{dy}{dx} = \dfrac{-3x^2 - 2xy}{x^2 + 8y} = -\dfrac{x(3x + 2y)}{x^2 + 8y}.$

**7.** $\dfrac{d}{dx}(x^2y + xy^2) = \dfrac{d}{dx}(3x) \ \Rightarrow \ \left(x^2 \cdot \dfrac{dy}{dx} + y \cdot 2x\right) + \left(x \cdot 2y\dfrac{dy}{dx} + y^2 \cdot 1\right) = 3 \ \Rightarrow$

$x^2\dfrac{dy}{dx} + 2xy\dfrac{dy}{dx} = 3 - 2xy - y^2 \ \Rightarrow \ (x^2 + 2xy)\dfrac{dy}{dx} = 3 - 2xy - y^2 \ \Rightarrow \ \dfrac{dy}{dx} = \dfrac{3 - 2xy - y^2}{x^2 + 2xy}$

**9.** $\dfrac{d}{dx}(e^{x^2y}) = \dfrac{d}{dx}(x + y) \ \Rightarrow \ e^{x^2y}\dfrac{d}{dx}(x^2y) = 1 + \dfrac{dy}{dx} \ \Rightarrow \ e^{x^2y}\left(x^2 \cdot \dfrac{dy}{dx} + y \cdot 2x\right) = 1 + \dfrac{dy}{dx} \ \Rightarrow$

$x^2e^{x^2y}\dfrac{dy}{dx} - \dfrac{dy}{dx} = 1 - 2xye^{x^2y} \ \Rightarrow \ (x^2e^{x^2y} - 1)\dfrac{dy}{dx} = 1 - 2xye^{x^2y} \ \Rightarrow \ \dfrac{dy}{dx} = \dfrac{1 - 2xye^{x^2y}}{x^2e^{x^2y} - 1}$

**11.** $\dfrac{d}{dx}\{f(x) + x^2[f(x)]^3\} = \dfrac{d}{dx}(10) \ \Rightarrow \ f'(x) + \left(x^2 \cdot 3[f(x)]^2 f'(x) + [f(x)]^3 \cdot 2x\right) = 0.$ If $x = 1$, we have

$f'(1) + 1^2 \cdot 3[f(1)]^2 f'(1) + [f(1)]^3 \cdot 2(1) = 0 \ \Rightarrow \ f'(1) + 1 \cdot 3 \cdot 2^2 \cdot f'(1) + 2^3 \cdot 2 = 0 \ \Rightarrow$

$f'(1) + 12f'(1) = -16 \ \Rightarrow \ 13f'(1) = -16 \ \Rightarrow \ f'(1) = -\frac{16}{13}.$

**13.** When $C = 0$, $C + L^3 = e^{2C} \ \Rightarrow \ 0 + L^3 = e^0 \ \Rightarrow \ L^3 = 1 \ \Rightarrow \ L = 1.$ To find $\dfrac{dC}{dL}$, we differentiate both sides of

$C + L^3 = e^{2C}$ with respect to $L$, treating $C$ as a function of $L$: $\dfrac{d}{dL}(C + L^3) = \dfrac{d}{dL}(e^{2C}) \ \Rightarrow$

$\dfrac{dC}{dL} + 3L^2 = e^{2C} \cdot 2\dfrac{dC}{dL} \ \Rightarrow \ 3L^2 = 2e^{2C}\dfrac{dC}{dL} - \dfrac{dC}{dL} \ \Rightarrow \ (2e^{2C} - 1)\dfrac{dC}{dL} = 3L^2 \ \Rightarrow \ \dfrac{dC}{dL} = \dfrac{3L^2}{2e^{2C} - 1}.$

Substituting $C = 0$ and $L = 1$ gives $\dfrac{dC}{dL} = \dfrac{3(1)^2}{2e^0 - 1} = \dfrac{3}{2 - 1} = 3.$

**15.** $\dfrac{d}{dx}(x^2 + xy + y^2) = \dfrac{d}{dx}(3) \ \Rightarrow \ 2x + \left(x \cdot \dfrac{dy}{dx} + y \cdot 1\right) + 2y\dfrac{dy}{dx} = 0 \ \Rightarrow \ x\dfrac{dy}{dx} + 2y\dfrac{dy}{dx} = -2x - y \ \Rightarrow$

$(x + 2y)\dfrac{dy}{dx} = -2x - y \ \Rightarrow \ \dfrac{dy}{dx} = \dfrac{-2x - y}{x + 2y}.$ When $x = 1$ and $y = 1$, we have $\dfrac{dy}{dx} = \dfrac{-2(1) - 1}{1 + 2(1)} = \dfrac{-3}{3} = -1,$

so an equation of the tangent line is $y - 1 = -1(x - 1)$ or $y = -x + 2.$

**17.** $\dfrac{d}{dx}\left[2(x^2+y^2)^2\right]=\dfrac{d}{dx}\left[25(x^2-y^2)\right]$ $\Rightarrow$ $4(x^2+y^2)\left(2x+2y\dfrac{dy}{dx}\right)=25\left(2x-2y\dfrac{dy}{dx}\right)$ $\Rightarrow$

$8y(x^2+y^2)\dfrac{dy}{dx}+50y\dfrac{dy}{dx}=50x-8x(x^2+y^2)$ $\Rightarrow$ $\left[8y(x^2+y^2)+50y\right]\dfrac{dy}{dx}=50x-8x(x^2+y^2)$ $\Rightarrow$

$\dfrac{dy}{dx}=\dfrac{50x-8x(x^2+y^2)}{8y(x^2+y^2)+50y}$. When $x=3$ and $y=1$, we have $\dfrac{dy}{dx}=\dfrac{150-24(10)}{8(10)+50}=\dfrac{-90}{130}=-\dfrac{9}{13}$,

so an equation of the tangent line is $y-1=-\dfrac{9}{13}(x-3)$ or $y=-\dfrac{9}{13}x+\dfrac{40}{13}$.

**19.** We differentiate both sides of $2px^2+3px=58{,}000$ with respect to $p$, treating $x$ as a function of $p$:

$\dfrac{d}{dp}\left(2px^2+3px\right)=\dfrac{d}{dp}(58{,}000)$ $\Rightarrow$ $2\left(p\cdot 2x\dfrac{dx}{dp}+x^2\cdot 1\right)+3\left(p\cdot\dfrac{dx}{dp}+x\cdot 1\right)=0$ $\Rightarrow$

$(4px+3p)\dfrac{dx}{dp}=-2x^2-3x$ $\Rightarrow$ $\dfrac{dx}{dp}=\dfrac{-2x^2-3x}{4px+3p}$. When $p=40$ we have $2(40)x^2+3(40)x=58{,}000$ $\Rightarrow$

$80x^2+120x-58{,}000=0$ $\Rightarrow$ $2x^2+3x-1450=0$. The quadratic formula gives

$x=\dfrac{-3\pm\sqrt{3^2-4(2)(-1450)}}{2(2)}=\dfrac{-3\pm\sqrt{11609}}{4}\approx 26.186$ or $-27.686$. Taking the positive solution, we have

$\dfrac{dx}{dp}=\dfrac{-2x^2-3x}{4px+3p}\approx\dfrac{-2(26.186)^2-3(26.186)}{4(40)(26.186)+3(40)}\approx-0.336$. Thus when the price increases from \$40, the number of copies

that will sell decreases at a rate of 0.336 thousand copies per dollar price increase, or 336 copies/dollar.

**21.** $f(x)=3x-2\ln x$ $\Rightarrow$ $f'(x)=3-2\cdot\dfrac{1}{x}=3-\dfrac{2}{x}$

**23.** $y=1.5x+\ln x$ $\Rightarrow$ $\dfrac{dy}{dx}=1.5+\dfrac{1}{x}$

**25.** $y=(\ln x)^5$ $\Rightarrow$ $\dfrac{dy}{dx}=5(\ln x)^4\cdot\dfrac{d}{dx}(\ln x)=5(\ln x)^4\cdot\dfrac{1}{x}=\dfrac{5(\ln x)^4}{x}$

**27.** $f(x)=\sqrt[5]{\ln x}=(\ln x)^{1/5}$ $\Rightarrow$ $f'(x)=\tfrac{1}{5}(\ln x)^{-4/5}\dfrac{d}{dx}(\ln x)=\dfrac{1}{5(\ln x)^{4/5}}\cdot\dfrac{1}{x}=\dfrac{1}{5x\,(\ln x)^{4/5}}$

**29.** $y=(\ln x+1)^2+(e^x+1)^2$ $\Rightarrow$

$$\dfrac{dy}{dx}=2(\ln x+1)^1\cdot\dfrac{d}{dx}(\ln x+1)+2(e^x+1)^1\cdot\dfrac{d}{dx}(e^x+1)$$

$$=2(\ln x+1)\left(\dfrac{1}{x}+0\right)+2(e^x+1)(e^x+0)=\dfrac{2(\ln x+1)}{x}+2e^x(e^x+1)$$

**31.** It is easier to first simplify the function using properties of logarithms:

$F(t)=\ln\dfrac{(2t+1)^3}{(3t-1)^4}=\ln(2t+1)^3-\ln(3t-1)^4=3\ln(2t+1)-4\ln(3t-1)$. Then

$F'(t)=3\cdot\dfrac{1}{2t+1}(2)-4\cdot\dfrac{1}{3t-1}(3)=\dfrac{6}{2t+1}-\dfrac{12}{3t-1}$, or combined, $\dfrac{6(3t-1)-12(2t+1)}{(2t+1)(3t-1)}=\dfrac{-6t-18}{(2t+1)(3t-1)}$.

**33.** $f(u)=\dfrac{\ln u}{1+\ln(2u)}$ $\Rightarrow$

$f'(u)=\dfrac{\left[1+\ln(2u)\right]\cdot\frac{1}{u}-\ln u\cdot\frac{1}{2u}(2)}{\left[1+\ln(2u)\right]^2}=\dfrac{\frac{1}{u}\left[1+\ln(2u)-\ln u\right]}{\left[1+\ln(2u)\right]^2}=\dfrac{1+(\ln 2+\ln u)-\ln u}{u\left[1+\ln(2u)\right]^2}=\dfrac{1+\ln 2}{u\left[1+\ln(2u)\right]^2}$

**35.** Use properties of logarithms to first simplify the function:

$y = \ln(e^{-x} + xe^{-x}) = \ln(e^{-x}(1+x)) = \ln(e^{-x}) + \ln(1+x) = -x + \ln(1+x)$. Then

$y' = -1 + \dfrac{1}{1+x}(1) = \dfrac{-(1+x)}{1+x} + \dfrac{1}{1+x} = \dfrac{-1-x+1}{1+x} = -\dfrac{x}{1+x}$.

**37.** $y = \ln(\ln x) \;\Rightarrow\; y' = \dfrac{1}{\ln x} \cdot \dfrac{d}{dx}(\ln x) = \dfrac{1}{\ln x} \cdot \dfrac{1}{x} = \dfrac{1}{x \ln x}$

**39.** $f(x) = \ln(x^2 - 5) \;\Rightarrow\; f'(x) = \dfrac{1}{x^2 - 5}(2x) = \dfrac{2x}{x^2 - 5} \;\Rightarrow\; f''(x) = \dfrac{(x^2 - 5) \cdot 2 - 2x \cdot 2x}{(x^2 - 5)^2} = \dfrac{-2x^2 - 10}{(x^2 - 5)^2}$

**41.** $f(x) = e^x \ln x \;\Rightarrow\; f'(x) = e^x \cdot \dfrac{1}{x} + (\ln x) \cdot e^x = e^x\left(\dfrac{1}{x} + \ln x\right) = e^x\left(x^{-1} + \ln x\right) \;\Rightarrow\;$

$f''(x) = e^x\left(-x^{-2} + \dfrac{1}{x}\right) + (x^{-1} + \ln x)e^x = e^x\left(-\dfrac{1}{x^2} + \dfrac{1}{x} + \dfrac{1}{x} + \ln x\right) = e^x\left(\ln x + \dfrac{2}{x} - \dfrac{1}{x^2}\right)$

**43.** $y = \ln(x^2 - 3) \;\Rightarrow\; \dfrac{dy}{dx} = \dfrac{1}{x^2 - 3} \cdot 2x = \dfrac{2x}{x^2 - 3}$. The slope of the tangent line at $(2, 0)$ is $\dfrac{dy}{dx} = \dfrac{2(2)}{2^2 - 3} = 4$ and its

equation is $y - 0 = 4(x - 2)$  or  $y = 4x - 8$.

**45.** $f(x) = 4 \ln x - x \;\Rightarrow\; f'(x) = 4 \cdot \dfrac{1}{x} - 1 = \dfrac{4}{x} - 1$.

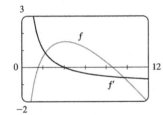

Notice that $f'(x) = 0$ where $f$ has a horizontal tangent, $f'$ is positive where

$f$ is increasing, and $f'$ is negative where $f$ is decreasing.

**47.** $\dfrac{d}{dx} \log_a x = \dfrac{d}{dx} \dfrac{\ln x}{\ln a} = \dfrac{d}{dx}\left(\dfrac{1}{\ln a} \ln x\right) = \dfrac{1}{\ln a} \cdot \dfrac{1}{x} = \dfrac{1}{x \ln a}$

**49.** $B = 120 + 10 \log_{10} I$, so a sound with intensity $I = 100$ W/m² has a perceived loudness of

$B = 120 + 10 \log_{10}(100) = 120 + 10(2) = 140$ dB. Using the rule in Exercise 47, $\dfrac{dB}{dI} = 0 + 10 \cdot \dfrac{1}{I \ln 10} = \dfrac{10}{I \ln 10}$.

Then $\dfrac{dB}{dI}\bigg|_{I=100} = \dfrac{10}{100 \ln 10} = \dfrac{1}{10 \ln 10} \approx 0.0434$ dB/(W/m²).

**51.** Using the method of implicit differentiation, $y = \ln(x^2 + y^2) \;\Rightarrow\; \dfrac{dy}{dx} = \dfrac{1}{x^2 + y^2} \dfrac{d}{dx}(x^2 + y^2) \;\Rightarrow\;$

$\dfrac{dy}{dx} = \dfrac{1}{x^2 + y^2}\left(2x + 2y\dfrac{dy}{dx}\right) \;\Rightarrow\; \dfrac{dy}{dx} - \dfrac{2y}{x^2 + y^2}\dfrac{dy}{dx} = \dfrac{2x}{x^2 + y^2} \;\Rightarrow\; \left(1 - \dfrac{2y}{x^2 + y^2}\right)\dfrac{dy}{dx} = \dfrac{2x}{x^2 + y^2} \;\Rightarrow\;$

$\dfrac{x^2 + y^2 - 2y}{x^2 + y^2}\dfrac{dy}{dx} = \dfrac{2x}{x^2 + y^2} \;\Rightarrow\; \dfrac{dy}{dx} = \dfrac{2x/(x^2 + y^2)}{(x^2 + y^2 - 2y)/(x^2 + y^2)} = \dfrac{2x}{x^2 + y^2 - 2y}$.

**53.** (a) Using implicit differentiation, we have $\dfrac{d}{dx}\left(y^2\right) = \dfrac{d}{dx}\left(5x^4 - x^2\right)$ $\Rightarrow$ $2y\dfrac{dy}{dx} = 20x^3 - 2x$ $\Rightarrow$

$\dfrac{dy}{dx} = \dfrac{20x^3 - 2x}{2y} = \dfrac{10x^3 - x}{y}$. At the point $(1, 2)$, $\dfrac{dy}{dx} = \dfrac{10(1)^3 - 1}{2} = \dfrac{9}{2}$, and an equation of the tangent line is

$y - 2 = \frac{9}{2}(x - 1)$ or $y = \frac{9}{2}x - \frac{5}{2}$.

(b)

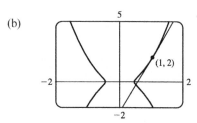

**55.** (a)

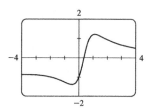

Looking at the graph of $f(x) = \ln(x^2 + e^{-x})$, we see that at first the slope appears to be relatively constant, about $-1$, after which the curve falls a little more steeply. Just to the right of the origin, the curve has a local minimum where the slope changes from negative to positive. The curve rises from there, getting steeper until about $x = 2$ and then getting less steep. We use this information to sketch a graph of $f'$, which appears similar to the graph shown in part (b).

(b) $f'(x) = \dfrac{1}{x^2 + e^{-x}}\left[2x + e^{-x}(-1)\right] = \dfrac{2x - e^{-x}}{x^2 + e^{-x}}$

**57.** $\dfrac{x^2}{a^2} - \dfrac{y^2}{b^2} = 1$ $\Rightarrow$ $\dfrac{d}{dx}\left(\dfrac{x^2}{a^2} - \dfrac{y^2}{b^2}\right) = \dfrac{d}{dx}(1)$ $\Rightarrow$ $\dfrac{1}{a^2}(2x) - \dfrac{1}{b^2}\left(2y\dfrac{dy}{dx}\right) = 0$ $\Rightarrow$ $\dfrac{2y}{b^2}\dfrac{dy}{dx} = \dfrac{2x}{a^2}$ $\Rightarrow$

$\dfrac{dy}{dx} = \dfrac{2x}{a^2}\cdot\dfrac{b^2}{2y} = \dfrac{b^2 x}{a^2 y}$. Thus the slope of the tangent line at $(x_0, y_0)$ is $\dfrac{b^2 x_0}{a^2 y_0}$ and its equation is

$y - y_0 = \dfrac{b^2 x_0}{a^2 y_0}(x - x_0)$ $\Rightarrow$ $a^2 y_0(y - y_0) = b^2 x_0(x - x_0)$ $\Rightarrow$ $a^2 y_0 y - a^2 y_0^2 = b^2 x_0 x - b^2 x_0^2$ or

$b^2 x_0 x - a^2 y_0 y = b^2 x_0^2 - a^2 y_0^2$. Note that we can simplify the equation further by dividing both sides by $a^2 b^2$, giving

$\dfrac{x_0 x}{a^2} - \dfrac{y_0 y}{b^2} = \dfrac{x_0^2}{a^2} - \dfrac{y_0^2}{b^2}$, and since $(x_0, y_0)$ lies on the hyperbola, we have $\dfrac{x_0^2}{a^2} - \dfrac{y_0^2}{b^2} = 1$, so the equation becomes

$\dfrac{x_0 x}{a^2} - \dfrac{y_0 y}{b^2} = 1$.

**59.** To find the points at which the ellipse $x^2 - xy + y^2 = 3$ crosses the $x$-axis, let $y = 0$ and solve for $x$:

$y = 0$ $\Rightarrow$ $x^2 - x(0) + 0^2 = 3$ $\Leftrightarrow$ $x = \pm\sqrt{3}$. So the graph of the ellipse crosses the $x$-axis at the points $\left(\pm\sqrt{3}, 0\right)$.

Using implicit differentiation to find $\dfrac{dy}{dx}$, we get $\dfrac{d}{dx}\left(x^2 - xy + y^2\right) = \dfrac{d}{dx}(3)$ $\Rightarrow$

$2x - \left(x\cdot\dfrac{dy}{dx} + y\cdot 1\right) + 2y\dfrac{dy}{dx} = 0$ $\Rightarrow$ $(2y - x)\dfrac{dy}{dx} = y - 2x$ $\Rightarrow$ $\dfrac{dy}{dx} = \dfrac{y - 2x}{2y - x}$. At $\left(\sqrt{3}, 0\right)$,

$\dfrac{dy}{dx} = \dfrac{0 - 2\sqrt{3}}{2(0) - \sqrt{3}} = 2$, and at $\left(-\sqrt{3}, 0\right)$, $\dfrac{dy}{dx} = \dfrac{0 + 2\sqrt{3}}{2(0) + \sqrt{3}} = 2$. Because the slopes are equal, the tangent lines at these

points are parallel.

**61.** We first take the natural logarithm of both sides of the equation $y = x^x$, giving $\ln y = \ln(x^x) \;\Rightarrow\; \ln y = x \ln x$. Then use

implicit differentiation to find $\dfrac{dy}{dx}$: $\dfrac{d}{dx}(\ln y) = \dfrac{d}{dx}(x \ln x) \;\Rightarrow\; \dfrac{1}{y}\dfrac{dy}{dx} = x \cdot \dfrac{1}{x} + (\ln x) \cdot 1 \;\Rightarrow\; \dfrac{dy}{dx} = y(1 + \ln x)$. Since

$y = x^x$, we have $\dfrac{dy}{dx} = x^x(1 + \ln x)$.

**63.** $y = \dfrac{e^{x^2}\sqrt{x^3 + x}}{(2x + 3)^4} \;\Rightarrow\;$

$\ln y = \ln \dfrac{e^{x^2}\sqrt{x^3 + x}}{(2x + 3)^4} = \ln e^{x^2} + \ln(x^3 + x)^{1/2} - \ln(2x + 3)^4 = x^2 + \tfrac{1}{2}\ln(x^3 + x) - 4\ln(2x + 3)$. Differentiating

implicitly with respect to $x$ gives $\dfrac{1}{y}\dfrac{dy}{dx} = 2x + \dfrac{1}{2}\dfrac{1}{x^3 + x}(3x^2 + 1) - 4 \cdot \dfrac{1}{2x + 3}(2) = 2x + \dfrac{3x^2 + 1}{2(x^3 + x)} - \dfrac{8}{2x + 3}$. Thus

$\dfrac{dy}{dx} = y\left(2x + \dfrac{3x^2 + 1}{2(x^3 + x)} - \dfrac{8}{2x + 3}\right) = \dfrac{e^{x^2}\sqrt{x^3 + x}}{(2x + 3)^4}\left(2x + \dfrac{3x^2 + 1}{2(x^3 + x)} - \dfrac{8}{2x + 3}\right)$.

## 3.6 Exponential Growth and Decay

---

**Prepare Yourself**

**1.** (a) $f(t) = 14e^{-0.2t} \;\Rightarrow\; f(4.5) = 14e^{-0.2(4.5)} \approx 5.69$

(b) $f(t) = 4.5 \;\Rightarrow\; 14e^{-0.2t} = 4.5 \;\Rightarrow\; e^{-0.2t} = \frac{4.5}{14} \;\Rightarrow\; \ln e^{-0.2t} = \ln \frac{4.5}{14} \;\Rightarrow\; -0.2t = \ln \frac{4.5}{14} \;\Rightarrow\;$

$t = \frac{1}{-0.2}\ln \frac{4.5}{14} \approx 5.67$

**2.** (a) $100(1.08)^x = 160 \;\Rightarrow\; (1.08)^x = \frac{160}{100} = 1.6 \;\Rightarrow\; \ln(1.08)^x = \ln 1.6 \;\Rightarrow\; x \ln 1.08 = \ln 1.6 \;\Rightarrow\;$

$x = \dfrac{\ln 1.6}{\ln 1.08} \approx 6.11$

(b) $5(0.8)^x = 3.1 \;\Rightarrow\; (0.8)^x = \frac{3.1}{5} = 0.62 \;\Rightarrow\; \ln(0.8)^x = \ln 0.62 \;\Rightarrow\; x \ln 0.8 = \ln 0.62 \;\Rightarrow\;$

$x = \dfrac{\ln 0.62}{\ln 0.8} \approx 2.14$

(c) $45e^{0.3x} = 85 \;\Rightarrow\; e^{0.3x} = \frac{85}{45} \;\Rightarrow\; \ln e^{0.3x} = \ln \frac{85}{45} \;\Rightarrow\; 0.3x = \ln \frac{85}{45} \;\Rightarrow\; x = \frac{1}{0.3}\ln \frac{85}{45} \approx 2.12$

**3.** (a) $g(x) = \dfrac{240}{1 + 2e^{-0.4x}} \;\Rightarrow\; g(7.3) = \dfrac{240}{1 + 2e^{-0.4(7.3)}} \approx 216.63$

(b) $g(x) = 100 \;\Rightarrow\; \dfrac{240}{1 + 2e^{-0.4x}} = 100 \;\Rightarrow\; \dfrac{240}{100} = 1 + 2e^{-0.4x} \;\Rightarrow\; 2e^{-0.4x} = 2.4 - 1 = 1.4 \;\Rightarrow\;$

$e^{-0.4x} = \frac{1.4}{2} = 0.7 \;\Rightarrow\; \ln(e^{-0.4x}) = \ln 0.7 \;\Rightarrow\; -0.4x = \ln 0.7 \;\Rightarrow\; x = \frac{1}{-0.4}\ln 0.7 = -2.5\ln 0.7 \approx 0.89$

**Exercises**

**1.** (a) Each year the population increases 2.4%, so it is $100\% + 2.4\% = 102.4\%$ of the previous year's population. Thus the population is multiplied by 1.024 each year, and after $t$ years, the population is $46{,}500(1.024)^t$.

(b) If the population decreases by 2.4% each year, then each year the population is $100\% - 2.4\% = 97.6\%$ of the previous year. After $t$ years, the population is $46{,}500(0.976)^t$.

**3.** (a) By Equation 1, the value of the investment after $t$ years is $A(t) = P(1+r)^t = 5000(1+0.062)^t = 5000(1.062)^t$ dollars. Notice that this means each year the value is 106.2% of the previous year's value.

(b) After 7.5 years the investment is worth $A(7.5) = 5000(1.062)^{7.5} \approx 7850.62$ dollars.

(c) $A(t) = 8000 \Rightarrow 5000(1.062)^t = 8000 \Rightarrow (1.062)^t = \frac{8000}{5000} = 1.6 \Rightarrow \ln(1.062)^t = \ln 1.6 \Rightarrow$

$t \ln 1.062 = \ln 1.6 \Rightarrow t = \dfrac{\ln 1.6}{\ln 1.062} \approx 7.81$. Thus the investment will be worth \$8000 after about 7.81 years.

**5.** (a) The value decreases by 15% each year, so 85% of the value is retained. Thus the value after $t$ years is $V(t) = 28.6(0.85)^t$ million dollars.

(b) $V(t) = \frac{1}{2}(28.6) = 14.3 \Rightarrow 28.6(0.85)^t = 14.3 \Rightarrow (0.85)^t = \frac{1}{2} \Rightarrow \ln(0.85)^t = \ln\frac{1}{2} \Rightarrow$

$t \ln 0.85 = \ln\frac{1}{2} \Rightarrow t = \dfrac{\ln(1/2)}{\ln 0.85} \approx 4.27$, so the equipment is reduced to half its value after about 4.27 years.

(c) $V'(t) = 28.6(0.85)^t \ln 0.85$, so $V'(5) = 28.6(0.85)^5 \ln 0.85 \approx -2.06$. Thus after 5 years, the value of the equipment is decreasing at a rate of about 2.06 million dollars per year.

(d) $V'(t) = -1 \Rightarrow 28.6(0.85)^t \ln 0.85 = -1 \Rightarrow (0.85)^t = \dfrac{-1}{28.6\ln 0.85} \Rightarrow$

$t \ln 0.85 = \ln\left(\dfrac{-1}{28.6\ln 0.85}\right) \Rightarrow t = \dfrac{1}{\ln 0.85}\ln\left(\dfrac{-1}{28.6\ln 0.85}\right) \approx 9.45$. Thus the value will be decreasing at a rate of \$1 million per year after about 9.45 years.

**7.** (a) Using Equations 1, 2, and 3 with $P = 3000$, $r = 0.05$, and $t = 5$, we have

    (i) Annually: $\qquad\qquad A = 3000(1 + 0.05)^5 = \$3828.84$

    (ii) Quarterly: $n = 4$; $\quad A = 3000\left(1 + \frac{0.05}{4}\right)^{4\cdot 5} = \$3846.11$

    (iii) Monthly: $n = 12$; $\quad A = 3000\left(1 + \frac{0.05}{12}\right)^{12\cdot 5} = \$3850.08$

    (iv) Weekly: $n = 52$; $\quad A = 3000\left(1 + \frac{0.05}{52}\right)^{52\cdot 5} = \$3851.61$

    (v) Daily: $n = 365$; $\quad A = 3000\left(1 + \frac{0.05}{365}\right)^{365\cdot 5} = \$3852.01$

    (vi) Continuously: $\qquad A = 3000e^{(0.05)5} = \$3852.08$

(b) $3000\left(1 + \frac{0.05}{4}\right)^{4t} = 6000 \Rightarrow (1.0125)^{4t} = \frac{6000}{3000} = 2 \Rightarrow 4t\ln 1.0125 = \ln 2 \Rightarrow t = \dfrac{\ln 2}{4\ln 1.0125} \approx 13.95$,

so it will take about 13.95 years for the balance to double.

(c) $3000e^{0.05t} = 6000 \Rightarrow e^{0.05t} = 2 \Rightarrow 0.05t = \ln 2 \Rightarrow t = \frac{1}{0.05}\ln 2 \approx 13.86$ years.

**9.** (a) By Equation 2, $V(t) = 16{,}000\left(1 + \frac{0.043}{12}\right)^{12t}$ dollars.

(b) $V'(t) = 16{,}000\left(1 + \frac{0.043}{12}\right)^{12t} \ln\left(1 + \frac{0.043}{12}\right) \cdot 12 \quad\Rightarrow$

$V'(3.5) = 16{,}000\left(1 + \frac{0.043}{12}\right)^{12(3.5)} \ln\left(1 + \frac{0.043}{12}\right) \cdot 12 \approx 798.10$. Thus after 3.5 years, the value of the account is

increasing at a rate of about $798.10/year.

**11.** If we invest, say, $100 at 5.1% interest compounded continuously, then after one year the value is $100e^{0.051(1)} \approx \$105.23$.

At 5.25% interest compounded quarterly, the value is $100\left(1 + \frac{0.0525}{4}\right)^{4(1)} \approx \$105.35$, so this is the better option.

**13.** Using Equation 4 with $C = 1300$ and $k = 0.172$, the population after $t$ years is $P(t) = 1300e^{0.172t}$. After 7.5 years, the

population is $P(7.5) = 1300e^{0.172(7.5)} \approx 4723$.

**15.** Using Equation 4, the population after $t$ days is $P(t) = 2e^{0.7944t}$, so after 6 days, the population is

$P(6) = 2e^{0.7944(6)} \approx 235$.

**17.** (a) The population after $t$ hours is $P(t) = 100e^{kt}$. After one hour the population is 420, so $P(1) = 420 \quad\Rightarrow$

$100e^{k(1)} = 420 \quad\Rightarrow\quad e^k = 4.2 \quad\Rightarrow\quad k = \ln 4.2$. Thus the number of bacteria after $t$ hours is

$P(t) = 100e^{(\ln 4.2)t} = 100(e^{\ln 4.2})^t = 100(4.2)^t$.

(b) $P(3) = 100(4.2)^3 = 7408.8 \approx 7409$ bacteria

(c) $P'(t) = 100(4.2)^t \ln 4.2 \quad\Rightarrow\quad P'(3) = 100(4.2)^3 \ln 4.2 \approx 10{,}632$ bacteria per hour

(d) $P(t) = 10{,}000 \quad\Rightarrow\quad 100(4.2)^t = 10{,}000 \quad\Rightarrow\quad (4.2)^t = 100 \quad\Rightarrow\quad t \ln 4.2 = \ln 100 \quad\Rightarrow\quad t = \dfrac{\ln 100}{\ln 4.2} \approx 3.2$ hours

**19.** (a) The population after $t$ days is $P(t) = 260e^{kt}$, and $P(3) = 1720 \quad\Rightarrow\quad 260e^{3k} = 1720 \quad\Rightarrow\quad e^{3k} = \frac{1720}{260} \quad\Rightarrow$

$3k = \ln\frac{1720}{260} \quad\Rightarrow\quad k = \frac{1}{3}\ln\frac{1720}{260} \approx 0.6298$, so a model for the virus population is $P(t) = 260e^{0.6298t}$.

(b) After 7 days, the model predicts that there will be $P(7) = 260e^{0.6298(7)} \approx 21{,}360$ viruses.

(c) $P'(t) = 260e^{0.6298t}(0.6298) = 163.748e^{0.6298t}$, so after one week the population is increasing at a rate of

$P'(7) = 163.748e^{0.6298(7)} \approx 13{,}453$ viruses/day.

**21.** (a) Let $P(t)$ be the population (in millions) in the year $t$, where $t = 0$ corresponds to 1750. Then $P(t) = 790e^{kt}$, and

$P(50) = 790e^{k(50)} = 980 \quad\Rightarrow\quad e^{50k} = \frac{980}{790} \quad\Rightarrow\quad 50k = \ln\frac{980}{790} \quad\Rightarrow\quad k = \frac{1}{50}\ln\frac{980}{790} \approx 0.004310$. According to this

model, the population in 1900 was $P(150) = 790e^{0.004310(150)} \approx 1508$ million, and in 1950 it was

$P(200) = 790e^{0.004310(200)} \approx 1871$ million. Both of these estimates are much too low.

(b) Let $t = 0$ correspond to the year 1850. Then the population in millions is $P(t) = 1260e^{kt}$, and

$P(50) = 1260e^{k(50)} = 1650 \quad\Rightarrow\quad e^{50k} = \frac{1650}{1260} \quad\Rightarrow\quad 50k = \ln\frac{1650}{1260} \quad\Rightarrow\quad k = \frac{1}{50}\ln\frac{1650}{1260} \approx 0.005393$. So with this

model, we estimate the population in 1950 to be $P(100) = 1260e^{0.005393(100)} \approx 2161$ million. This is still too low, but

closer than the estimate for 1950 in part (a).

(c) Let $t = 0$ correspond to the year 1900. Then the population in millions is $P(t) = 1650e^{kt}$, and

$$P(50) = 1650e^{k(50)} = 2560 \quad \Rightarrow \quad e^{50k} = \tfrac{2560}{1650} \quad \Rightarrow \quad 50k = \ln \tfrac{2560}{1650} \quad \Rightarrow \quad k = \tfrac{1}{50} \ln \tfrac{2560}{1650} \approx 0.008785.$$ With this

model, we estimate the population in 2000 to be $P(100) = 1650e^{0.008785(100)} \approx 3972$ million. This is much too low. The

discrepancy is explained by the fact that the world birth rate (average yearly number of births per person) is about the same

as always, whereas the mortality rate (especially the infant mortality rate) is much lower, owing mostly to advances in

medical science and to the wars in the first part of the twentieth century. The exponential model assumes, among other

things, that the birth and mortality rates will remain constant.

**23.** (a) If $m(t)$ is the mass (in mg) remaining after $t$ years, then $m(t) = Ce^{kt} = 100e^{kt}$. The half-life is 30 years, so

$$m(30) = \tfrac{1}{2}(100) = 50 \quad \Rightarrow \quad 100e^{k(30)} = 50 \quad \Rightarrow \quad e^{30k} = \tfrac{1}{2} \quad \Rightarrow \quad 30k = \ln \tfrac{1}{2} \quad \Rightarrow \quad k = \tfrac{1}{30} \ln \tfrac{1}{2} \approx -0.02310$$ and

$m(t) = 100e^{-0.02310t}$ or, since the mass is halved every 30 years, we can equivalently write $m(t) = 100\left(\tfrac{1}{2}\right)^{t/30}$.

(b) $m(100) = 100\left(\tfrac{1}{2}\right)^{100/30} \approx 9.92$ mg

(c) $m(t) = 1 \quad \Rightarrow \quad 100\left(\tfrac{1}{2}\right)^{t/30} = 1 \quad \Rightarrow \quad \left(\tfrac{1}{2}\right)^{t/30} = \tfrac{1}{100} \quad \Rightarrow \quad (t/30) \ln \tfrac{1}{2} = \ln \tfrac{1}{100} \quad \Rightarrow \quad t = \dfrac{30 \ln(1/100)}{\ln(1/2)} \approx 199.3$,

so only 1 mg remains after about 199.3 years.

(d) $m(t) = 100\left(\tfrac{1}{2}\right)^{t/30} \quad \Rightarrow \quad m'(t) = 100\left(\tfrac{1}{2}\right)^{t/30}\left(\ln \tfrac{1}{2}\right) \cdot \tfrac{1}{30}$  [or $m(t) = 100e^{-0.02310t} \quad \Rightarrow$

$m'(t) = 100e^{-0.02310t}(-0.02310) = -2.31e^{-0.02310t}$ ], so $m'(100) = \tfrac{100}{30}\left(\ln \tfrac{1}{2}\right)\left(\tfrac{1}{2}\right)^{100/30} \approx -0.229$. Thus after

100 years, the mass is decreasing at a rate of about 0.229 mg/year.

**25.** Let $A(t)$ be the level of radioactivity. Thus, $A(t) = Ce^{kt}$ and $k$ is determined by using the half-life:

$$A(5730) = \tfrac{1}{2}C \quad \Rightarrow \quad Ce^{k(5730)} = \tfrac{1}{2}C \quad \Rightarrow \quad e^{5730k} = \tfrac{1}{2} \quad \Rightarrow \quad 5730k = \ln \tfrac{1}{2} \quad \Rightarrow \quad k = \tfrac{1}{5730} \ln \tfrac{1}{2} \approx -0.00012097.$$

Thus $A(t) = Ce^{-0.00012097t}$, or equivalently, since $t$ is halved every 5730 years, $A(t) = C\left(\tfrac{1}{2}\right)^{t/5730}$. If 74% of the $^{14}$C

remains, then we know that $A(t) = 0.74C \quad \Rightarrow \quad Ce^{-0.00012097t} = 0.74C \quad \Rightarrow \quad e^{-0.00012097t} = 0.74 \quad \Rightarrow$

$-0.00012097t = \ln 0.74 \quad \Rightarrow$

$t = \dfrac{1}{-0.00012097} \ln 0.74 \approx 2489$, so the parchment is about 2500 years old.

**27.** (a) The mass, in grams, of the sample that remains after $t$ hours is $m(t) = Ce^{kt} = 100e^{kt}$. We are given that

$$m(26) = 30 \quad \Rightarrow \quad 100e^{k(26)} = 30 \quad \Rightarrow \quad e^{26k} = 0.3 \quad \Rightarrow \quad 26k = \ln 0.3 \quad \Rightarrow \quad k = \tfrac{1}{26} \ln 0.3 \approx -0.046307.$$ Thus

$m(t) = 100e^{-0.046307t}$. The half-life is the time required for half of the sample to remain, so we solve

$$m(t) = \tfrac{1}{2}(100) = 50 \quad \Rightarrow \quad 100e^{-0.046307t} = 50 \quad \Rightarrow \quad e^{-0.046307t} = \tfrac{1}{2} \quad \Rightarrow \quad -0.046307t = \ln \tfrac{1}{2} \quad \Rightarrow$$

$t = \dfrac{1}{-0.046307} \ln \tfrac{1}{2} \approx 14.97$. Thus the half-life is about 14.97 hours.

(b) $m(2) = 100e^{-0.046307(2)} \approx 91.2$ g

(c) $m(t) = 5 \quad \Rightarrow \quad 100e^{-0.046307t} = 5 \quad \Rightarrow \quad e^{-0.046307t} = \tfrac{1}{20} \quad \Rightarrow \quad -0.046307t = \ln \tfrac{1}{20} \quad \Rightarrow$

$t = \dfrac{1}{-0.046307} \ln \tfrac{1}{20} \approx 64.7$ hours.

**29.** (a) The difference between the turkey temperature $T$ and the room temperature follows exponential decay: $T - 75 = Ce^{kt}$, where $t$ is the time in minutes. The initial temperature difference is $C = 185 - 75 = 110$, and after 30 minutes the temperature is $150°\text{F}$, so $150 - 75 = 110e^{k(30)}$ $\Rightarrow$ $\frac{75}{110} = e^{30k}$ $\Rightarrow$ $\ln\frac{75}{110} = 30k$ $\Rightarrow$

$k = \frac{1}{30}\ln\frac{75}{110} \approx -0.012766$. Thus $T = 75 + 110e^{-0.012766t}$, and after 45 minutes the temperature of the turkey is

$T = 75 + 110e^{-0.012766(45)} \approx 137°\text{F}$.

(b) $T = 100$ $\Rightarrow$ $75 + 110e^{-0.012766t} = 100$ $\Rightarrow$ $e^{-0.012766t} = \frac{25}{110}$ $\Rightarrow$ $-0.012766t = \ln\frac{25}{110}$ $\Rightarrow$

$t = \frac{1}{-0.012766t}\ln\frac{25}{110} \approx 116$ min.

(c) $\frac{dT}{dt} = 0 + 110e^{-0.012766t}(-0.012766) = -1.40426e^{-0.012766t}$ $\Rightarrow$ $\frac{dT}{dt}\bigg|_{t=30} = -1.40426e^{-0.012766(30)} \approx -0.957$,

thus after half an hour the turkey is cooling at a rate of approximately $0.96\,°\text{F/min}$.

**31.** (a) $P$ is a logistic function, and comparing to Equation 6 we see that the carrying capacity is $M = 1680$ lions. The number of mountain lions on January 1, 2010, is $P(0) = \dfrac{1680}{1 + 4.2e^{-0.11(0)}} = \dfrac{1680}{1 + 4.2(1)} = \dfrac{1680}{5.2} \approx 323$.

(b) $P(15) = \dfrac{1680}{1 + 4.2e^{-0.11(15)}} \approx 930$ lions

(c) $P(t) = 1500$ $\Rightarrow$ $\dfrac{1680}{1 + 4.2e^{-0.11t}} = 1500$ $\Rightarrow$ $\dfrac{1680}{1500} = 1 + 4.2e^{-0.11t}$ $\Rightarrow$ $1.12 - 1 = 4.2e^{-0.11t}$ $\Rightarrow$

$\frac{0.12}{4.2} = e^{-0.11t}$ $\Rightarrow$ $\ln\frac{0.12}{4.2} = -0.11t$ $\Rightarrow$ $t = \frac{1}{-0.11}\ln\frac{0.12}{4.2} \approx 32.3$. Thus the population reaches 1500 about

32.3 years after January 1, 2010, or April, 2042.

(d) To find $P'$ we can use the Quotient Rule or we can rewrite $P$ as $P(t) = 1680(1 + 4.2e^{-0.11t})^{-1}$. Then, using the Chain Rule,

$$P'(t) = 1680(-1)(1 + 4.2e^{-0.11t})^{-2} \cdot \frac{d}{dt}(1 + 4.2e^{-0.11t})$$

$$= 1680(-1)(1 + 4.2e^{-0.11t})^{-2}[4.2e^{-0.11t}(-0.11)]$$

$$= 776.16e^{-0.11t}(1 + 4.2e^{-0.11t})^{-2} = \frac{776.16e^{-0.11t}}{(1 + 4.2e^{-0.11t})^2}$$

Then $P'(12) = \dfrac{776.16e^{-0.11(12)}}{(1 + 4.2e^{-0.11(12)})^2} \approx 46.0$, so on January 1, 2022, the number of mountain lions is increasing at a rate

of about 46 lions/year.

**33.** $P(t) = 23.7(1 + 4.8e^{-0.2t})^{-1}$ $\Rightarrow$ $P'(t) = 23.7(-1)(1 + 4.8e^{-0.2t})^{-2}[4.8e^{-0.2t}(-0.2)] = \dfrac{22.752e^{-0.2t}}{(1 + 4.8e^{-0.2t})^2}$, and

$P'(8) = \dfrac{22.752e^{-0.2(8)}}{(1 + 4.8e^{-0.2(8)})^2} \approx 1.18$. Thus on January 1, 2008, the animal population was increasing at a rate of about

1.18 thousand per year, or 1180 animals/year.

**35.** (a) $B$ is a logistic function with carrying capacity $M = 8 \times 10^7$, $k = 0.71$, and the initial biomass is $B_0 = 2 \times 10^7$, so

$$A = \frac{M - B_0}{B_0} = \frac{(8 \times 10^7) - (2 \times 10^7)}{2 \times 10^7} = \frac{6 \times 10^7}{2 \times 10^7} = 3. \text{ By Equation 6, } B(t) = \frac{M}{1 + Ae^{-kt}} = \frac{8 \times 10^7}{1 + 3e^{-0.71t}} \text{ kg after}$$

$t$ years. After one year, the biomass is $B(1) = \dfrac{8 \times 10^7}{1 + 3e^{-0.71(1)}} \approx 3.23 \times 10^7$ kg.

(b) $B(t) = 4 \times 10^7 \ \Rightarrow \ \dfrac{8 \times 10^7}{1 + 3e^{-0.71t}} = 4 \times 10^7 \ \Rightarrow \ \dfrac{8 \times 10^7}{4 \times 10^7} = 1 + 3e^{-0.71t} \ \Rightarrow \ 2 - 1 = 3e^{-0.71t} \ \Rightarrow$

$\frac{1}{3} = e^{-0.71t} \ \Rightarrow \ \ln\frac{1}{3} = -0.71t \ \Rightarrow \ t = \frac{1}{-0.71}\ln\frac{1}{3} \approx 1.55$ years

**37.** (a) Let $P$ be the world population, in billions, and let $t$ be the year, with $t = 0$ corresponding to 1990. The relative growth rate

is $\dfrac{dP/dt}{P} \approx \dfrac{0.02}{5.3} \approx 0.00377$.   (20 million is 0.02 billion.)

(b) $A = \dfrac{M - P_0}{P_0} = \dfrac{100 - 5.3}{5.3} = \dfrac{94.7}{5.3} \approx 17.87$, so by Equation 6,   $P(t) = \dfrac{M}{1 + Ae^{-kt}} \approx \dfrac{100}{1 + 17.87e^{-0.00377t}}$. The

model estimates the population in 2000 to be $P(10) = \dfrac{100}{1 + 17.87e^{-0.00377(10)}} \approx 5.49$ billion. This is a little lower (10%)

than the actual population.

(c) According to the model, the population in 2100 will be $P(110) = \dfrac{100}{1 + 17.87e^{-0.00377(110)}} \approx 7.81$ billion, and in 2500,

$P(510) = \dfrac{100}{1 + 17.87e^{-0.00377(510)}} \approx 27.68$ billion.

**39.** (a)

From the graph of $p(t) = 1/(1 + 10e^{-0.5t})$, it appears that $p(t) = 0.8$

(indicating that 80% of the population has heard the rumor) when

$t \approx 7.4$ hours.

(b) $p(t) = (1 + 10e^{-0.5t})^{-1} \ \Rightarrow \ p'(t) = -1(1 + 10e^{-0.5t})^{-2} \cdot 10e^{-0.5t}(-0.5) = \dfrac{5e^{-0.5t}}{(1 + 10e^{-0.5t})^2}$ and

$p'(6) = \dfrac{5e^{-0.5(6)}}{(1 + 10e^{-0.5(6)})^2} \approx 0.111$. Thus after six hours the rumor is spreading at a rate of about 11.1 percent of the

population per hour.

**41.** Following the hint, we choose $t = 0$ to correspond to 1960 and subtract

94,000 from each of the population figures. (This shifts the scatter plot

downward so that the horizontal asymptote is approximately the

$x$-axis.) We then use a calculator to obtain the models and add 94,000

to get the exponential function $P_E(t) \approx 1578.3(1.0933)^t + 94{,}000$ and

the logistic function $P_L(t) \approx \dfrac{32{,}658.5}{1 + 12.75e^{-0.1706t}} + 94{,}000$. $P_L$ is a

reasonably accurate model, while $P_E$ does not fit the data closely.

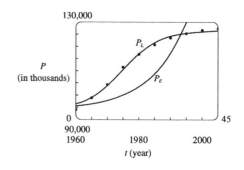

**43.** If the population decreases at a constant percentage rate, then the rate of decrease is proportional to the population. Thus as the population declines, the rate of change becomes smaller negative, and $B'(4) < B'(8)$ [$B'(4)$ is larger negative]. Also, the graph of $B$ follows exponential decay, so it is decreasing and gets less steep as we look to the right.

**45.** (a) $R(t) = R_0[R_0 + (1 - R_0)e^{-kt}]^{-1}$ $\Rightarrow$

$$R'(t) = -R_0[R_0 + (1 - R_0)e^{-kt}]^{-2}(1 - R_0)e^{-kt}(-k) = \frac{kR_0(1 - R_0)e^{-kt}}{[R_0 + (1 - R_0)e^{-kt}]^2}. \text{ We have}$$

$$1 - R(t) = 1 - \frac{R_0}{R_0 + (1 - R_0)e^{-kt}} = \frac{R_0 + (1 - R_0)e^{-kt} - R_0}{R_0 + (1 - R_0)e^{-kt}} = \frac{(1 - R_0)e^{-kt}}{R_0 + (1 - R_0)e^{-kt}}, \text{ so}$$

$$k \cdot R(t) \cdot [1 - R(t)] = k \cdot \frac{R_0}{R_0 + (1 - R_0)e^{-kt}} \cdot \frac{(1 - R_0)e^{-kt}}{R_0 + (1 - R_0)e^{-kt}}$$

$$= \frac{kR_0(1 - R_0)e^{-kt}}{[R_0 + (1 - R_0)e^{-kt}]^2} = R'(t)$$

as desired.

(b) Let $t$ be the number of hours since 8 AM. Then $R_0 = \frac{80}{1000} = 0.08$ and $R(4) = \frac{1}{2}$, so $\dfrac{0.08}{0.08 + (1 - 0.08)e^{-k(4)}} = \dfrac{1}{2}$ $\Rightarrow$

$2(0.08) = 0.08 + 0.92e^{-4k}$ $\Rightarrow$ $0.08 = 0.92e^{-4k}$ $\Rightarrow$ $e^{-4k} = \frac{0.08}{0.92}$ $\Rightarrow$ $-4k = \ln\frac{8}{92}$ $\Rightarrow$

$k = -\frac{1}{4}\ln\frac{8}{92} \approx 0.6106$. Thus $R(t) = \dfrac{0.08}{0.08 + 0.92e^{-0.6106t}}$. $R(t) = 0.9$ $\Rightarrow$ $\dfrac{0.08}{0.08 + 0.92e^{-0.6106t}} = 0.9$ $\Rightarrow$

$\frac{0.08}{0.9} = 0.08 + 0.92e^{-0.6106t}$ $\Rightarrow$ $\frac{8}{90} - 0.08 = 0.92e^{-0.6106t}$ $\Rightarrow$ $e^{-0.6106t} = \frac{(8/90) - 0.08}{0.92}$ $\Rightarrow$

$-0.6106t = \ln\frac{(8/90) - 0.08}{0.92}$ $\Rightarrow$ $t = \frac{1}{-0.6106}\ln\frac{(8/90) - 0.08}{0.92} \approx 7.60$ h or 7 h 36 min. Thus, 90% of the population will have heard the rumor by 3:36 PM.

**47.** (a) We graph $P(t) = 1000e^{-\ln(1000/100)e^{-0.05t}} = 1000e^{-(\ln 10)e^{-0.05t}}$.

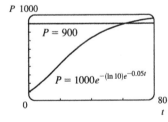

The graphs look very similar. For the Gompertz function, $P(40) \approx 732$, nearly the same as the logistic function. The Gompertz function reaches $P = 900$ at $t \approx 61.7$, and its value at $t = 80$ is about 959, so it doesn't increase quite as fast as the logistic curve.

(b) The population in 40 years is $P(40) = 1000e^{-\ln(10)e^{-0.05(40)}} \approx 732$. We have

$$P'(t) = 1000e^{-(\ln 10)e^{-0.05t}}\left[-(\ln 10)e^{-0.05t}(-0.05)\right] = 50(\ln 10)e^{-0.05t}e^{-(\ln 10)e^{-0.05t}}, \text{ so the rate of change of the}$$

population after 40 years is $P'(40) = 50(\ln 10)e^{-0.05(40)}e^{-(\ln 10)e^{-0.05(40)}} \approx 11.41$ members/year.

## 3 Review

**1.** $f(x) = 5x^3 - 7x + 13$ $\Rightarrow$ $f'(x) = 5(3x^2) - 7(1) + 0 = 15x^2 - 7$

**3.** $q = \sqrt[3]{r} + 6/r = r^{1/3} + 6r^{-1}$ $\Rightarrow$ $\dfrac{dq}{dr} = \frac{1}{3}r^{-2/3} + 6(-1)r^{-2} = \frac{1}{3r^{2/3}} - \frac{6}{r^2} = \frac{1}{3\sqrt[3]{r^2}} - \frac{6}{r^2}$

**5.** $h(u) = 3e^u + 1/\sqrt{u} = 3e^u + u^{-1/2} \;\Rightarrow\; h'(u) = 3e^u + \left(-\tfrac{1}{2}u^{-3/2}\right) = 3e^u - \dfrac{1}{2u^{3/2}}$ or $3e^u - \dfrac{1}{2u\sqrt{u}}$

**7.** $E(x) = 2.3(1.06)^x \;\Rightarrow\; E'(x) = 2.3(1.06)^x \ln 1.06 = 2.3(\ln 1.06)(1.06)^x$

**9.** $B(t) = 1 + 4\ln t \;\Rightarrow\; B'(t) = 0 + 4(1/t) = 4/t$

**11.** $C(a) = \sqrt{a}(e^a + 1) = a^{1/2}(e^a + 1) \;\Rightarrow$

$C'(a) = a^{1/2}\cdot(e^a + 0) + (e^a + 1)\cdot\tfrac{1}{2}a^{-1/2} = e^a\sqrt{a} + \dfrac{e^a + 1}{2\sqrt{a}}$ by the Product Rule.

**13.** $y = \dfrac{t}{1-t^2} \;\Rightarrow\; y' = \dfrac{(1-t^2)(1) - t(0-2t)}{(1-t^2)^2} = \dfrac{1 - t^2 + 2t^2}{(1-t^2)^2} = \dfrac{t^2 + 1}{(1-t^2)^2}$ by the Quotient Rule.

**15.** $y = (x^4 - 3x^2 + 5)^3 \;\Rightarrow\; y' = 3(x^4 - 3x^2 + 5)^2 \cdot \dfrac{d}{dx}(x^4 - 3x^2 + 5) = 3(x^4 - 3x^2 + 5)^2(4x^3 - 6x)$ by the Chain Rule.

**17.** $A = 16e^{-2t} \;\Rightarrow\; A' = 16\cdot e^{-2t}(-2) = -32e^{-2t}$

**19.** $y = \ln(x^3 + 5) \;\Rightarrow\; \dfrac{dy}{dx} = \dfrac{1}{x^3 + 5}(3x^2 + 0) = \dfrac{3x^2}{x^3 + 5}$

**21.** $y = 2x\sqrt{x^2 + 1} = 2x(x^2 + 1)^{1/2} \;\Rightarrow$

$$y' = 2x\cdot\tfrac{1}{2}(x^2 + 1)^{-1/2}(2x) + (x^2 + 1)^{1/2}\cdot 2 = \dfrac{2x^2}{\sqrt{x^2 + 1}} + 2\sqrt{x^2 + 1}$$

$$= \dfrac{2x^2}{\sqrt{x^2 + 1}} + 2\sqrt{x^2 + 1}\cdot\dfrac{\sqrt{x^2 + 1}}{\sqrt{x^2 + 1}} = \dfrac{2x^2 + 2(x^2 + 1)}{\sqrt{x^2 + 1}} = \dfrac{4x^2 + 2}{\sqrt{x^2 + 1}} = \dfrac{2(2x^2 + 1)}{\sqrt{x^2 + 1}}$$

**23.** $z = \sqrt{\dfrac{t}{t^2 + 4}} = \left[\dfrac{t}{t^2 + 4}\right]^{1/2} \;\Rightarrow$

$$\dfrac{dz}{dt} = \dfrac{1}{2}\left[\dfrac{t}{t^2 + 4}\right]^{-1/2}\cdot\dfrac{d}{dt}\left(\dfrac{t}{t^2 + 4}\right) = \dfrac{1}{2}\cdot\dfrac{(t^2 + 4)^{1/2}}{t^{1/2}}\cdot\dfrac{(t^2 + 4)(1) - t(2t)}{(t^2 + 4)^2}$$

$$= \dfrac{t^2 + 4 - 2t^2}{2t^{1/2}(t^2 + 4)^{3/2}} = \dfrac{4 - t^2}{2\sqrt{t}(t^2 + 4)^{3/2}}$$

**25.** $y = xe^{-1/x} = xe^{-x^{-1}} \;\Rightarrow\; y' = x\cdot e^{-x^{-1}}[-(-1)x^{-2}] + e^{-x^{-1}}(1) = xe^{-1/x}(1/x^2) + e^{-1/x} = e^{-1/x}(1/x + 1)$

**27.** $f(x) = 10^{x\sqrt{x-1}} \;\Rightarrow$

$$f'(x) = 10^{x\sqrt{x-1}}(\ln 10)\cdot\dfrac{d}{dx}[x(x-1)^{1/2}] = 10^{x\sqrt{x-1}}(\ln 10)[x\cdot\tfrac{1}{2}(x-1)^{-1/2}(1) + (x-1)^{1/2}\cdot 1]$$

$$= 10^{x\sqrt{x-1}}(\ln 10)\left(\dfrac{x}{2\sqrt{x-1}} + \sqrt{x-1}\right)$$

**29.** $A(r) = 6(\ln r)^4 \;\Rightarrow\; A'(r) = 6\cdot 4(\ln r)^3\cdot\dfrac{1}{r} = \dfrac{24(\ln r)^3}{r}$

**31.** $y = 3^{x\ln x} \;\Rightarrow\; y' = 3^{x\ln x}(\ln 3)\cdot\dfrac{d}{dx}(x\ln x) = 3^{x\ln x}(\ln 3)\left(x\cdot\dfrac{1}{x} + \ln x\cdot 1\right) = 3^{x\ln x}(\ln 3)(1 + \ln x)$

**33.** $y = [\ln(x^2 + 1)]^3 \quad \Rightarrow$

$$\frac{dy}{dx} = 3[\ln(x^2 + 1)]^2 \cdot \frac{d}{dx}[\ln(x^2 + 1)] = 3[\ln(x^2 + 1)]^2 \cdot \frac{1}{x^2 + 1}(2x) = \frac{6x[\ln(x^2 + 1)]^2}{x^2 + 1}$$

**35.** $f(t) = 500e^{0.65t} \quad \Rightarrow \quad f'(t) = 500e^{0.65t}(0.65) = 325e^{0.65t} \quad \Rightarrow$

$f''(t) = 325e^{0.65t}(0.65) = 211.25e^{0.65t}$

**37.** $f(x) = 2x - 5x^{3/4} \quad \Rightarrow \quad f'(x) = 2 - 5 \cdot \frac{3}{4}x^{-1/4} = 2 - \frac{15}{4}x^{-1/4} \quad \Rightarrow$

$f''(x) = 0 - \frac{15}{4}\left(-\frac{1}{4}x^{-5/4}\right) = \frac{15}{16}x^{-5/4}$

Note that $f'$ is negative when $f$ is decreasing and positive when $f$ is
increasing. $f''$ is always positive since $f'$ is always increasing.

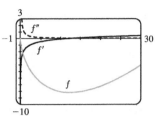

**39.** $\dfrac{d}{dx}(xy^4 + x^2y) = \dfrac{d}{dx}(x + 3y) \quad \Rightarrow \quad \left(x \cdot 4y^3 \dfrac{dy}{dx} + y^4 \cdot 1\right) + \left(x^2 \cdot \dfrac{dy}{dx} + y \cdot 2x\right) = 1 + 3\dfrac{dy}{dx} \quad \Rightarrow$

$4xy^3 \dfrac{dy}{dx} + x^2 \dfrac{dy}{dx} - 3\dfrac{dy}{dx} = 1 - y^4 - 2xy \quad \Rightarrow \quad (4xy^3 + x^2 - 3)\dfrac{dy}{dx} = 1 - y^4 - 2xy \quad \Rightarrow \quad \dfrac{dy}{dx} = \dfrac{1 - y^4 - 2xy}{4xy^3 + x^2 - 3}$

**41.** $f(t) = \sqrt{10 + 3t} = (10 + 3t)^{1/2} \quad \Rightarrow \quad f'(t) = \frac{1}{2}(10 + 3t)^{-1/2}(3) = \frac{3}{2}(10 + 3t)^{-1/2} \quad \Rightarrow$

$f''(t) = \frac{3}{2}\left(-\frac{1}{2}\right)(10 + 3t)^{-3/2}(3) = -\frac{9}{4}(10 + 3t)^{-3/2}$, so

$f''(2) = -\frac{9}{4}(10 + 3(2))^{-3/2} = -\frac{9}{4}(16)^{-3/2} = -\dfrac{9}{4(\sqrt{16})^3} = -\dfrac{9}{4(4)^3} = -\dfrac{9}{256}.$

**43.** $y = (2 + x)e^{-x} \quad \Rightarrow \quad dy/dx = (2 + x) \cdot e^{-x}(-1) + e^{-x} \cdot 1 = e^{-x}[-(2 + x) + 1] = e^{-x}(-x - 1).$

At $(0, 2)$, $dy/dx = e^0(0 - 1) = 1(-1) = -1$, so an equation of the tangent line is $y - 2 = -1(x - 0)$, or $y = -x + 2$.

**45.** $y = (3x - 2)^5 \quad \Rightarrow \quad dy/dx = 5(3x - 2)^4(3) = 15(3x - 2)^4.$ At $(1, 1)$, $dy/dx = 15(1)^4 = 15$, so an equation of the
tangent line is $y - 1 = 15(x - 1)$, or $y = 15x - 14.$

**47.** (a) $f(x) = x\sqrt{5 - x} = x(5 - x)^{1/2} \quad \Rightarrow$

$$f'(x) = x\left[\frac{1}{2}(5 - x)^{-1/2}(-1)\right] + (5 - x)^{1/2}(1) = \frac{-x}{2\sqrt{5 - x}} + \sqrt{5 - x} \cdot \frac{2\sqrt{5 - x}}{2\sqrt{5 - x}}$$

$$= \frac{-x}{2\sqrt{5 - x}} + \frac{2(5 - x)}{2\sqrt{5 - x}} = \frac{-x + 10 - 2x}{2\sqrt{5 - x}} = \frac{10 - 3x}{2\sqrt{5 - x}}$$

(b) At $(1, 2)$: $f'(1) = \frac{10 - 3}{2\sqrt{5 - 1}} = \frac{7}{4}.$

So an equation of the tangent line is $y - 2 = \frac{7}{4}(x - 1)$ or $y = \frac{7}{4}x + \frac{1}{4}.$

At $(4, 4)$: $f'(4) = \frac{10 - 12}{2\sqrt{5 - 4}} = \frac{-2}{2} = -1.$

So an equation of the tangent line is $y - 4 = -1(x - 4)$ or $y = -x + 8.$

(c)

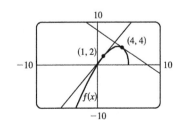

(d)

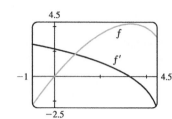

The graphs look reasonable, since $f'$ is positive where $f$ is increasing, and $f'$ is negative where $f$ is decreasing.

**49.** (a) The average cost function is $\dfrac{C(q)}{q} = \dfrac{920 + 2q - 0.02q^2 + 0.00007q^3}{q} = \dfrac{920}{q} + 2 - 0.02q + 0.00007q^2$, and the

average cost after producing 1500 units is $\dfrac{C(1500)}{1500} = \dfrac{195{,}170}{1500} \approx \$130.11/\text{unit}.$

(b) $C(q) = 920 + 2q - 0.02q^2 + 0.00007q^3$, so the marginal cost is

$C'(q) = 0 + 2 - 0.02(2q) + 0.00007(3q^2) = 2 - 0.04q + 0.00021q^2$ dollars per unit.

(c) $C'(100) = 2 - 0.04(100) + 0.00021(100)^2 = 2 - 4 + 2.1 = \$0.10/\text{unit}$. This value represents the rate at which costs are increasing as the hundredth unit is produced, and is the approximate cost of producing the 101st unit.

(d) The actual cost of producing the 101st item is $C(101) - C(100) = 990.10107 - 990 = \$0.10107$, slightly higher than $C'(100)$.

**51.** (a) $P(q) = R(q) - C(q) = (1.36q - 0.0001q^2) - (380 + 0.32q + 0.0002q^2) = -380 + 1.04q - 0.0003q^2$

(b) The average revenue is $\dfrac{R(q)}{q} = \dfrac{1.36q - 0.0001q^2}{q} = 1.36 - 0.0001q$ dollars per can, and the marginal revenue is

$R'(q) = 1.36 - 0.0001(2q) = 1.36 - 0.0002q$ dollars per can.

(c) $C'(q) = 0.32 + 0.004q$, and $R'(q) > C'(q)$ initially, so profit is maximized when $R'(q) = C'(q)$ $\Rightarrow$

$1.36 - 0.0002q = 0.32 + 0.0004q$ $\Rightarrow$ $1.04 = 0.0006q$ $\Rightarrow$ $q = \frac{1.04}{0.0006} \approx 1733$ cans.

**53.** (a) Each year the value is 104.6% of the previous year's value, so after $t$ years, the value is

$V(t) = 2.6(1.046)^t$ million dollars. After 3.5 years, the property is worth $V(3.5) = 2.6(1.046)^{3.5} \approx \$3.04$ million.

(b) The value decreases 4.6% each year, so 95.4% of the value is maintained. After $t$ years, the value is

$V(t) = 2.6(0.954)^t$ million dollars. When $V(t) = 2$ we have $2.6(0.954)^t = 2$ $\Rightarrow$ $(0.954)^t = \frac{2}{2.6}$ $\Rightarrow$

$t \ln 0.954 = \ln \frac{2}{2.6}$ $\Rightarrow$ $t = \frac{1}{\ln 0.954} \ln \frac{2}{2.6} \approx 5.57$, so the value declines to \$2 million after about 5.57 years.

**55.** (a) The population after $t$ hours is $P(t) = 200e^{kt}$, and in half an hour there are 360 cells, so $P(\frac{1}{2}) = 200e^{k/2} = 360$ $\Rightarrow$

$e^{k/2} = \frac{360}{200} = 1.8$ $\Rightarrow$ $k/2 = \ln 1.8$ $\Rightarrow$ $k = 2 \ln 1.8 \approx 1.1756$. Thus $P(t) = 200e^{1.1756t}$.

(b) $P(4) = 200e^{1.1756(4)} \approx 22{,}040$ bacteria

(c) $P'(t) = 200e^{1.1756t}(1.1756) = 235.12e^{1.1756t}$, so $P'(4) = 235.12e^{1.1756(4)} \approx 25{,}910$ bacteria per hour

(d) The relative growth rate is $k = 2 \ln 1.8 \approx 1.1756$ percent/hour.

(e) $P(t) = 10{,}000$ $\Rightarrow$ $200e^{1.1756t} = 10{,}000$ $\Rightarrow$ $e^{1.1756t} = 50$ $\Rightarrow$ $1.1756t = \ln 50$ $\Rightarrow$

$t = \frac{1}{1.1756} \ln 50 \approx 3.33$ hours

**57.** Let $C_0$ be the initial concentration of the drug; then the concentration after $t$ hours is $C(t) = C_0 e^{kt}$.  $C(30) = \frac{1}{2}C_0$ since

the concentration is reduced by half after 30 hours, so $C_0 e^{30k} = \frac{1}{2}C_0$  $\Rightarrow$  $30k = \ln\frac{1}{2}$  $\Rightarrow$  $k = \frac{1}{30}\ln\frac{1}{2} \approx -0.02310$.

Since 10% of the original concentration remains if 90% is eliminated, we want the value of $t$ such that $C(t) = \frac{1}{10}C_0$.

Therefore, $C_0 e^{-0.02310t} = \frac{1}{10}C_0$  $\Rightarrow$  $-0.02310t = \ln\frac{1}{10}$  $\Rightarrow$  $t = \frac{1}{-0.02310}\ln\frac{1}{10} \approx 99.68$. Thus it takes about

100 hours to eliminate 90% of the drug.

**59. (a)** The mass (in ounces) remaining after $t$ years is $m(t) = 2e^{kt}$. We are given that $m(6.3) = 0.8$  $\Rightarrow$  $2e^{k(6.3)} = 0.8$  $\Rightarrow$

$e^{6.3k} = 0.4$  $\Rightarrow$  $6.3k = \ln 0.4$  $\Rightarrow$  $k = \frac{1}{6.3}\ln 0.4 \approx -0.1454$. Thus $m(t) = 2e^{-0.1454t}$, and the half-life is the time

required for half of the mass to remain: $m(t) = 1$  $\Rightarrow$  $2e^{-0.1454t} = 1$  $\Rightarrow$  $e^{-0.1454t} = \frac{1}{2}$  $\Rightarrow$

$-0.1454t = \ln\frac{1}{2}$  $\Rightarrow$  $t = \frac{1}{-0.1454}\ln\frac{1}{2} \approx 4.77$ years.

**(b)** To find the time required for just 10% of the material to remain, we solve $m(t) = 0.1(2) = 0.2$  $\Rightarrow$

$2e^{-0.1454t} = 0.2$  $\Rightarrow$  $e^{-0.1454t} = 0.1$  $\Rightarrow$  $-0.1454t = \ln 0.1$  $\Rightarrow$  $t = \frac{1}{-0.1454}\ln 0.1 \approx 15.84$ years.

**61. (a)** Comparing to Equation 6 in Section 3.6, we see that $M = 285$, so the carrying capacity is 285,000. The population on

January 1, 2010, is $P(0) = \dfrac{285}{1 + 3.8e^0} = 59.375$ thousand, or 59,375.

**(b)** $P(25) = \dfrac{285}{1 + 3.8e^{-0.08(25)}} \approx 188$ thousand, or 188,000.

**(c)** $P(t) = 200$  $\Rightarrow$  $\dfrac{285}{1 + 3.8e^{-0.08t}} = 200$  $\Rightarrow$  $1 + 3.8e^{-0.08t} = \frac{285}{200} = 1.425$  $\Rightarrow$  $3.8e^{-0.08t} = 0.425$  $\Rightarrow$

$e^{-0.08t} = \frac{0.425}{3.8}$  $\Rightarrow$  $-0.08t = \ln\frac{0.425}{3.8}$  $\Rightarrow$  $t = \frac{1}{-0.08}\ln\frac{0.425}{3.8} \approx 27.4$. Thus the population reaches 200,000 about

27.4 years after January 1, 2010, or May 2037.

**(d)** $P(t) = 285(1 + 3.8e^{-0.08t})^{-1}$  $\Rightarrow$  $P'(t) = 285(-1)(1 + 3.8e^{-0.08t})^{-2}(3.8e^{-0.08t})(-0.08) = \dfrac{86.64e^{-0.08t}}{(1 + 3.8e^{-0.08t})^2}$,

so $P'(30) = \dfrac{86.64e^{-0.08(30)}}{(1 + 3.8e^{-0.08(30)})^2} \approx 4.35$. This means that on January 1, 2040, the population will be increasing at a rate

of about 4350 animals per year.

**63.** $y = [\ln(x + 4)]^2$  $\Rightarrow$  $\dfrac{dy}{dx} = 2[\ln(x + 4)]^1 \cdot \dfrac{1}{x + 4}(1) = \dfrac{2\ln(x + 4)}{x + 4}$ and $\dfrac{dy}{dx} = 0$  $\Leftrightarrow$  $\ln(x + 4) = 0$  $\Leftrightarrow$

$x + 4 = e^0$  $\Rightarrow$  $x + 4 = 1$  $\Rightarrow$  $x = -3$. When $x = -3$, $y = [\ln(1)]^2 = 0$, so the tangent is horizontal at the point

$(-3, 0)$.

**65.** We find $\dfrac{dy}{dx}$ by differentiating implicitly: $\dfrac{d}{dx}(x^2 + 2y^2) = \dfrac{d}{dx}(1)$  $\Rightarrow$  $2x + 4y\dfrac{dy}{dx} = 0$  $\Rightarrow$  $\dfrac{dy}{dx} = \dfrac{-2x}{4y} = -\dfrac{x}{2y}$. Thus

$\dfrac{dy}{dx} = 1$  $\Leftrightarrow$  $-\dfrac{x}{2y} = 1$  $\Leftrightarrow$  $x = -2y$. Because the points lie on the ellipse, we must have $(-2y)^2 + 2y^2 = 1$  $\Rightarrow$

$6y^2 = 1$  $\Rightarrow$  $y = \pm\frac{1}{\sqrt{6}}$. Since $x = -2y$, the points are $\left(-\frac{2}{\sqrt{6}}, \frac{1}{\sqrt{6}}\right)$ and $\left(\frac{2}{\sqrt{6}}, -\frac{1}{\sqrt{6}}\right)$.

**67.** (a) $h(x) = f(x) g(x) \Rightarrow h'(x) = f(x) g'(x) + g(x) f'(x) \Rightarrow$

$h'(2) = f(2) g'(2) + g(2) f'(2) = (3)(4) + (5)(-2) = 12 - 10 = 2$

(b) $F(x) = f(g(x)) \Rightarrow F'(x) = f'(g(x)) g'(x) \Rightarrow F'(2) = f'(g(2)) g'(2) = f'(5)(4) = 11 \cdot 4 = 44$

# 4 APPLICATIONS OF DIFFERENTIATION

## 4.1 Related Rates

**1.**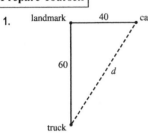

Let $d$ be the distance between the car and the truck. Looking at the figure, we see that we have a right triangle, so by the Pythagorean Theorem, $40^2 + 60^2 = d^2$ $\Rightarrow$ $d^2 = 1600 + 3600 = 5200$ $\Rightarrow$ $d = \sqrt{5200} = 20\sqrt{13} \approx 72.1$ miles.

**2.** The distance $d$ between the planes corresponds to the hypotenuse of a right triangle whose legs are the paths followed by the planes. Thus $x^2 + y^2 = d^2$ $\Rightarrow$ $d = \sqrt{x^2 + y^2}$ miles.

**3.** (a) $y = [f(x)]^4$ $\Rightarrow$ $y' = 4[f(x)]^3 f'(x)$ by the Chain Rule.

   (b) $y = x^2 + x\, f(x)$ $\Rightarrow$ $y' = 2x + [x \cdot f'(x) + f(x) \cdot 1] = 2x + x\, f'(x) + f(x)$

**4.** (a) $y = A(t)\, B(t)$ $\Rightarrow$ $y' = A(t) \cdot B'(t) + B(t) \cdot A'(t) = A(t)B'(t) + A'(t)B(t)$ by the Product Rule.

   (b) $y = [A(t)]^2 + [B(t)]^2$ $\Rightarrow$ $y' = 2A(t)\, A'(t) + 2B(t)\, B'(t)$   by the Chain Rule (twice).

   (c) $y = [A(t)]^2 B(t)$ $\Rightarrow$ $y' = [A(t)]^2 \cdot B'(t) + B(t) \cdot 2[A(t)]^1 A'(t) = [A(t)]^2 B'(t) + 2A(t)\, A'(t)B(t)$

---

**Exercises**

**1.** The volume of the cube is $V = x^3$. Then using the Chain Rule we have $\dfrac{dV}{dt} = \dfrac{dV}{dx}\dfrac{dx}{dt} = 3x^2\,\dfrac{dx}{dt}$.

**3.** Let $s$ denote the side of a square. Then the square's area $A$ is given by $A = s^2$, and differentiating with respect to $t$ gives $\dfrac{dA}{dt} = 2s\,\dfrac{ds}{dt}$. We are given that $\dfrac{ds}{dt} = 6$, and when $A = 16$, $s = 4$. Thus at that time the area of the square is increasing at the rate $\dfrac{dA}{dt} = 2(4)(6) = 48$ cm$^2$/s.

**5.** $y = x^3 + 2x$, and differentiating with respect to $t$ gives $\dfrac{dy}{dt} = 3x^2\,\dfrac{dx}{dt} + 2\,\dfrac{dx}{dt}$. We are given that $\dfrac{dx}{dt} = 5$, and when $x = 2$,
$\dfrac{dy}{dt} = 3(2)^2(5) + 2(5) = 70$.

**7.** $z^2 = x^2 + y^2$ $\Rightarrow$ $2z\,\dfrac{dz}{dt} = 2x\,\dfrac{dx}{dt} + 2y\,\dfrac{dy}{dt}$. When $x = 5$ and $y = 12$,
$z^2 = 5^2 + 12^2$ $\Rightarrow$ $z^2 = 169$ $\Rightarrow$ $z = 13$ (since $z$ is positive). Also $\dfrac{dx}{dt} = 2$ and $\dfrac{dy}{dt} = 3$, so
$2(13)\,\dfrac{dz}{dt} = 2(5)(2) + 2(12)(3)$ $\Rightarrow$ $\dfrac{dz}{dt} = \dfrac{20 + 72}{26} = \dfrac{46}{13}$.

**9.** $C = 2200 + 16q - 0.01q^2$ and $C$ and $q$ are changing with time, so differentiating with respect to $t$ gives

$\dfrac{dC}{dt} = 16\dfrac{dq}{dt} - 0.02q\dfrac{dq}{dt}$. When $q = 600$ and $\dfrac{dq}{dt} = 40$, the rate of change of cost is

$\dfrac{dC}{dt} = 16(40) - 0.02(600)(40) = \$160/\text{week}$.

**11.** Both $q$ and $p$ are changing with respect to time $t$, so $qe^{0.03p} = 5000 \Rightarrow q = 5000e^{-0.03p} \Rightarrow$

$\dfrac{dq}{dt} = 5000e^{-0.03p}\left(-0.03\dfrac{dp}{dt}\right) = -150e^{-0.03p}\dfrac{dp}{dt}$. When $p = 14$ and $\dfrac{dp}{dt} = 1.20$, we have

$\dfrac{dq}{dt} = -150e^{-0.03(14)}(1.20) = -180e^{-0.42} \approx -118.27$. Thus the demand is decreasing at a rate of about 118 bottles/year.

**13.** (a) Given: the rate of decrease of the surface area is $1\ \text{cm}^2/\text{min}$. If we let $t$ be time (in minutes) and $S$ the surface

area (in $\text{cm}^2$), then we are given that $dS/dt = -1\ \text{cm}^2/\text{s}$.

(b) Unknown: the rate of decrease of the diameter when the diameter is 10 cm.     (c)

If we let $x$ be the diameter, then we want to find $dx/dt$ when $x = 10$ cm.

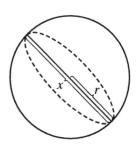

(d) If the radius is $r$, then $x = 2r \Rightarrow r = \frac{1}{2}x$ and the surface area is $S = 4\pi r^2 = 4\pi\left(\frac{1}{2}x\right)^2 = \pi x^2$. Differentiating with

respect to $t$, we have $\dfrac{dS}{dt} = 2\pi x\dfrac{dx}{dt}$.

(e) When $x = 10$, $\dfrac{dS}{dt} = 2\pi x\dfrac{dx}{dt} \Rightarrow -1 = 2\pi(10)\dfrac{dx}{dt} \Rightarrow \dfrac{dx}{dt} = -\dfrac{1}{20\pi}$. So the rate of decrease

is $\dfrac{1}{20\pi} \approx 0.0159$ cm/min.

**15.** (a) Given: a plane flying horizontally at an altitude of 1 mi and a speed of 500 mi/h passes directly over a radar station.

If we let $t$ be time (in hours) and $x$ the horizontal distance traveled by the plane (in mi), then we are given

that $dx/dt = 500$ mi/h.

(b) Unknown: the rate at which the distance from the plane to the station is     (c)

increasing when the plane is 2 mi from the station. If we let $y$ be the distance

from the plane to the station, then we want to find $dy/dt$ when $y = 2$ mi.

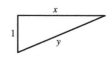

(d) By the Pythagorean Theorem, $y^2 = x^2 + 1 \Rightarrow 2y\,(dy/dt) = 2x\,(dx/dt)$.

(e) $y = 2 \Rightarrow 4 = x^2 + 1 \Rightarrow x = \sqrt{3}$. Then $2(2)\dfrac{dy}{dt} = 2\sqrt{3}(500) \Rightarrow \dfrac{dy}{dt} = \dfrac{1000\sqrt{3}}{4} = 250\sqrt{3} \approx 433$, so the

distance from the plane to the station is increasing at a rate of about 433 mi/h.

**17.**  Let $x$ be the distance the first car travels, $y$ the distance of the second car, and $z$ the

distance between the cars. We are given that $\dfrac{dx}{dt} = 60$ mi/h and $\dfrac{dy}{dt} = 25$ mi/h.

$$z^2 = x^2 + y^2 \quad\Rightarrow\quad 2z\,\frac{dz}{dt} = 2x\,\frac{dx}{dt} + 2y\,\frac{dy}{dt} \quad\Rightarrow\quad z\,\frac{dz}{dt} = x\,\frac{dx}{dt} + y\,\frac{dy}{dt}.$$

After 2 hours, $x = (60\text{ mi/h})(2\text{ h}) = 120$ mi and $y = (25\text{ mi/h})(2\text{ h}) = 50$ mi $\;\Rightarrow\;$ $z = \sqrt{120^2 + 50^2} = 130$ mi, so

$(130)\dfrac{dz}{dt} = (120)(60) + (50)(25) \;\Rightarrow\; \dfrac{dz}{dt} = \dfrac{120(60) + 50(25)}{130} = 65$. Thus the distance between the cars is increasing at

a rate of 65 mi/h.

**19.** 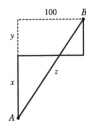 Let $x$ be the distance ship $A$ sails and $y$ the distance for ship $B$. We are given that

$\dfrac{dx}{dt} = 35$ km/h and $\dfrac{dy}{dt} = 25$ km/h. If $z$ is the distance between the ships, then by the

Pythagorean Theorem, $z^2 = (x+y)^2 + 100^2 \;\Rightarrow\; 2z\,\dfrac{dz}{dt} = 2(x+y)\left(\dfrac{dx}{dt} + \dfrac{dy}{dt}\right)$.

At 4:00 PM, $x = 4(35) = 140$ and $y = 4(25) = 100 \;\Rightarrow\;$

$z = \sqrt{(x+y)^2 + 100^2} = \sqrt{(140+100)^2 + 100^2} = \sqrt{67{,}600} = 260$, so $2(260)\dfrac{dz}{dt} = 2(140+100)(35+25) \;\Rightarrow\;$

$\dfrac{dz}{dt} = \dfrac{2(240)(60)}{2(260)} = \dfrac{720}{13} \approx 55.4$ km/h.

**21.** Differentiating both sides of $PV = C$ with respect to $t$ and using the Product Rule gives $P\dfrac{dV}{dt} + V\dfrac{dP}{dt} = 0$. When

$V = 600$, $P = 150$, and $\dfrac{dP}{dt} = 20$, we have $(150)\dfrac{dV}{dt} + (600)(20) = 0 \;\Rightarrow\; \dfrac{dV}{dt} = -\dfrac{(600)(20)}{150} = -80$. Thus, the

volume is decreasing at a rate of 80 cm$^3$/min.

**23.** Let $y$ be the height of the rocket and $\ell$ the distance from the camera to the rocket. By the Pythagorean Theorem,

$4000^2 + y^2 = \ell^2$. Differentiating with respect to $t$, we obtain $2y\,\dfrac{dy}{dt} = 2\ell\,\dfrac{d\ell}{dt}$.

We know that $\dfrac{dy}{dt} = 600$ ft/s, and when $y = 3000$ ft we have

$\ell = \sqrt{4000^2 + y^2} = \sqrt{4000^2 + 3000^2} = \sqrt{25{,}000{,}000} = 5000$ ft. Thus

$2(3000)(600) = 2(5000)\dfrac{d\ell}{dt} \;\Rightarrow\; \dfrac{d\ell}{dt} = \dfrac{2(3000)(600)}{2(5000)} = \dfrac{1800}{5} = 360$ ft/s.

**25.** 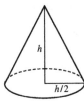 If $V$ is the volume, then we are given that $\dfrac{dV}{dt} = 30$ ft$^3$/min. The diameter = the height,

so $h = 2r \;\Rightarrow\; r = \dfrac{h}{2}$ and $V = \dfrac{1}{3}\pi r^2 h = \dfrac{1}{3}\pi\left(\dfrac{h}{2}\right)^2 h = \dfrac{\pi h^3}{12}$. Differentiating with

respect to $t$, we have $\dfrac{dV}{dt} = \dfrac{\pi}{12}\cdot 3h^2\dfrac{dh}{dt}$.

When $h = 10$ ft, $30 = \dfrac{\pi}{4}(10)^2\dfrac{dh}{dt} \;\Rightarrow\; \dfrac{dh}{dt} = \dfrac{30}{10^2(\pi/4)} = \dfrac{6}{5\pi} \approx 0.38$ ft/min.

**27.** $W = 13.12 + 0.6215T - 11.37v^{0.16} + 0.3965Tv^{0.16}$ $\Rightarrow$

$\dfrac{dW}{dt} = 0.6215\dfrac{dT}{dt} - 11.37(0.16v^{-0.84})\dfrac{dv}{dt} + 0.3965\left[T(0.16v^{-0.84})\dfrac{dv}{dt} + v^{0.16}\dfrac{dT}{dt}\right]$. When

$T = -15$, $v = 30$, $\dfrac{dT}{dt} = 2$, and $\dfrac{dv}{dt} = 4$, the apparent temperature is changing at a rate of

$\dfrac{dW}{dt} = 0.6215(2) - 11.37[0.16(30)^{-0.84}](4) + 0.3965\left[0.16(-15)(30)^{-0.84}(4) + (30)^{0.16}(2)\right] \approx 1.97\,°\mathrm{C/h}$.

**29.** The area of a triangle is $A = \frac{1}{2}bh$, where $b$ is the base and $h$ is the altitude. We are given that $\dfrac{dh}{dt} = 1$ cm/min and

$\dfrac{dA}{dt} = 2$ cm$^2$/min. Using the Product Rule, we have $\dfrac{dA}{dt} = \dfrac{1}{2}\left(b\dfrac{dh}{dt} + h\dfrac{db}{dt}\right)$. When $h = 10$ and $A = 100$, we have

$100 = \frac{1}{2}b(10)$ $\Rightarrow$ $\frac{1}{2}b = 10$ $\Rightarrow$ $b = 20$, so $2 = \dfrac{1}{2}\left(20\cdot 1 + 10\dfrac{db}{dt}\right)$ $\Rightarrow$ $4 = 20 + 10\dfrac{db}{dt}$ $\Rightarrow$

$\dfrac{db}{dt} = \dfrac{4-20}{10} = -1.6$ cm/min.

**31.** 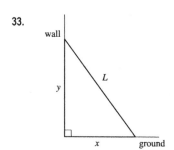 Let $h$ be the depth of the water (in feet). By similar triangles, $\dfrac{3}{b} = \dfrac{1}{h}$, so $b = 3h$.

The water has volume

$$V = (\text{area of end})\,(\text{length}) = \tfrac{1}{2}bh \cdot 10 = \tfrac{1}{2}(3h)h \cdot 10 = 15h^2 \quad\Rightarrow\quad \dfrac{dV}{dt} = 30h\dfrac{dh}{dt}.$$

We are given that $\dfrac{dV}{dt} = 12$, and when the water depth is 6 inches or a half foot, we have $h = \frac{1}{2}$, so $12 = 30\left(\dfrac{1}{2}\right)\dfrac{dh}{dt}$ $\Rightarrow$

$\dfrac{dh}{dt} = \dfrac{12}{15} = \dfrac{4}{5}$ ft/min.

**33.** 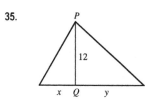 From the figure and given information, we have $x^2 + y^2 = L^2$, $\dfrac{dy}{dt} = -0.15$ m/s, and

$\dfrac{dx}{dt} = 0.2$ m/s when $x = 3$ m. Differentiating implicitly with respect to $t$, we get

$x^2 + y^2 = L^2$ $\Rightarrow$ $2x\dfrac{dx}{dt} + 2y\dfrac{dy}{dt} = 0$, and substituting the given information gives

$2(3)(0.2) + 2y(-0.15) = 0$ $\Rightarrow$ $y = \dfrac{2(3)(0.2)}{2(0.15)} = 4$. Thus

$L^2 = x^2 + y^2 = 3^2 + 4^2 = 25$ $\Rightarrow$ $L = 5$, so the ladder is 5 m long.

**35.** We are given $\dfrac{dx}{dt} = 2$ ft/s and need to find $\dfrac{dy}{dt}$ when $x = 5$. Using the Pythagorean

Theorem, the length of rope to the left of $P$ is $\sqrt{x^2 + 12^2}$ and the length to the right of

$P$ is $\sqrt{y^2 + 12^2}$; the rope is 39 feet long, so $\sqrt{x^2 + 144} + \sqrt{y^2 + 144} = 39$.

Differentiating with respect to $t$, we have $\frac{1}{2}(x^2 + 144)^{-1/2}\cdot 2x\dfrac{dx}{dt} + \frac{1}{2}(y^2 + 144)^{-1/2}\cdot 2y\dfrac{dy}{dt} = 0$ $\Rightarrow$

$\dfrac{x}{\sqrt{x^2 + 144}}\dfrac{dx}{dt} + \dfrac{y}{\sqrt{y^2 + 144}}\dfrac{dy}{dt} = 0$. When $x = 5$, $\sqrt{5^2 + 144} + \sqrt{y^2 + 144} = 39$ $\Rightarrow$ $\sqrt{y^2 + 144} = 39 - 13$ $\Rightarrow$

$y^2 = 26^2 - 144 = 532 \ \Rightarrow \ y = \sqrt{532}$, so we have $\dfrac{5}{\sqrt{25+144}}(2) + \dfrac{\sqrt{532}}{\sqrt{532+144}}\dfrac{dy}{dt} = 0 \ \Rightarrow$

$\dfrac{10}{13} + \dfrac{\sqrt{532}}{26}\dfrac{dy}{dt} = 0 \ \Rightarrow \ \dfrac{dy}{dt} = \dfrac{-10/13}{\sqrt{532}/26} = \dfrac{10}{\sqrt{133}} \approx -0.87$ ft/s. Thus cart $B$ is moving towards $Q$ at about $0.87$ ft/s.

## 4.2 Maximum and Minimum Values

> **Prepare Yourself**

**1.** (a) $4x^2 + x - 3 = 0 \ \Rightarrow \ (4x-3)(x+1) = 0 \ \Rightarrow \ 4x - 3 = 0 \ $ or $\ x + 1 = 0 \ \Rightarrow \ x = \frac{3}{4} \ $ or $\ x = -1$

(b) $\dfrac{a^2 - 2a}{(a+1)^2} = 0$ only when $a^2 - 2a = 0 \ \Rightarrow \ a(a-2) = 0 \ \Rightarrow \ a = 0 \ $ or $\ a - 2 = 0 \ \Rightarrow \ a = 0 \ $ or $\ a = 2$.

(c) $5t^2 - 5 = 0 \ \Rightarrow \ 5(t^2 - 1) = 0 \ \Rightarrow \ 5(t+1)(t-1) = 0 \ \Rightarrow \ t + 1 = 0 \ $ or $\ t - 1 = 0 \ \Rightarrow$

$t = -1 \ $ or $\ t = 1$

(d) $3 + \dfrac{1}{x} = 0 \ \Rightarrow \ \dfrac{1}{x} = -3 \ \Rightarrow \ x = -\dfrac{1}{3}$

**2.** $2x^2 - x - 5 = 0$ is a quadratic equation of form $ax^2 + bx + c = 0$ with $a = 2$, $b = -1$, and $c = -5$, so the solutions are

$x = \dfrac{-b \pm \sqrt{b^2 - 4ac}}{2a} = \dfrac{-(-1) \pm \sqrt{(-1)^2 - 4(2)(-5)}}{2(2)} = \dfrac{1 \pm \sqrt{41}}{4} = \dfrac{1}{4} \pm \dfrac{1}{4}\sqrt{41}$.

**3.** Since $(1 + x^2)^{-1/2} \cdot (1 + x^2) = (1 + x^2)^{(-1/2)+1} = (1 + x^2)^{1/2}$, we can write

$x^2(1 + x^2)^{-1/2} + 3(1 + x^2)^{1/2} = (1 + x^2)^{-1/2}\left[x^2 + 3(1 + x^2)\right] = (1 + x^2)^{-1/2}(4x^2 + 3)$.

**4.** We factor out the lowest powers of $t$ and $(t + 2)$:

$t^{2/3}(t+2) - 5t^{-1/3}(t+2)^2 = t^{-1/3}(t+2)\left[t^1 - 5(t+2)\right] = t^{-1/3}(t+2)(-4t - 10)$. We can now factor out $-2$ from

the last factor to obtain $-2t^{-1/3}(t+2)(2t+5)$.

**5.** (a) $3xe^x + 2e^x = 0 \ \Rightarrow \ e^x(3x + 2) = 0 \ \Rightarrow \ 3x + 2 = 0 \ $ [since $e^x \neq 0$ for all $x$] $\ \Rightarrow \ 3x = -2 \ \Rightarrow \ x = -\frac{2}{3}$

(b) $\ln x - 2 = 0 \ \Rightarrow \ \ln x = 2 \ \Rightarrow \ e^{\ln x} = e^2 \ \Rightarrow \ x = e^2$

(c) $2x(x^2 - 9)^3 = 0 \ \Rightarrow \ 2x[(x+3)(x-3)]^3 = 0 \ \Rightarrow \ 2x(x+3)^3(x-3)^3 = 0 \ \Rightarrow$

$2x = 0 \ $ or $\ x + 3 = 0 \ $ or $\ x - 3 = 0 \ \Rightarrow \ x = 0 \ $ or $\ x = -3 \ $ or $\ x = 3$

**6.** (a) $f(x) = xe^{5x} \ \Rightarrow \ f'(x) = x \cdot e^{5x}(5) + e^{5x} \cdot 1 = (5x + 1)e^{5x}$ by the Product Rule (and the Chain Rule).

(b) $f(t) = \dfrac{t+3}{t^2+4} \ \Rightarrow \ f'(t) = \dfrac{(t^2+4)(1) - (t+3)(2t)}{(t^2+4)^2} = \dfrac{t^2 + 4 - 2t^2 - 6t}{(t^2+4)^2} = \dfrac{-t^2 - 6t + 4}{(t^2+4)^2}$ by the Quotient Rule.

(c) $y = \sqrt{1 + \ln x} = (1 + \ln x)^{1/2} \ \Rightarrow \ \dfrac{dy}{dx} = \dfrac{1}{2}(1 + \ln x)^{-1/2} \cdot \dfrac{1}{x} = \dfrac{1}{2x\sqrt{1 + \ln x}}$

(d) $y = x\sqrt{1 + \ln x} = x(1 + \ln x)^{1/2}$ $\Rightarrow$

$$\frac{dy}{dx} = x \cdot \frac{1}{2}(1 + \ln x)^{-1/2}\left(\frac{1}{x}\right) + (1 + \ln x)^{1/2} \cdot 1 = \frac{1}{2\sqrt{1 + \ln x}} + \sqrt{1 + \ln x}$$

## Exercises

**1.** A function $f$ has an absolute minimum at $x = c$ if $f(c)$ is the smallest function value on the entire domain of $f$, whereas $f$ has a local minimum at $c$ if $f(c)$ is the smallest function value when $x$ is near $c$.

**3.** Absolute maximum at $b$; absolute minimum at $d$; local maximum at both $b$ and $e$; local minimum at both $d$ and $s$; neither a maximum nor a minimum at $a$, $c$, $r$, and $t$.

**5.** Absolute maximum value is $f(4) = 4$; absolute minimum value is $f(7) = 0$; local maximum values are $f(4) = 4$ and $f(6) = 3$; local minimum values are $f(2) = 1$ and $f(5) = 2$.

**7.** Absolute minimum at 2, absolute maximum at 3, local minimum at 4

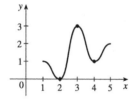

**9.** Absolute maximum at 5, absolute minimum at 2, local maximum at 3, local minima at 2 and 4

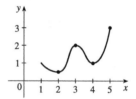

**11.** (a)

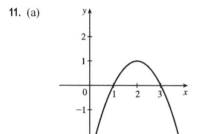

(b)

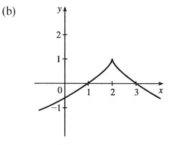

(c)

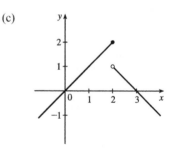

**13.** (a) *Note:* By the Extreme Value Theorem, $f$ must *not* be continuous; if it were, it would attain an absolute minimum.

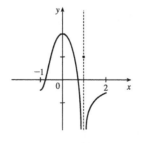

(b)

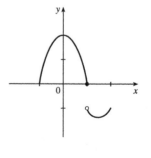

**15.** $f(x) = 8 - 3x$, $x \geq 1$. Absolute maximum $f(1) = 5$; no local maximum. No absolute or local minimum.

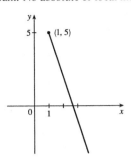

**17.** $f(x) = x^2$, $0 < x < 2$. No absolute or local maximum or minimum value.

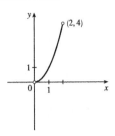

**19.** $f(x) = \ln x$, $0 < x \leq 2$. Absolute maximum $f(2) = \ln 2 \approx 0.69$; no local maximum. No absolute or local minimum.

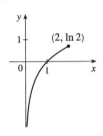

**21.** $f(x) = 1 - \sqrt{x}$. Absolute maximum $f(0) = 1$; no local maximum. No absolute or local minimum.

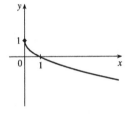

**23.** $f(x) = 5x^2 + 4x \;\Rightarrow\; f'(x) = 10x + 4$. $f'(x) = 0 \;\Rightarrow\; 10x = -4 \;\Rightarrow\; x = -\frac{2}{5}$, and $f'(x)$ is always defined, so $-\frac{2}{5}$ is the only critical number.

**25.** $f(x) = x^3 + 3x^2 - 24x \;\Rightarrow\; f'(x) = 3x^2 + 6x - 24 = 3(x^2 + 2x - 8)$.
$f'(x) = 0 \;\Rightarrow\; 3(x + 4)(x - 2) = 0 \;\Rightarrow\; x + 4 = 0$ or $x - 2 = 0 \;\Rightarrow\; x = -4, 2$. Because $f'(x)$ is defined for all values of $x$, these are the only critical numbers.

**27.** $s(t) = 3t^4 + 4t^3 - 6t^2 \;\Rightarrow\; s'(t) = 12t^3 + 12t^2 - 12t$. $s'(t) = 0 \;\Rightarrow\; 12t(t^2 + t - 1) \;\Rightarrow\;$
$t = 0$ or $t^2 + t - 1 = 0$. Using the quadratic formula to solve the latter equation gives
$t = \dfrac{-1 \pm \sqrt{1^2 - 4(1)(-1)}}{2(1)} = \dfrac{-1 \pm \sqrt{5}}{2}$. Thus the three critical numbers are $0$, $-\frac{1}{2} + \frac{1}{2}\sqrt{5} \approx 0.618$, and
$-\frac{1}{2} - \frac{1}{2}\sqrt{5} \approx -1.618$.

**29.** $f(x) = x \ln x \;\Rightarrow\; f'(x) = x(1/x) + (\ln x) \cdot 1 = \ln x + 1$. $f'(x) = 0 \;\Rightarrow\; \ln x = -1 \;\Rightarrow\; x = e^{-1} = 1/e$.
$f'(x)$ is defined for all values of $x$ in the domain of $f$ ($x > 0$), so the only critical number is $x = 1/e$.

**31.** $g(y) = \dfrac{y - 1}{y^2 - y + 1} \;\Rightarrow\;$
$g'(y) = \dfrac{(y^2 - y + 1)(1) - (y - 1)(2y - 1)}{(y^2 - y + 1)^2} = \dfrac{y^2 - y + 1 - (2y^2 - 3y + 1)}{(y^2 - y + 1)^2} = \dfrac{-y^2 + 2y}{(y^2 - y + 1)^2} = \dfrac{y(2 - y)}{(y^2 - y + 1)^2}$.

[continued]

$g'(y) = 0$ only when $y(2 - y) = 0 \Rightarrow y = 0, 2$. Also, the expression $y^2 - y + 1$ is never equal to 0, so $g'(y)$ exists for all real numbers. The critical numbers are 0 and 2.

**33.** $F(x) = x^{4/5}(x - 4)^2 \Rightarrow$

$F'(x) = x^{4/5} \cdot 2(x - 4)(1) + (x - 4)^2 \cdot \frac{4}{5}x^{-1/5} = x^{-1/5}(x - 4)\left[2x^1 + \frac{4}{5}(x - 4)\right]$

$= x^{-1/5}(x - 4) \cdot \frac{1}{5}[5 \cdot 2x + 4(x - 4)] = \frac{1}{5}x^{-1/5}(x - 4)(14x - 16)$

$= \frac{(x - 4)(14x - 16)}{5x^{1/5}} = \frac{2(x - 4)(7x - 8)}{5\sqrt[5]{x}}$

$F'(x) = 0$ when $2(x - 4)(7x - 8) \Rightarrow x = 4, \frac{8}{7}$. Also, $F'(x)$ does not exist when $x = 0$, so the three critical numbers are $0, \frac{8}{7}$, and 4.

**35.** $f(x) = 3x^2 - 12x + 5$, $[0, 3]$. $f'(x) = 0$ when $6x - 12 = 0 \Rightarrow x = 2$ and $f'(x)$ is always defined, so 2 is the only critical number (and it lies within the given interval). Applying the Closed Interval Method, we compute the function values at the critical number and the endpoints of the interval: $f(0) = 5$, $f(2) = -7$, and $f(3) = -4$. The largest of these, $f(0) = 5$, is the absolute maximum value and the smallest, $f(2) = -7$, is the absolute minimum value.

**37.** $f(x) = 2x^3 - 3x^2 - 12x + 1$, $[-2, 3]$. $f'(x) = 6x^2 - 6x - 12$ and $f'(x) = 0 \Rightarrow$

$6(x^2 - x - 2) = 6(x - 2)(x + 1) = 0 \Rightarrow x = 2, -1$. Note that both of these critical numbers are within the given interval. We compute $f(-2) = -3$, $f(-1) = 8$, $f(2) = -19$, and $f(3) = -8$. Then $f(-1) = 8$ is the absolute maximum value and $f(2) = -19$ is the absolute minimum value.

**39.** $f(x) = x^4 - 2x^2 + 3$, $[-2, 3]$. $f'(x) = 4x^3 - 4x$ and $f'(x) = 0 \Rightarrow 4x(x^2 - 1) = 4x(x + 1)(x - 1) = 0 \Rightarrow$ $x = 0, -1, 1$. All three of these critical numbers are within the given interval, so we compute $f(-2) = 11$, $f(-1) = 2$, $f(0) = 3$, $f(1) = 2$, $f(3) = 66$. Then $f(3) = 66$ is the absolute maximum value and $f(\pm 1) = 2$ is the absolute minimum value.

**41.** $f(t) = t\sqrt{4 - t^2} = t(4 - t^2)^{1/2}$, $[-1, 2]$.

$$f'(t) = t \cdot \frac{1}{2}(4 - t^2)^{-1/2}(-2t) + (4 - t^2)^{1/2} \cdot 1 = \frac{-t^2}{\sqrt{4 - t^2}} + \sqrt{4 - t^2}$$

$$= \frac{-t^2}{\sqrt{4 - t^2}} + \sqrt{4 - t^2} \cdot \frac{\sqrt{4 - t^2}}{\sqrt{4 - t^2}} = \frac{-t^2 + (4 - t^2)}{\sqrt{4 - t^2}} = \frac{4 - 2t^2}{\sqrt{4 - t^2}}$$

$f'(t) = 0 \Rightarrow 4 - 2t^2 = 0 \Rightarrow t^2 = 2 \Rightarrow t = \pm\sqrt{2}$, but $t = -\sqrt{2}$ is not in the interval $(-1, 2)$.

$f'(t)$ does not exist if $4 - t^2 = 0 \Rightarrow t = \pm 2$, but neither of these values is in the interval $(-1, 2)$. (Note that $t = 2$ is an endpoint of the interval.) Thus the only critical point we need to check is $\sqrt{2}$. We compute $f(-1) = -\sqrt{3}$, $f(\sqrt{2}) = 2$, and $f(2) = 0$. Then $f(\sqrt{2}) = 2$ is the absolute maximum value and $f(-1) = -\sqrt{3}$ is the absolute minimum value.

**43.** $f(x) = xe^{-x^2/8}$, $[-1, 4]$. $f'(x) = x \cdot e^{-x^2/8}(-\frac{1}{4}x) + e^{-x^2/8} \cdot 1 = e^{-x^2/8}(1 - \frac{1}{4}x^2)$. Since $e^{-x^2/8}$ is never 0,

$f'(x) = 0 \Rightarrow 1 - \frac{1}{4}x^2 = 0 \Rightarrow 1 = \frac{1}{4}x^2 \Rightarrow x^2 = 4 \Rightarrow x = \pm 2$, but $-2$ is not in the interval $(-1, 4)$, so 2 is the only critical number to check. $f(-1) = -e^{-1/8} \approx -0.88$, $f(2) = 2e^{-1/2} \approx 1.21$, and $f(4) = 4e^{-2} \approx 0.54$. Thus

$f(2) = 2e^{-1/2} = 2/\sqrt{e}$ is the absolute maximum value and $f(-1) = -e^{-1/8} = -1/\sqrt[8]{e}$ is the absolute minimum value.

**45.** $f(x) = \ln(x^2 + x + 1)$, $[-1, 1]$. $f'(x) = \dfrac{1}{x^2 + x + 1} \cdot (2x + 1) = \dfrac{2x + 1}{x^2 + x + 1}$. $f'(x) = 0$ only when $2x + 1 = 0$ $\Rightarrow$

$x = -\frac{1}{2}$. Since $x^2 + x + 1 > 0$ for all $x$, the domain of $f$ and $f'$ is $\mathbb{R}$. Thus the only critical number is $-\frac{1}{2}$, which lies within

the given interval. We compute $f(-1) = \ln 1 = 0$, $f(-\frac{1}{2}) = \ln\frac{3}{4} \approx -0.29$, and $f(1) = \ln 3 \approx 1.10$. Thus $f(1) = \ln 3$ is

the absolute maximum value and $f(-\frac{1}{2}) = \ln\frac{3}{4}$ is the absolute minimum value.

**47.** We use a graphing calculator (or graphing software) to graph

$f(x) = |x^2 - 8| - 3x$. $f'(x) = 0$ where the graph has a horizontal

tangent line, at $x \approx -1.50$, and we observe cusps (corners) on the curve,

where the derivative does not exist, at $x \approx -2.83$ and $x \approx 2.83$. Thus the

critical numbers are approximately $-2.83$, $-1.50$, and $2.83$.

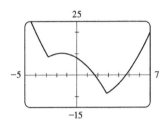

**49. (a)**

From the graph, it appears that the absolute maximum value is about

$f(-0.77) = 2.19$, and the absolute minimum value is about $f(0.77) = 1.81$.

**(b)** $f(x) = x^5 - x^3 + 2$ $\Rightarrow$ $f'(x) = 5x^4 - 3x^2 = x^2(5x^2 - 3)$. So $f'(x) = 0$ $\Rightarrow$ $x = 0$ or $5x^2 - 3 = 0$ $\Rightarrow$

$x^2 = \frac{3}{5}$ $\Rightarrow$ $x = \pm\sqrt{\frac{3}{5}}$. From the graph, we know that $f(0)$ is neither a minimum nor a maximum.

$$f\left(-\sqrt{\tfrac{3}{5}}\right) = \left(-\sqrt{\tfrac{3}{5}}\right)^5 - \left(-\sqrt{\tfrac{3}{5}}\right)^3 + 2 = -\left(\tfrac{3}{5}\right)^2 \sqrt{\tfrac{3}{5}} + \tfrac{3}{5}\sqrt{\tfrac{3}{5}} + 2$$

$$= \left(\tfrac{3}{5} - \tfrac{9}{25}\right)\sqrt{\tfrac{3}{5}} + 2 = \tfrac{6}{25}\sqrt{\tfrac{3}{5}} + 2 \approx 2.1859 \quad \text{(maximum)}$$

and similarly, $f\left(\sqrt{\tfrac{3}{5}}\right) = -\tfrac{6}{25}\sqrt{\tfrac{3}{5}} + 2 \approx 1.8141$ (minimum).

**51. (a)**

From the graph, it appears that the absolute maximum value is about

$g(1) = 0.37$, and the absolute minimum value is about $g(4.5) = 0.05$ (at

the right endpoint).

**(b)** $g(t) = te^{-t}$ $\Rightarrow$ $g'(t) = t \cdot e^{-t}(-1) + e^{-t} \cdot 1 = e^{-t}(1 - t)$. $e^{-t}$ is never 0, so $g'(t) = 0$ when $1 - t = 0$ $\Rightarrow$

$t = 1$. The maximum is $g(1) = e^{-1} = 1/e \approx 0.3679$, and the minimum is at the right endpoint,

$g(4.5) = 4.5e^{-4.5} = 4.5/e^{4.5} \approx 0.04999$.

**53.** $f(x) = \dfrac{e^x}{1 + \sqrt{x}}, \quad 0 \le x \le 1$

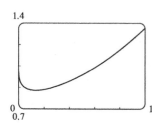

Using a graphing calculator, the absolute minimum is approximately

$f(0.134) = 0.837$, and the absolute maximum occurs at the right endpoint:

$f(1) \approx 1.359$.

**55.** $p(t) = \sqrt[3]{t^2 - t - 1}, \quad -1 \le t \le 3$

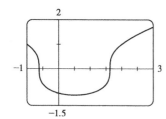

Using a graphing calculator, the absolute minimum is approximately

$p(0.500) = -1.077$, and the absolute maximum occurs at the right

endpoint: $p(3) \approx 1.710$.

**57.** $C(t) = 1.2te^{-2.6t} \quad \Rightarrow \quad C'(t) = 1.2\left[t \cdot e^{-2.6t}(-2.6) + e^{-2.6t} \cdot 1\right] = 1.2e^{-2.6t}(1 - 2.6t). \quad C'(t) = 0$ when

$1 - 2.6t = 0 \quad \Rightarrow \quad t = \frac{1}{2.6} \approx 0.385$, so we compute the function values at this critical number as well as at the endpoints of

the interval $[0, 3]$:   $C(0) = 0, C\left(\frac{1}{2.6}\right) = \frac{1.2}{2.6}e^{-2.6/2.6} = \frac{12}{26}e^{-1} = \frac{6}{13e} \approx 0.170$, and $C(3) = 3.6e^{-7.8} \approx 0.0015$. Thus the

maximum BAC for $0 \le t \le 3$ is $\frac{6}{13e} \approx 0.170$ mg/mL, which occurs after $\frac{1}{2.6} \approx 0.385$ hours (about 23 minutes).

**59.** $V(T) = 999.87 - 0.06426T + 0.0085043T^2 - 0.0000679T^3 \quad \Rightarrow \quad V'(T) = -0.06426 + 0.0170086T - 0.0002037T^2$.

Setting this equal to 0 and using the quadratic formula to find $T$, we get

$$T = \frac{-0.0170086 \pm \sqrt{0.0170086^2 - 4(-0.0002037)(-0.06426)}}{2(-0.0002037)} \approx 3.9665 \text{ or } 79.5318.$$ Only the critical number 3.9665 is

in the interval $0 < T < 30$, so we compare the function value there to the values at the endpoints of the interval:

$V(0) = 999.87$, $V(3.9665) \approx 999.745$, $V(30) \approx 1003.763$. Thus between $0°$C and $30°$C, water has its minimum volume at

about $3.9665°$C.

**61.** $S(t) = -0.00003237t^5 + 0.0009037t^4 - 0.008956t^3 + 0.03629t^2 - 0.04458t + 0.4074 \quad \Rightarrow$

$S'(t) = -0.00003237(5t^4) + 0.0009037(4t^3) - 0.008956(3t^2) + 0.03629(2t) - 0.04458$.

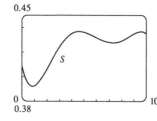

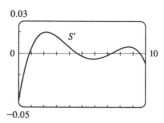

We can use a graphing device to esimate the maximum and minimum values directly from the graph of $S$, or we can use the graph of $S'$ to locate the critical numbers and use the Closed Interval Method for $0 \le t \le 10$. Since $S'$ exists for all $t$, the only critical numbers of $S$ occur when $S'(t) = 0$. The graph of $S'$ has $t$-intercepts at $t_1 \approx 0.855$, $t_2 \approx 4.618$, $t_3 \approx 7.292$, and $t_4 \approx 9.570$. The values of $S$ at these critical numbers are $S(t_1) \approx 0.39$, $S(t_2) \approx 0.43645$, $S(t_3) \approx 0.427$, and $S(t_4) \approx 0.43641$, and the values of $S$ at the endpoints of the interval are $S(0) \approx 0.41$ and $S(10) \approx 0.435$. Comparing the six numbers, we see that sugar was most expensive at $t_2 \approx 4.618$ (corresponding roughly to March 1998) and cheapest at $t_1 \approx 0.855$ (June 1994).

**63.** $f(x) = x^a(1-x)^b$, $0 \le x \le 1$, $a > 0$, $b > 0$.

$$f'(x) = x^a \cdot b(1-x)^{b-1}(-1) + (1-x)^b \cdot ax^{a-1} = x^{a-1}(1-x)^{b-1}[x \cdot b(-1) + (1-x) \cdot a]$$
$$= x^{a-1}(1-x)^{b-1}(a - ax - bx)$$

At the endpoints, we have $f(0) = f(1) = 0$   [the minimum value of $f$]. In the interval $(0, 1)$, $f'(x) = 0$ only when

$$a - ax - bx = 0 \quad \Rightarrow \quad a = (a+b)x \quad \Rightarrow \quad x = \frac{a}{a+b}.$$

$$f\left(\frac{a}{a+b}\right) = \left(\frac{a}{a+b}\right)^a \left(1 - \frac{a}{a+b}\right)^b = \frac{a^a}{(a+b)^a}\left(\frac{a+b-a}{a+b}\right)^b = \frac{a^a}{(a+b)^a} \cdot \frac{b^b}{(a+b)^b} = \frac{a^a b^b}{(a+b)^{a+b}}.$$

So $f\left(\dfrac{a}{a+b}\right) = \dfrac{a^a b^b}{(a+b)^{a+b}}$ is the absolute maximum value.

**65.** $f(x) = x^{101} + x^{51} + x + 1 \quad \Rightarrow \quad f'(x) = 101x^{100} + 51x^{50} + 1$. But $101x^{100} \ge 0$ and $51x^{50} \ge 0$ for all values of $x$, so $101x^{100} + 51x^{50} + 1 \ge 1$ for all $x$, thus $f'(x) = 0$ has no solution. $f'(x)$ is never undefined, so $f$ has no critical numbers. Thus $f(x)$ can have no local maximum or minimum.

## 4.3   Derivatives and the Shapes of Curves

**Prepare Yourself**

**1.** (a) First $r^2 - 3r - 18 = (r+3)(r-6) = 0$ when $r = -3$ or $r = 6$. These numbers divide the real number line into three intervals: $(-\infty, -3)$, $(-3, 6)$, and $(6, \infty)$. On each of these intervals we determine the signs of the factors.

| Interval | $r + 3$ | $r - 6$ | $r^2 - 3r - 18 = (r+3)(r-6)$ |
|---|---|---|---|
| $r < -3$ | $-$ | $-$ | $+$ |
| $-3 < r < 6$ | $+$ | $-$ | $-$ |
| $r > 6$ | $+$ | $+$ | $+$ |

Thus $r^2 - 3r - 18 < 0$ on the interval $(-3, 6)$.

Alternatively, we can use test values in each interval. For instance, $0$ is in the interval $(-3, 6)$, and when $r = 0$ we have $r^2 - 3r - 18 = -18 < 0$, so we know that $r^2 - 3r - 18 < 0$ for all values of $r$ in that interval. Test values for the other two intervals give positive numbers, so those intervals are not part of the solution.

(b) $x^3 - 9x = x(x^2 - 9) = x(x+3)(x-3) = 0$ when $x = 0, -3,$ or $3$.

| Interval | $x$ | $x + 3$ | $x - 3$ | $x^3 - 9x = x(x+3)(x-3)$ |
|----------|-----|---------|---------|--------------------------|
| $x < -3$ | $-$ | $-$ | $-$ | $-$ |
| $-3 < x < 0$ | $-$ | $+$ | $-$ | $+$ |
| $0 < x < 3$ | $+$ | $+$ | $-$ | $-$ |
| $x > 3$ | $+$ | $+$ | $+$ | $+$ |

Thus $x^3 - 9x > 0$ on the intervals $(-3, 0)$ and $(3, \infty)$.

(c) $x^2 e^x - 4xe^x = xe^x(x - 4) = 0$ when $x = 0$ or $x = 4$. ($e^x > 0$ always.)

| Interval | $x$ | $x - 4$ | $xe^x(x-4)$ |
|----------|-----|---------|-------------|
| $x < 0$ | $-$ | $-$ | $+$ |
| $0 < x < 4$ | $+$ | $-$ | $-$ |
| $x > 4$ | $+$ | $+$ | $+$ |

Thus $x^2 e^x - 4xe^x > 0$ on the intervals $(-\infty, 0)$ and $(4, \infty)$.

(d) Because $(t^2 + 2)^2$ is always positive, $\dfrac{t - 4}{(t^2 + 2)^2} < 0$ when $t - 4 < 0 \;\Rightarrow\; t < 4$.

(e) The domain of $\dfrac{x \ln x - 6x}{x^2}$ is $(0, \infty)$, thus $x$ is always positive, so $\dfrac{x \ln x - 6x}{x^2} = \dfrac{x(\ln x - 6)}{x^2} = \dfrac{\ln x - 6}{x} > 0$ when

$\ln x - 6 > 0 \;\Rightarrow\; \ln x > 6 \;\Rightarrow\; x > e^6$.

**2.** (a) $f(x) = \dfrac{2 + \ln x}{x} \;\Rightarrow\; f'(x) = \dfrac{x(1/x) - (2 + \ln x)(1)}{x^2} = \dfrac{-1 - \ln x}{x^2}$ by the Quotient Rule. The domain of $f$ and $f'$ is

$(0, \infty)$, where the denominator $x^2$ is never $0$ and $f'$ is always defined. $f'(x) = 0$ when $-1 - \ln x = 0 \;\Rightarrow$

$\ln x = -1 \;\Rightarrow\; x = e^{-1}$, so $e^{-1} = 1/e$ is the only critical number.

(b) $g(t) = (t^2 + 2t)e^t \;\Rightarrow\; g'(t) = (t^2 + 2t) \cdot e^t + e^t \cdot (2t + 2) = (t^2 + 4t + 2)e^t$ by the Product Rule. $e^t$ is never $0$, so

$g'(t) = 0$ when $t^2 + 4t + 2 = 0 \;\Rightarrow\; t = \dfrac{-4 \pm \sqrt{4^2 - 4(1)(2)}}{2(1)} = \dfrac{-4 \pm 2\sqrt{2}}{2} = \dfrac{-4}{2} \pm \dfrac{2\sqrt{2}}{2} = -2 \pm \sqrt{2}$ by the

quadratic formula, and these are the only two critical numbers.

**3.** (a) $B(t) = 3te^{-2t} \;\Rightarrow\; B'(t) = 3[t \cdot e^{-2t}(-2) + e^{-2t} \cdot 1] = 3(1 - 2t)e^{-2t} \;\Rightarrow$

$B''(t) = 3[(1 - 2t) \cdot e^{-2t}(-2) + e^{-2t}(-2)] = 3(-4 + 4t)e^{-2t} = (12t - 12)e^{-2t}$ or $12(t-1)e^{-2t}$

(b) $y = \ln(x^3 + x) \;\Rightarrow\; \dfrac{dy}{dx} = \dfrac{1}{x^3 + x}(3x^2 + 1) = \dfrac{3x^2 + 1}{x^3 + x} \;\Rightarrow$

$\dfrac{d^2 y}{dx^2} = \dfrac{(x^3 + x)(6x) - (3x^2 + 1)(3x^2 + 1)}{(x^3 + x)^2} = \dfrac{6x^4 + 6x^2 - (9x^4 + 6x^2 + 1)}{(x^3 + x)^2} = -\dfrac{3x^4 + 1}{(x^3 + x)^2}$

**4.** $f(x) = \dfrac{x^2}{2x^2 + 1}$ $\Rightarrow$ $f'(x) = \dfrac{(2x^2 + 1)(2x) - x^2(4x)}{(2x^2 + 1)^2} = \dfrac{2x}{(2x^2 + 1)^2}$ $\Rightarrow$

$f''(x) = \dfrac{(2x^2 + 1)^2(2) - 2x(2)(2x^2 + 1)(4x)}{[(2x^2 + 1)^2]^2} = \dfrac{2(2x^2 + 1)[(2x^2 + 1) - 8x^2]}{(2x^2 + 1)^4} = \dfrac{2(1 - 6x^2)}{(2x^2 + 1)^3}$, so

$f''(3) = \dfrac{2(1 - 6 \cdot 9)}{(2 \cdot 9 + 1)^3} = \dfrac{2(-53)}{(19)^3}$ is negative.

---

### Exercises

**1.** (a) $f$ is increasing on $(0, 6)$ and $(8, 9)$.

(b) $f$ is decreasing on $(6, 8)$.

(c) $f$ is concave upward on $(2, 4)$ and $(7, 9)$.

(d) $f$ is concave downward on $(0, 2)$ and $(4, 7)$.

(e) The points of inflection are $(2, 3)$, $(4, \approx 4.5)$, and $(7, 4)$ (where the concavity changes direction).

**3.** (a) Use the Increasing/Decreasing (I/D) Test.

(b) Use the Concavity Test.

(c) At any value of $x$ at which the concavity changes direction, we have an inflection point at $(x, f(x))$.

**5.** $y = 3x^2 - 11x + 4$ $\Rightarrow$ $y' = 6x - 11$. Then $y' = 0$ when $6x - 11 = 0$ $\Rightarrow$ $x = \frac{11}{6}$, so $\frac{11}{6}$ is the only critical number.

We have $y' < 0$ for $x < \frac{11}{6}$ and $y' > 0$ for $x > \frac{11}{6}$. Because $y'$ changes from negative to positive at $x = \frac{11}{6}$, we have a local

minimum value there, where $y = -\frac{73}{12}$. (There is no local maximum value.)

**7.** $M(t) = 4t^3 - 11t^2 - 20t + 7$ $\Rightarrow$ $M'(t) = 12t^2 - 22t - 20 = 2(6t^2 - 11t - 10) = 2(2t - 5)(3t + 2)$. Thus $M'(t) = 0$

when $2t - 5 = 0$ $\Rightarrow$ $t = \frac{5}{2}$ or $3t + 2 = 0$ $\Rightarrow$ $t = -\frac{2}{3}$. So the critical numbers are $-\frac{2}{3}$ and $\frac{5}{2}$, and we determine the

sign of $M'(t)$ in the intervals determined by the critical numbers.

[We don't need to include the "2" in the chart to determine the sign of $M'(t)$.]

| Interval | $2t - 5$ | $3t + 2$ | $M'(t)$ |
|---|---|---|---|
| $x < -\frac{2}{3}$ | $-$ | $-$ | $+$ |
| $-\frac{2}{3} < x < \frac{5}{2}$ | $-$ | $+$ | $-$ |
| $x > \frac{5}{2}$ | $+$ | $+$ | $+$ |

$M'(t)$ changes from positive to negative at $-\frac{2}{3}$, so $M\left(-\frac{2}{3}\right) = \frac{385}{27} \approx 14.26$ is a local maximum value. $M'(t)$ changes from

negative to positive at $\frac{5}{2}$, so $M\left(\frac{5}{2}\right) = -\frac{197}{4} = -49.25$ is a local minimum value.

**9.** $f(x) = \dfrac{\ln x}{\sqrt{x}}$. (Note that $f$ is defined only for $x > 0$.)

$f'(x) = \dfrac{\sqrt{x}(1/x) - \ln x\left(\frac{1}{2}x^{-1/2}\right)}{x} = \dfrac{\dfrac{\sqrt{x}}{x} - \dfrac{\ln x}{2\sqrt{x}}}{x} \cdot \dfrac{2\sqrt{x}}{2\sqrt{x}} = \dfrac{2 - \ln x}{2x^{3/2}}$. The denominator is always positive for $x > 0$, so

the only critical number occurs when $2 - \ln x = 0$ $\Rightarrow$ $\ln x = 2$ $\Rightarrow$ $x = e^2$. When $0 < x < e^2$ we have $f'(x) > 0$, and

$f'(x) < 0$ for $x > e^2$. Because $f'(x)$ changes from positive to negative at $x = e^2$, $f(e^2) = \dfrac{\ln e^2}{\sqrt{e^2}} = \dfrac{2}{e}$ is a local

maximum value.

**11.** $f(x) = x^4 - 4x^3 + 6x^2 - 1 \;\Rightarrow\; f'(x) = 4x^3 - 12x^2 + 12x \;\Rightarrow$

$f''(x) = 12x^2 - 24x + 12 = 12(x^2 - 2x + 1) = 12(x-1)^2$. Then $f''(x) = 0$ at $x = 1$, but $f''(x) > 0$ elsewhere. Because

the concavity does not change at 1, there is no inflection point, and since $f'$ is increasing for all $x$, we can say that $f$ is concave

upward on $(-\infty, \infty)$.

**13.** $h(t) = -1.6t^3 + 0.9t^2 + 2.2t - 6.4 \;\Rightarrow\; h'(t) = -4.8t^2 + 1.8t + 2.2 \;\Rightarrow\; h''(t) = -9.6t + 1.8$. $h''(t) = 0$ when

$t = \frac{1.8}{9.6} = 0.1875$, and $h''(t) > 0$ for $t < 0.1875$, $h''(t) < 0$ for $t > 0.1875$. Thus $h$ is concave upward on $(-\infty, 0.1875)$

and concave downward on $(0.1875, \infty)$. $h''(t)$ changes sign at $t = 0.1875$, where $h(0.1875) \approx -5.9664$, so

$(0.1875, \approx -5.9664)$ is an inflection point.

**15.** (a) $f(x) = x^3 - 12x + 1 \;\Rightarrow\; f'(x) = 3x^2 - 12 = 3(x^2 - 4) = 3(x+2)(x-2)$. Then $f'(x) = 0$ when $x = -2$ or

$x = 2$, so these are the critical numbers.

[We don't need to include "3" in the chart to determine the sign of $f'(x)$.]

| Interval | $x + 2$ | $x - 2$ | $f'(x)$ | $f$ |
|----------|---------|---------|---------|-----|
| $x < -2$ | $-$ | $-$ | $+$ | increasing on $(-\infty, -2)$ |
| $-2 < x < 2$ | $+$ | $-$ | $-$ | decreasing on $(-2, 2)$ |
| $x > 2$ | $+$ | $+$ | $+$ | increasing on $(2, \infty)$ |

So $f$ is increasing on $(-\infty, -2)$ and $(2, \infty)$, and $f$ is decreasing on $(-2, 2)$.

(b) $f'$ changes from positive to negative at $x = -2$ and from negative to positive at $x = 2$. Thus, $f(-2) = 17$ is a local

maximum value and $f(2) = -15$ is a local minimum value.

(c) $f''(x) = 6x$. $f''(x) = 0$ when $x = 0$, $f''(x) > 0$ when $x > 0$, and $f''(x) < 0$ when $x < 0$. Thus, $f$ is concave upward

on $(0, \infty)$ and concave downward on $(-\infty, 0)$. Concavity changes direction at $x = 0$, and $f(0) = 1$, so we have the

inflection point $(0, 1)$.

**17.** (a) $f(x) = x^4 - 2x^2 + 2 \;\Rightarrow\; f'(x) = 4x^3 - 4x = 4x(x^2 - 1) = 4x(x+1)(x-1)$. We have critical numbers where

$f'(x) = 0$, at $x = 0$, $x = -1$, and $x = 1$.

| Interval | $x$ | $x + 1$ | $x - 1$ | $f'(x)$ | $f$ |
|----------|-----|---------|---------|---------|-----|
| $x < -1$ | $-$ | $-$ | $-$ | $-$ | decreasing on $(-\infty, -1)$ |
| $-1 < x < 0$ | $-$ | $+$ | $-$ | $+$ | increasing on $(-1, 0)$ |
| $0 < x < 1$ | $+$ | $+$ | $-$ | $-$ | decreasing on $(0, 1)$ |
| $x > 1$ | $+$ | $+$ | $+$ | $+$ | increasing on $(1, \infty)$ |

(b) $f'$ changes from positive to negative at $x = 0$, so $f(0) = 2$ is a local maximum value. $f'$ changes from negative to positive

at $x = -1$ and $x = 1$, so $f(-1) = 1$ and $f(1) = 1$ are local minimum values.

(c) $f''(x) = 12x^2 - 4 = 4(3x^2 - 1)$. Thus $f''(x) = 0$ when $3x^2 - 1 = 0$ $\Rightarrow$ $x^2 = \frac{1}{3}$ $\Rightarrow$ $x = \pm\sqrt{1/3} = \pm 1/\sqrt{3}$.

We have $f''(x) < 0$ when $-1/\sqrt{3} < x < 1/\sqrt{3}$, and $f''(x) > 0$ when $x < -1/\sqrt{3}$ or $x > 1/\sqrt{3}$, so $f$ is concave upward on $(-\infty, -1/\sqrt{3})$ and $(1/\sqrt{3}, \infty)$, and $f$ is concave downward on $(-1/\sqrt{3}, 1/\sqrt{3})$. Concavity changes direction at $x = \pm 1/\sqrt{3}$, so inflection points are $(-1/\sqrt{3}, 13/9)$ and $(1/\sqrt{3}, 13/9)$.

**19.** (a) $f(x) = 5xe^{-0.2x}$ $\Rightarrow$ $f'(x) = 5\left[x \cdot e^{-0.2x}(-0.2) + e^{-0.2x} \cdot 1\right] = (5-x)e^{-0.2x}$. Because $e^{-0.2x} > 0$, $f'(x) = 0$ only for $x = 5$, so 5 is the only critical number. We have $f'(x) > 0$ for $x < 5$ and $f'(x) < 0$ for $x > 5$, so $f$ is increasing on $(-\infty, 5)$ and decreasing on $(5, \infty)$.

(b) $f'$ changes from positive to negative at $x = 5$, so $f(5) = 25e^{-1} = 25/e$ is a local maximum value.

(c) $f''(x) = (5-x) \cdot e^{-0.2x}(-0.2) + e^{-0.2x} \cdot (-1) = (0.2x - 2)e^{-0.2x}$, and $f''(x) = 0$ when $0.2x - 2 = 0$ $\Rightarrow$ $x = 10$. $f''(x) < 0$ when $x < 10$ and $f''(x) > 0$ when $x > 10$, so $f$ is concave upward on $(10, \infty)$ and concave downward on $(-\infty, 10)$. Concavity changes direction at $x = 10$, so the point $(10, 50e^{-2}) = (10, 50/e^2)$ is an inflection point.

**21.** (a) $f(x) = xe^x$ $\Rightarrow$ $f'(x) = x \cdot e^x + e^x \cdot 1 = (x+1)e^x$. Now $e^x$ is always positive, so $f'(x) > 0$ when $x + 1 > 0$ $\Rightarrow$ $x > -1$ and $f'(x) < 0$ when $x < -1$. Thus, $f$ is increasing on $(-1, \infty)$ and decreasing on $(-\infty, -1)$.

(b) $f'$ changes from negative to positive at its only critical number, $x = -1$. Thus, $f(-1) = -e^{-1} = -1/e$ is a local minimum value.

(c) $f''(x) = (x+1) \cdot e^x + e^x \cdot 1 = (x+2)e^x$. So $f''(x) > 0$ when $x + 2 > 0$ $\Rightarrow$ $x > -2$ and $f''(x) < 0$ when $x < -2$. Thus, $f$ is concave upward on $(-2, \infty)$ and concave downward on $(-\infty, -2)$. Since the concavity changes direction at $x = -2$, the point $(-2, -2e^{-2})$ is an inflection point.

**23.** (a) $f(x) = \dfrac{\ln x}{x}$ $\Rightarrow$ $f'(x) = \dfrac{x(1/x) - (\ln x)(1)}{x^2} = \dfrac{1 - \ln x}{x^2}$. The domain of $f$ and $f'$ is $(0, \infty)$, and $f'(x) = 0$ when $1 - \ln x = 0$ $\Rightarrow$ $\ln x = 1$ $\Rightarrow$ $x = e$. $f'(x) > 0$ for $0 < x < e$ and $f'(x) < 0$ for $x > e$, so $f$ is increasing on $(0, e)$ and is decreasing on $(e, \infty)$.

(b) $f'$ changes from positive to negative at $x = e$, so $f(e) = 1/e$ is a local maximum value.

(c) $f''(x) = \dfrac{x^2(-1/x) - (1 - \ln x)(2x)}{(x^2)^2} = \dfrac{-3x + 2x \ln x}{x^4} = \dfrac{x(2\ln x - 3)}{x^4} = \dfrac{2\ln x - 3}{x^3}$. For $x > 0$, $f''(x) = 0$ when $2\ln x - 3 = 0$ $\Rightarrow$ $\ln x = \frac{3}{2}$ $\Rightarrow$ $x = e^{3/2}$. We have $f''(x) > 0$ for $x > e^{3/2}$ and $f''(x) < 0$ for $0 < x < e^{3/2}$, so $f$ is concave upward on $(e^{3/2}, \infty)$ and is concave downward on $(0, e^{3/2})$. There is an inflection point at $(e^{3/2}, 3/(2e^{3/2}))$.

**25.** (a) $f(x) = 2x^3 - 3x^2 - 12x$ $\Rightarrow$ $f'(x) = 6x^2 - 6x - 12 = 6(x^2 - x - 2) = 6(x-2)(x+1)$. $f'(x) = 0$ when $x = 2$ or $x = -1$. $f'(x) > 0$ when $x < -1$ or $x > 2$, and $f'(x) < 0$ when $-1 < x < 2$. Thus $f$ is increasing on $(-\infty, -1)$ and $(2, \infty)$, and $f$ is decreasing on $(-1, 2)$.

(b) Since $f'$ changes from positive to negative at $x = -1$, $f(-1) = 7$ is a local maximum value. Since $f'$ changes from negative to positive at $x = 2$, $f(2) = -20$ is a local minimum value.

(c) $f''(x) = 12x - 6 = 6(2x - 1)$, so $f''(x) = 0$ when $2x - 1 = 0$ $\Rightarrow$ $x = \frac{1}{2}$, and $f''(x) > 0$ for $x > \frac{1}{2}$, $f''(x) < 0$ for $x < \frac{1}{2}$. Thus $f$ is concave upward on $\left(\frac{1}{2}, \infty\right)$ and concave downward on $\left(-\infty, \frac{1}{2}\right)$. Concavity changes direction at $x = \frac{1}{2}$, so we have an inflection point at $\left(\frac{1}{2}, -\frac{13}{2}\right)$.

(d)
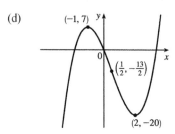

**27.** (a) $h(x) = 3x^5 - 5x^3 + 3$ $\Rightarrow$ $h'(x) = 15x^4 - 15x^2 = 15x^2(x^2 - 1) = 15x^2(x + 1)(x - 1)$, so $h'(x) = 0$ when $x = 0, -1,$ or $1$.

| Interval | $x^2$ | $x + 1$ | $x - 1$ | $h'(x)$ | $h$ |
|----------|-------|---------|---------|---------|-----|
| $x < -1$ | $+$ | $-$ | $-$ | $+$ | increasing on $(-\infty, -1)$ |
| $-1 < x < 0$ | $+$ | $+$ | $-$ | $-$ | decreasing on $(-1, 0)$ |
| $0 < x < 1$ | $+$ | $+$ | $-$ | $-$ | decreasing on $(0, 1)$ |
| $x > 1$ | $+$ | $+$ | $+$ | $+$ | increasing on $(1, \infty)$ |

Notice that $h$ has a horizontal tangent line at $x = 0$, but the derivative does not change sign there, so $h$ is increasing on $(-\infty, -1)$ and $(1, \infty)$ and decreasing on $(-1, 1)$.

(b) $h'$ changes from positive to negative at $x = -1$ and from negative to positive at $x = 1$, so $h(-1) = 5$ is a local maximum value and $h(1) = 1$ is a local minimum value.

(c) $h''(x) = 60x^3 - 30x = 30x(2x^2 - 1)$. Then $h''(x) = 0$ when $x = 0$ or when $2x^2 - 1 = 0$ $\Rightarrow$ $x^2 = \frac{1}{2}$ $\Rightarrow$ $x = \pm\frac{1}{\sqrt{2}}$.

| Interval | $x$ | $2x^2 - 1$ | $h''(x)$ | $h$ |
|----------|-----|------------|----------|-----|
| $x < -\frac{1}{\sqrt{2}}$ | $-$ | $+$ | $-$ | concave downward on $\left(-\infty, -\frac{1}{\sqrt{2}}\right)$ |
| $-\frac{1}{\sqrt{2}} < x < 0$ | $-$ | $-$ | $+$ | concave upward on $\left(-\frac{1}{\sqrt{2}}, 0\right)$ |
| $0 < x < \frac{1}{\sqrt{2}}$ | $+$ | $-$ | $-$ | concave downward on $\left(0, \frac{1}{\sqrt{2}}\right)$ |
| $x > \frac{1}{\sqrt{2}}$ | $+$ | $+$ | $+$ | concave upward on $\left(\frac{1}{\sqrt{2}}, \infty\right)$ |

There are inflection points at $(0, 3)$, $\left(-\frac{1}{\sqrt{2}}, 3 + \frac{7}{8}\sqrt{2}\right) \approx (-0.71, 4.24)$, and $\left(\frac{1}{\sqrt{2}}, 3 - \frac{7}{8}\sqrt{2}\right) \approx (0.71, 1.76)$.

(d)
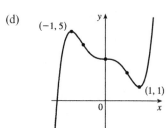

**29.** (a) $A(x) = x\sqrt{x+3} = x(x+3)^{1/2} \quad \Rightarrow$

$$A'(x) = x \cdot \tfrac{1}{2}(x+3)^{-1/2}(1) + \sqrt{x+3} \cdot 1 = \frac{x}{2\sqrt{x+3}} + \sqrt{x+3} \cdot \frac{2\sqrt{x+3}}{2\sqrt{x+3}}$$

$$= \frac{x + 2(x+3)}{2\sqrt{x+3}} = \frac{3x+6}{2\sqrt{x+3}}$$

The domain of $A$ is $[-3, \infty)$ and the domain of $A'$ is $(-3, \infty)$. $A'(x) > 0$ for $x > -2$ and $A'(x) < 0$ for $-3 < x < -2$, so $A$ is increasing on $(-2, \infty)$ and decreasing on $(-3, -2)$.

(b) Because $A'$ changes from negative to positive at $x = -2$, $A(-2) = -2$ is a local minimum value.

(c) $A''(x) = \dfrac{2\sqrt{x+3} \cdot 3 - (3x+6) \cdot 2\left[\tfrac{1}{2}(x+3)^{-1/2}(1)\right]}{\left(2\sqrt{x+3}\right)^2} = \dfrac{6\sqrt{x+3} - (3x+6) \cdot \dfrac{1}{\sqrt{x+3}}}{4(x+3)}$

$= \dfrac{6\sqrt{x+3} - (3x+6) \cdot \dfrac{1}{\sqrt{x+3}}}{4(x+3)} \cdot \dfrac{\sqrt{x+3}}{\sqrt{x+3}} = \dfrac{6(x+3) - (3x+6)}{4(x+3)^{3/2}} = \dfrac{3x+12}{4(x+3)^{3/2}} = \dfrac{3(x+4)}{4(x+3)^{3/2}}$

$A''(x) > 0$ for all $x > -3$, so $A$ is concave upward on $(-3, \infty)$. There are no inflection points.

(d)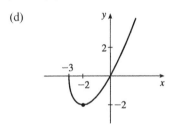

**31.** (a) There is an inflection point at $x = 3$ because the graph of $f$ changes from concave downward to concave upward there. There is an inflection point at $x = 5$ because the graph of $f$ changes from concave upward to concave downward there.

(b) There is an inflection point at $x = 2$ and at $x = 6$ because $f'$ changes from increasing to decreasing [$f'(x)$ has a local maximum value] there, and so $f''(x)$ changes from positive to negative there. There is an inflection point at $x = 4$ because $f'$ changes from decreasing to increasing [$f'(x)$ has a local minimum value] there and so $f''(x)$ changes from negative to positive there.

(c) There is an inflection point at $x = 1$ because $f''(x)$ changes from negative to positive there, and so the graph of $f$ changes from concave downward to concave upward. There is an inflection point at $x = 7$ because $f''(x)$ changes from positive to negative there, and so the graph of $f$ changes from concave upward to concave downward.

**33.** (a) $f$ is increasing where $f'$ is positive, that is, on $(0, 2)$, $(4, 6)$, and $(8, \infty)$; and decreasing where $f'$ is negative, that is, on $(2, 4)$ and $(6, 8)$.

(b) $f$ has local maximum values where $f'$ changes from positive to negative, at $x = 2$ and at $x = 6$, and local minimum values where $f'$ changes from negative to positive, at $x = 4$ and at $x = 8$.

(c) $f$ is concave upward where $f'$ is increasing, that is, on $(3, 6)$ and $(6, \infty)$, and concave downward where $f'$ is decreasing, on $(0, 3)$.

(d) There is a point of inflection where $f$ changes direction of concavity, that is, at $x = 3$.

(e)

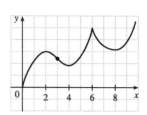

**35.** $f(x) = x^5 - 5x + 3 \quad \Rightarrow \quad f'(x) = 5x^4 - 5 = 5(x^4 - 1) = 5(x^2 + 1)(x^2 - 1) = 5(x^2 + 1)(x + 1)(x - 1)$. $f'(x) = 0$ when $x = \pm 1$, so these are the critical numbers.

*First Derivative Test:* $f'(x) > 0$ when $x < -1$ or $x > 1$, and $f'(x) < 0$ when $-1 < x < 1$. Since $f'$ changes from positive to negative at $x = -1$, $f(-1) = 7$ is a local maximum value. Since $f'$ changes from negative to positive at $x = 1$, $f(1) = -1$ is a local minimum value.

*Second Derivative Test:* $f''(x) = 20x^3$. $f''(-1) = -20$; since $f''(-1) < 0$, $f(-1) = 7$ is a local maximum value. $f''(1) = 20 > 0$, so $f(1) = -1$ is a local minimum value.

*Preference:* The Second Derivative Test may be slightly easier to apply in this case.

**37.** (a) By the Second Derivative Test, if $f'(2) = 0$ and $f''(2) < 0$, $f$ has a local maximum at $x = 2$.

(b) If $f'(6) = 0$, we know that $f$ has a horizontal tangent at $x = 6$. Knowing that $f''(6) = 0$ does not provide any additional information since the Second Derivative Test fails. We could have a local maximum value at 6, a local minimum value, or neither.

**39.** (a) I'm very unhappy. It's uncomfortably hot and $f'(3) = 2$ indicates that the temperature is increasing, and $f''(3) = 4$ indicates that the rate of increase is increasing. (The temperature is rapidly getting warmer.)

(b) I'm still unhappy, but not as unhappy as in part (a). It's uncomfortably hot and $f'(3) = 2$ indicates that the temperature is increasing, but $f''(3) = -4$ indicates that the rate of increase is decreasing. (The temperature is slowly getting warmer.)

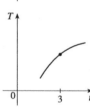

(c) I'm somewhat happy. It's uncomfortably hot and $f'(3) = -2$ indicates that the temperature is decreasing, but $f''(3) = 4$ indicates that the rate of change is increasing. (The rate of change is negative but it's becoming less negative. The temperature is slowly getting cooler.)

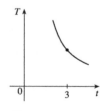

(d) I'm getting happier. It's uncomfortably hot but $f'(3) = -2$ indicates that the temperature is decreasing, and $f''(3) = -4$ indicates that the rate of change is decreasing, that is, becoming more negative. (The temperature is rapidly getting cooler.)

**41.** $P(t)$ is increasing, so $P'(t) > 0$. If the rate of increase is slowing, then $P'(t)$ is decreasing, so $P''(t) < 0$ (the graph of $P$ is concave downward).

**43.** (a)

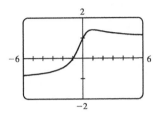

From the graph, we get an estimate of $f(1) \approx 1.41$ as a local maximum value, and no local minimum value.

$$f(x) = \frac{x+1}{\sqrt{x^2+1}} \quad \Rightarrow$$

$$f'(x) = \frac{\sqrt{x^2+1} \cdot 1 - (x+1) \cdot \frac{1}{2}(x^2-1)^{-1/2}(2x)}{\left(\sqrt{x^2+1}\right)^2} = \frac{\sqrt{x^2+1} - \dfrac{x(x+1)}{\sqrt{x^2+1}}}{x^2+1} \cdot \frac{\sqrt{x^2+1}}{\sqrt{x^2+1}}$$

$$= \frac{(x^2+1) - x(x+1)}{(x^2+1)^{3/2}} = \frac{1-x}{(x^2+1)^{3/2}}$$

The denominator is always positive, and $f'(x) = 0$ when $x = 1$, the only critical number. $f(1) = \frac{2}{\sqrt{2}} = \sqrt{2}$ is the exact local maximum value.

(b) From the graph in part (a), $f$ increases most rapidly somewhere between $x = -\frac{1}{2}$ and $x = -\frac{1}{4}$. (Note that the point where $f$ increases most rapidly is an inflection point of the graph.) To find the exact value, we need to find the maximum value of $f'$, which we can do by finding the critical numbers of $f'$.

$$\frac{d}{dx}[f'(x)] = f''(x) = \frac{(x^2+1)^{3/2}(-1) - (1-x) \cdot \frac{3}{2}(x^2+1)^{1/2}(2x)}{\left[(x^2+1)^{3/2}\right]^2} = \frac{-(x^2+1)^{3/2} - 3x(1-x)(x^2+1)^{1/2}}{(x^2+1)^3}$$

$$= \frac{(x^2+1)^{1/2}\left[-(x^2+1) - 3x(1-x)\right]}{(x^2+1)^3} = \frac{2x^2 - 3x - 1}{(x^2+1)^{5/2}}$$

The denominator is always positive, so $f''(x) = 0$ when $2x^2 - 3x - 1 = 0 \quad \Rightarrow$

$$x = \frac{-(-3) \pm \sqrt{(-3)^2 - 4(2)(-1)}}{2(2)} = \frac{3 \pm \sqrt{17}}{4} \text{ by the quadratic formula.}$$

$x = \dfrac{3 + \sqrt{17}}{4}$ corresponds to the *minimum* value of $f'$. The maximum value of $f'$ occurs at $x = \dfrac{3 - \sqrt{17}}{4} \approx -0.28$.

**45.** $f(x) = e^{-x} + x^3 \quad \Rightarrow \quad f'(x) = e^{-x}(-1) + 3x^2 = -e^{-x} + 3x^2 \quad \Rightarrow \quad f''(x) = e^{-x} + 6x$

(a)

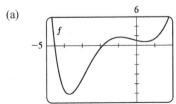

From the graph of $f$, it appears that $f$ has inflection points at about $(-2.8, -5.5)$ and $(-0.2, 1.2)$. We estimate that $f$ is concave upward on the intervals $(-\infty, -2.8)$, $(-0.2, \infty)$ and concave downward on $(-2.8, -0.2)$.

(b)

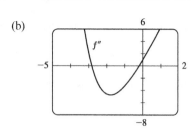

The graph of $f''$ has $x$-intercepts $x \approx -2.83$ and $x \approx -0.20$. We see that $f''(x) > 0$ for $x < -2.83$ and $x > -0.20$, and $f''(x) < 0$ for $-2.9 < x < -0.20$. Thus we estimate that $f$ is concave upward on $(-\infty, -2.83)$, $(-0.20, \infty)$ and concave downward on $(-2.83, -0.20)$. Concavity changes sign at $x \approx -2.83$ and $x \approx -0.20$, so more accurate estimates for the inflection points are $(-2.83, -5.72)$ and $(-0.20, 1.21)$.

**47.** $f(x) = \dfrac{8e^x}{1 + xe^x}$ $\Rightarrow$ $f'(x) = \dfrac{(1 + xe^x) \cdot 8e^x - 8e^x \cdot (x \cdot e^x + e^x \cdot 1)}{(1 + xe^x)^2} = \dfrac{8e^x(1 - e^x)}{(1 + xe^x)^2}$

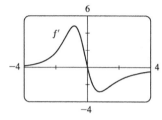

The graph of $f'$ has a local maximum at $x \approx -0.83$ and a local minimum at $x \approx 0.77$. $f$ is concave upward where the graph of $f'$ is increasing, on $(-\infty, -0.83)$ and $(0.77, \infty)$, and $f$ is concave downward where $f'$ is decreasing, on $(-0.83, 0.77)$.

**49.** Here $S(t) = 0.01t^4 e^{-0.07t}$ $\Rightarrow$ $S'(t) = 0.01 \left[ t^4 \cdot e^{-0.07t}(-0.07) + e^{-0.07t} \cdot 4t^3 \right] = 0.01t^3 e^{-0.07}(4 - 0.07t)$.

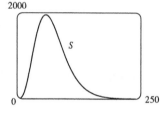

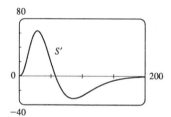

The graph of $S'$ has a local maximum at about $t \approx 28.57$, where $S'$ changes from increasing to decreasing, and a local minimum at about $t \approx 85.71$, where $S'$ changes from decreasing to increasing. Both of these times correspond to inflection points of $S$. At $\approx 28.57$ minutes, the rate of increase of the level of medication in the bloodstream is at its greatest, and at $\approx 85.71$ minutes, the rate of decrease is the greatest.

**51.** $f(t) = \dfrac{88.5}{1 + 17.7e^{-0.94t}}$ $\Rightarrow$ $f'(t) = \dfrac{0 - (88.5)[17.7e^{-0.94t}(-0.94)]}{(1 + 17.7e^{-0.94t})^2} = \dfrac{1472.463e^{-0.94t}}{(1 + 17.7e^{-0.94t})^2}$.

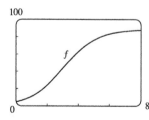

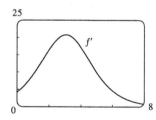

From the graph of $f$, we estimate that the graph is increasing most rapidly around $t = 3$, where the curve has an inflection point. We get a more accurate estimate by graphing the derivative. $f'$ has a local maximum at $t \approx 3.06$, corresponding to the greatest rate of change of $f$. [We could also look at a graph of $f''$, but the computation of $f''(t)$ is messy.] Thus we estimate that the percentage of households with DVD players was increasing most rapidly 3.06 years after midyear 2000.

**53.** The factors $(x+1)^2$ and $(x-6)^4$ are never negative, so they do not affect the sign of $f'(x) = (x+1)^2(x-3)^5(x-6)^4$.

Thus $f'(x) > 0$ when $(x-3)^5 > 0 \implies x - 3 > 0 \implies x > 3$, and $f$ is increasing on the interval $(3, \infty)$.

**55.** Let the cubic function be $f(x) = ax^3 + bx^2 + cx + d \implies f'(x) = 3ax^2 + 2bx + c \implies f''(x) = 6ax + 2b$.

$f$ is concave upward when $6ax + 2b > 0 \implies x > -b/(3a)$ and concave downward when $6ax + 2b < 0 \implies$

$x < -b/(3a)$. Thus the only point of inflection occurs when $x = -b/(3a)$.

**57.** $f(x) = ax^3 + bx^2 + cx + d \implies f'(x) = 3ax^2 + 2bx + c$.

We are given that $f(1) = 0 \implies a + b + c + d = 0$ and $f(-2) = 3 \implies -8a + 4b - 2c + d = 3$.

Also, $-2$ and $1$ must be critical numbers, so $f'(-2) = 0 \implies$

$12a - 4b + c = 0$ and $f'(1) = 0 \implies 3a + 2b + c = 0$. Solving this

system of four equations, we get $a = \frac{2}{9}, b = \frac{1}{3}, c = -\frac{4}{3}, d = \frac{7}{9}$, so the

function is $f(x) = \frac{2}{9}x^3 + \frac{1}{3}x^2 - \frac{4}{3}x + \frac{7}{9} = \frac{1}{9}\left(2x^3 + 3x^2 - 12x + 7\right)$.

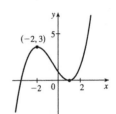

## 4.4  Asymptotes

**Prepare Yourself**

**1.** (a) If $a = 1/10$, then $\dfrac{1}{a^3} = \dfrac{1}{(1/10)^3} = \dfrac{1}{1/1000} = 1000$.

(b) If $b = -1/100$, then $\dfrac{1}{b^3} = \dfrac{1}{(-1/100)^3} = \dfrac{1}{-1/1,000,000} = -1,000,000$.

**2.** If $x$ becomes larger and larger, then

(a) $x^2$ becomes larger and larger, so $f(x) = -x^2$ gets larger and larger negative.

(b) $g(x) = 1/x$ gets closer and closer to 0.

(c) $h(x) = -1/x$ is negative and gets closer and closer to 0.

(d) $A(x) = e^x$ gets larger and larger (since $e^x$ is an increasing function).

(e) $e^x$ gets larger and larger, so $B(x) = e^{-x} = 1/e^x$ gets closer and closer to 0.

**3.** If $c$ gets closer and closer to 5, the numerator of $\dfrac{2c}{c-5}$ approaches 10, while the denominator approaches 0. If $c > 5$, the

denominator is a small positive number, while if $c < 5$, the denominator is a small negative number. Thus $\dfrac{2c}{c-5}$ grows larger

and larger as $c$ approaches 5 with $c > 5$, and if $c < 5$, $\dfrac{2c}{c-5}$ grows larger and larger negative as $c$ approaches 5.

**4.** As $a$ gets closer and closer to 0, the values of $\ln a$ get larger and larger negative. (See the graph of $y = \ln x$ in Figure 6.)

**5.** $\dfrac{4x^3 + x^2 - 2}{3x^3 + 3x^2} \cdot \dfrac{1/x^3}{1/x^3} = \dfrac{\dfrac{4x^3 + x^2 - 2}{x^3}}{\dfrac{3x^3 + 3x^2}{x^3}} = \dfrac{\dfrac{4x^3}{x^3} + \dfrac{x^2}{x^3} - \dfrac{2}{x^3}}{\dfrac{3x^3}{x^3} + \dfrac{3x^2}{x^3}} = \dfrac{4 + \dfrac{1}{x} - \dfrac{2}{x^3}}{3 + \dfrac{3}{x}}$

Exercises

**1.** (a) As $x$ gets closer and closer to 2, the values of $f(x)$ become larger and larger.

(b) As $x$ approaches 1 through numbers larger than 1, the values of $f(x)$ become larger and larger negative.

(c) As $x$ becomes larger and larger, the values of $f(x)$ get closer and closer to 5.

(d) As $x$ becomes larger and larger negative, the values of $f(x)$ get closer and closer to 3.

**3.** (a) $\lim\limits_{x \to 2} f(x) = \infty$

(b) $\lim\limits_{x \to -1^-} f(x) = \infty$

(c) $\lim\limits_{x \to -1^+} f(x) = -\infty$

(d) $\lim\limits_{x \to \infty} f(x) = 1$

(e) $\lim\limits_{x \to -\infty} f(x) = 2$

(f) Vertical: $x = -1$, $x = 2$; Horizontal: $y = 1$, $y = 2$

**5.** $\lim\limits_{x \to 0} f(x) = -\infty$, $\quad \lim\limits_{x \to -\infty} f(x) = 5$,

$\lim\limits_{x \to \infty} f(x) = -5$

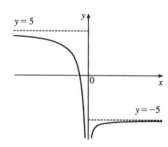

**7.** $\lim\limits_{x \to 2} f(x) = -\infty$, $\quad \lim\limits_{x \to \infty} f(x) = \infty$, $\quad \lim\limits_{x \to -\infty} f(x) = 0$,

$\lim\limits_{x \to 0^+} f(x) = \infty$, $\quad \lim\limits_{x \to 0^-} f(x) = -\infty$

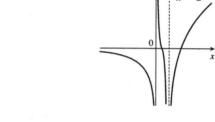

**9.** Vertical asymptotes: $x \approx -1.62$,

$x \approx 0.62$, $x = 1$;

Horizontal asymptote: $y = 1$

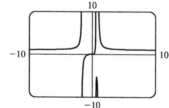

**11.** As $x$ approaches 4 while $x > 4$, $x - 4$ is positive and approaches 0. Thus $\dfrac{3}{x - 4}$ becomes larger and larger (positive), so

$$\lim_{x \to 4^+} \frac{3}{x - 4} = \infty.$$

**13.** As $x$ approaches $-3$ with $x > -3$, $x + 3$ is positive and approaches 0 while $x + 2$ approaches $-1$. Thus $\dfrac{x + 2}{x + 3}$ becomes

larger and larger negative, and $\lim\limits_{x \to -3^+} \dfrac{x + 2}{x + 3} = -\infty$.

**15.** As $x$ approaches 1, the numerator of $\dfrac{2 - x}{(x - 1)^2}$ approaches $2 - 1 = 1$ while the denominator is positive and approaches 0.

Thus $\dfrac{2 - x}{(x - 1)^2}$ grows arbitrarily large as $x \to 1$, and $\lim\limits_{x \to 1} \dfrac{2 - x}{(x - 1)^2} = \infty$.

**17.** As $x$ becomes larger and larger, $\dfrac{2}{x^3}$ gets closer to 0 (similarly as in Example 3). Thus $\lim\limits_{x \to \infty} \dfrac{2}{x^3} = 0$.

**19.** As in Example 4, we first divide both the numerator and denominator by $x^3$ (the highest power of $x$ that occurs in the denominator).

$$\lim_{x \to \infty} \frac{x^3 + 5x}{2x^3 - x^2 + 4} = \lim_{x \to \infty} \frac{\dfrac{x^3 + 5x}{x^3}}{\dfrac{2x^3 - x^2 + 4}{x^3}} = \lim_{x \to \infty} \frac{\dfrac{x^3}{x^3} + \dfrac{5x}{x^3}}{\dfrac{2x^3}{x^3} - \dfrac{x^2}{x^3} + \dfrac{4}{x^3}} = \lim_{x \to \infty} \frac{1 + \dfrac{5}{x^2}}{2 - \dfrac{1}{x} + \dfrac{4}{x^3}}$$

$$= \frac{\lim\limits_{x \to \infty} \left(1 + \dfrac{5}{x^2}\right)}{\lim\limits_{x \to \infty} \left(2 - \dfrac{1}{x} + \dfrac{4}{x^3}\right)} = \frac{\lim\limits_{x \to \infty} 1 + 5 \lim\limits_{x \to \infty} \dfrac{1}{x^2}}{\lim\limits_{x \to \infty} 2 - \lim\limits_{x \to \infty} \dfrac{1}{x} + 4 \lim\limits_{x \to \infty} \dfrac{1}{x^3}} = \frac{1 + 5(0)}{2 - 0 + 4(0)} = \frac{1}{2}$$

**21.** Divide both the numerator and denominator by $p^2$ (the highest power of $p$ that occurs in the denominator).

$$\lim_{p \to \infty} \frac{3p}{p^2 + 2p + 7} = \lim_{p \to \infty} \frac{\dfrac{3p}{p^2}}{\dfrac{p^2}{p^2} + \dfrac{2p}{p^2} + \dfrac{7}{p^2}} = \lim_{p \to \infty} \frac{\dfrac{3}{p}}{1 + \dfrac{2}{p} + \dfrac{7}{p^2}} = \frac{3 \lim\limits_{p \to \infty} \dfrac{1}{p}}{\lim\limits_{p \to \infty} 1 + 2 \lim\limits_{p \to \infty} \dfrac{1}{p} + 7 \lim\limits_{p \to \infty} \dfrac{1}{p^2}}$$

$$= \frac{3(0)}{1 + 2(0) + 7(0)} = 0$$

**23.** First, multiply the factors in the denominator. Then divide both the numerator and denominator by $u^4$.

$$\lim_{u \to \infty} \frac{4u^4 + 5}{(u^2 - 2)(2u^2 - 1)} = \lim_{u \to \infty} \frac{4u^4 + 5}{2u^4 - 5u^2 + 2} = \lim_{u \to \infty} \frac{\dfrac{4u^4 + 5}{u^4}}{\dfrac{2u^4 - 5u^2 + 2}{u^4}} = \lim_{u \to \infty} \frac{\dfrac{4u^4}{u^4} + \dfrac{5}{u^4}}{\dfrac{2u^4}{u^4} - \dfrac{5u^2}{u^4} + \dfrac{2}{u^4}}$$

$$= \lim_{u \to \infty} \frac{4 + \dfrac{5}{u^4}}{2 - \dfrac{5}{u^2} + \dfrac{2}{u^4}} = \frac{\lim\limits_{u \to \infty} 4 + 5 \lim\limits_{u \to \infty} \dfrac{1}{u^4}}{\lim\limits_{u \to \infty} 2 - 5 \lim\limits_{u \to \infty} \dfrac{1}{u^2} + 2 \lim\limits_{u \to \infty} \dfrac{1}{u^4}} = \frac{4 + 5(0)}{2 - 5(0) + 2(0)} = \frac{4}{2} = 2$$

**25.** First we factor the function $(x^4 + x^5)$ as $x^4(1 + x)$. Then $x^4$ becomes arbitrarily large (positive) as $x \to -\infty$, and $1 + x$ becomes arbitrarily large negative. Thus their product becomes larger and larger negative:

$$\lim_{x \to -\infty} (x^4 + x^5) = \lim_{x \to -\infty} x^4(1 + x) = -\infty.$$

**27.** $\lim\limits_{x \to \infty} \dfrac{x + x^3 + x^5}{1 - x^2 + x^4} = \lim\limits_{x \to \infty} \dfrac{(x + x^3 + x^5)/x^4}{(1 - x^2 + x^4)/x^4}$    [divide by the highest power of $x$ in the denominator]

$$= \lim_{x \to \infty} \frac{1/x^3 + 1/x + x}{1/x^4 - 1/x^2 + 1} = \infty$$

because $(1/x^3 + 1/x + x) \to \infty$ and $(1/x^4 - 1/x^2 + 1) \to 0 - 0 + 1 = 1$ as $x \to \infty$.

**29.** $\lim\limits_{x \to \infty} \dfrac{2x}{x^3 + 3} = \lim\limits_{x \to \infty} \dfrac{\dfrac{2x}{x^3}}{\dfrac{x^3 + 3}{x^3}} = \lim\limits_{x \to \infty} \dfrac{\dfrac{2}{x^2}}{1 + \dfrac{3}{x^3}} = \dfrac{\lim\limits_{x \to \infty} \dfrac{2}{x^2}}{\lim\limits_{x \to \infty} \left(1 + \dfrac{3}{x^3}\right)} = \dfrac{2 \lim\limits_{x \to \infty} \dfrac{1}{x^2}}{\lim\limits_{x \to \infty} 1 + 3 \lim\limits_{x \to \infty} \dfrac{1}{x^3}} = \dfrac{2(0)}{1 + 3(0)} = 0.$

Similarly, $\lim\limits_{x \to -\infty} \dfrac{2x}{x^3 + 3} = 0$. Thus $y = 0$ (the $x$-axis) is a horizontal asymptote.

**31.** $e^x + 1 \to \infty$ as $x \to \infty$, so its reciprocal $1/(e^x + 1) \to 0$. Thus $\lim\limits_{x \to \infty} \dfrac{1}{e^x + 1} = 0$, and $y = 0$ (the $x$-axis) is a horizontal

asymptote. As $x \to -\infty$, $e^x \to 0$ (for $x$ large negative, $e^x$ is the reciprocal of $e$ to a large positive power), so

$$\lim_{x \to -\infty} \frac{1}{e^x + 1} = \frac{\lim\limits_{x \to -\infty} 1}{\lim\limits_{x \to -\infty} e^x + \lim\limits_{x \to -\infty} 1} = \frac{1}{0 + 1} = 1, \text{ and } y = 1 \text{ is also a horizontal asymptote.}$$

**33.** $\lim\limits_{x \to \infty} \dfrac{2x^2 + x - 1}{x^2 + x - 2} = \lim\limits_{x \to \infty} \dfrac{\dfrac{2x^2 + x - 1}{x^2}}{\dfrac{x^2 + x - 2}{x^2}} = \lim\limits_{x \to \infty} \dfrac{2 + \dfrac{1}{x} - \dfrac{1}{x^2}}{1 + \dfrac{1}{x} - \dfrac{2}{x^2}} = \dfrac{\lim\limits_{x \to \infty}\left(2 + \dfrac{1}{x} - \dfrac{1}{x^2}\right)}{\lim\limits_{x \to \infty}\left(1 + \dfrac{1}{x} - \dfrac{2}{x^2}\right)}$

$$= \frac{\lim\limits_{x \to \infty} 2 + \lim\limits_{x \to \infty} \dfrac{1}{x} - \lim\limits_{x \to \infty} \dfrac{1}{x^2}}{\lim\limits_{x \to \infty} 1 + \lim\limits_{x \to \infty} \dfrac{1}{x} - 2\lim\limits_{x \to \infty} \dfrac{1}{x^2}} = \frac{2 + 0 - 0}{1 + 0 - 2(0)} = 2$$

Similarly, $\lim\limits_{x \to -\infty} \dfrac{2x^2 + x - 1}{x^2 + x - 2} = 2$, so $y = 2$ is the only horizontal asymptote.

$y = f(x) = \dfrac{2x^2 + x - 1}{x^2 + x - 2} = \dfrac{(2x - 1)(x + 1)}{(x + 2)(x - 1)}$, so $x = -2$ and $x = 1$ are not in the domain of $f$. We determine the

one-sided limits at these numbers. As $x \to -2^-$, the numerator of $f$ approaches $(-5)(-1) = 5$. In the denominator, the

factor $(x - 1) \to -3$, but $x + 2$ approaches $0$ through negative values. Thus the denominator of $f$ approaches $0$ through

positive values and $\lim\limits_{x \to -2^-} \dfrac{(2x - 1)(x + 1)}{(x + 2)(x - 1)} = \infty$. As $x \to -2^+$, $(x + 2)$ approaches $0$ through positive values, so the

denominator is negative and $\lim\limits_{x \to -2^+} f(x) = -\infty$. $\lim\limits_{x \to 1^-} f(x) = -\infty$ because the numerator of $f$ approaches $2$, and the

denominator approaches $0$ through negative values. Similarly,

$\lim\limits_{x \to 1^+} f(x) = \infty$ because the numerator approaches $2$ and the denominator

approaches $0$ through positive values. Thus, $x = -2$ and $x = 1$ are vertical

asymptotes. The graph confirms our work.

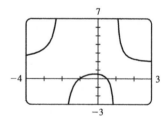

**35.** From the graph, it appears that $y = 1$ is a horizontal asymptote.

$$\lim_{x \to \infty} \frac{3x^3 + 500x^2}{x^3 + 500x^2 + 100x + 2000} = \lim_{x \to \infty} \frac{\dfrac{3x^3 + 500x^2}{x^3}}{\dfrac{x^3 + 500x^2 + 100x + 2000}{x^3}} = \lim_{x \to \infty} \frac{3 + (500/x)}{1 + (500/x) + (100/x^2) + (2000/x^3)}$$

$$= \frac{3 + 0}{1 + 0 + 0 + 0} = 3, \quad \text{so } y = 3 \text{ is a horizontal asymptote.}$$

The discrepancy can be explained by the choice of the viewing window. Try

$[-100{,}000, 100{,}000]$ by $[-1, 4]$ to get a graph that lends credibility to our

calculation that $y = 3$ is a horizontal asymptote.

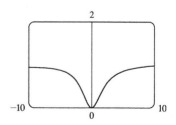

**37.** $y = f(x) = x/\sqrt{x^2+1}$, and using the hint from Exercise 36, we have

$\lim\limits_{x \to \infty} \dfrac{x}{\sqrt{x^2+1}} = \lim\limits_{x \to \infty} \dfrac{x}{\sqrt{x^2+1}} \cdot \dfrac{1/x}{1/\sqrt{x^2}} = \lim\limits_{x \to \infty} \dfrac{x/x}{\sqrt{(x^2+1)/x^2}} = \lim\limits_{x \to \infty} \dfrac{1}{\sqrt{1+1/x^2}} = \dfrac{1}{\sqrt{1+0}} = 1$. Thus $y = 1$ is a

horizontal asymptote. If $x \to -\infty$, then we can assume that $x$ is negative, so $\sqrt{x^2} = |x| = -x$:

$\lim\limits_{x \to -\infty} \dfrac{x}{\sqrt{x^2+1}} = \lim\limits_{x \to -\infty} \dfrac{x}{\sqrt{x^2+1}} \cdot \dfrac{1/x}{1/\left(-\sqrt{x^2}\right)} = \lim\limits_{x \to -\infty} \dfrac{x/x}{-\sqrt{(x^2+1)/x^2}} = \lim\limits_{x \to -\infty} \dfrac{1}{-\sqrt{1+1/x^2}}$

$= \dfrac{1}{-\sqrt{1+0}} = -1$, so $y = -1$ is also a horizontal asymptote.

$f'(x) = \dfrac{\sqrt{x^2+1} \cdot 1 - x \cdot \frac{1}{2}(x^2+1)^{-1/2}(2x)}{\left(\sqrt{x^2+1}\right)^2} = \dfrac{\sqrt{x^2+1} - x^2(x^2+1)^{-1/2}}{x^2+1} \cdot \dfrac{(x^2+1)^{1/2}}{(x^2+1)^{1/2}}$

$= \dfrac{x^2+1-x^2}{(x^2+1)^{3/2}} = \dfrac{1}{(x^2+1)^{3/2}}$

which is positive for all $x$, so $f$ is increasing on $\mathbb{R}$.

$f'(x) = (x^2+1)^{-3/2} \;\Rightarrow\; f''(x) = -\frac{3}{2}(x^2+1)^{-5/2} \cdot 2x = \dfrac{-3x}{(x^2+1)^{5/2}}.$

The denominator is always positive, so $f''(x) > 0$ for $x < 0$ and

$f''(x) < 0$ for $x > 0$. Thus, $f$ is concave upward on $(-\infty, 0)$ and concave

downward on $(0, \infty)$. We have an inflection point at $(0, 0)$.

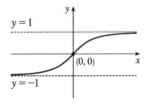

**39.** (a) In both viewing rectangles,

$\lim\limits_{x \to \infty} P(x) = \lim\limits_{x \to \infty} Q(x) = \infty$ and

$\lim\limits_{x \to -\infty} P(x) = \lim\limits_{x \to -\infty} Q(x) = -\infty.$

In the larger viewing rectangle, $P$ and $Q$

become less distinguishable.

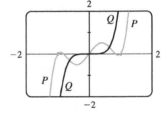
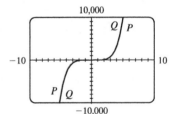

(b) $\lim\limits_{x \to \infty} \dfrac{P(x)}{Q(x)} = \lim\limits_{x \to \infty} \dfrac{3x^5 - 5x^3 + 2x}{3x^5} = \lim\limits_{x \to \infty} \left( \dfrac{3x^5}{3x^5} - \dfrac{5x^3}{3x^5} + \dfrac{2x}{3x^5} \right) = \lim\limits_{x \to \infty} \left( 1 - \dfrac{5}{3} \cdot \dfrac{1}{x^2} + \dfrac{2}{3} \cdot \dfrac{1}{x^4} \right)$

$= 1 - \frac{5}{3}(0) + \frac{2}{3}(0) = 1$, thus $P$ and $Q$ have the same end behavior.

**41.** (a) After $t$ minutes, $25t$ liters of brine with 30 g of salt per liter have been pumped into the tank, so it contains

$(5000 + 25t)$ liters of water and $25t \cdot 30 = 750t$ grams of salt. Therefore, the salt concentration at time $t$ will be

$C(t) = \dfrac{750t}{5000 + 25t} = \dfrac{30t}{200 + t} \dfrac{\text{g}}{\text{L}}.$

(b) $\lim\limits_{t \to \infty} C(t) = \lim\limits_{t \to \infty} \dfrac{30t}{200 + t} = \lim\limits_{t \to \infty} \dfrac{30t/t}{200/t + t/t} = \lim\limits_{t \to \infty} \dfrac{30}{200/t + 1} = \dfrac{30}{0 + 1} = 30$ g/L. So the salt concentration

approaches that of the brine being pumped into the tank.

## 4.5  Curve Sketching

Prepare Yourself

**1.** The domain of the rational function $y = \dfrac{x}{x^2 - 4} = \dfrac{x}{(x+2)(x-2)}$ is $\{x \mid x \neq \pm 2\}$.

**2.** $L(t) = t^3 - 3t^2 - 9t \;\Rightarrow\; L'(t) = 3t^2 - 6t - 9 = 3(t^2 - 2t - 3) = 3(t-3)(t+1)$. $L'(t)$ is defined everywhere, so the critical numbers are the numbers where $L'(t) = 0$, namely 3 and $-1$.

**3.** $g(x) = \dfrac{x^2}{x^2 + 1} \;\Rightarrow\; g'(x) = \dfrac{(x^2+1)(2x) - x^2(2x)}{(x^2+1)^2} = \dfrac{2x}{(x^2+1)^2}$. The denominator is always positive and $g'(x) = 0$ when $x = 0$, so 0 is the only critical number. $g'(x) > 0$ when $x > 0$ and $g'(x) < 0$ when $x < 0$, so $g$ is increasing on $(0, \infty)$ and decreasing on $(-\infty, 0)$. Because $g'$ changes from negative to positive at 0, $g(0) = 0$ is a local minimum value.

**4.** $f(w) = 3w^4 - 2w^3 + 1 \;\Rightarrow\; f'(w) = 12w^3 - 6w^2 \;\Rightarrow\; f''(w) = 36w^2 - 12w = 12w(3w - 1)$. $f''(w) = 0$ for $w = 0$ or $w = \frac{1}{3}$, $f''(w) > 0$ when $w < 0$ or $w > \frac{1}{3}$, and $f''(w) < 0$ when $0 < w < \frac{1}{3}$. Thus $f$ is concave upward on $(-\infty, 0)$ and $\left(\frac{1}{3}, \infty\right)$ and concave downward on $\left(0, \frac{1}{3}\right)$. Concavity changes direction at $w = 0$ and $w = \frac{1}{3}$, so $(0, 1)$ and $\left(\frac{1}{3}, \frac{26}{27}\right)$ are inflection points.

**5.** First we find the limits of $R(t)$ as $t \to \infty$ and $t \to -\infty$ by dividing numerator and denominator by $t^2$, the highest power of $t$ appearing in the denominator.

$$\lim_{t \to \infty} \frac{2t^2 + 1}{t^2 + 7t} = \lim_{t \to \infty} \frac{\dfrac{2t^2 + 1}{t^2}}{\dfrac{t^2 + 7t}{t^2}} = \lim_{t \to \infty} \frac{2 + \dfrac{1}{t^2}}{1 + \dfrac{7}{t}} = \frac{\lim_{t \to \infty}\left(2 + \dfrac{1}{t^2}\right)}{\lim_{t \to \infty}\left(1 + \dfrac{7}{t}\right)} = \frac{\lim_{t \to \infty} 2 + \lim_{t \to \infty} \dfrac{1}{t^2}}{\lim_{t \to \infty} 1 + 7 \lim_{t \to \infty} \dfrac{1}{t}} = \frac{2 + 0}{1 + 7(0)} = 2$$

Similarly, $\displaystyle\lim_{t \to -\infty} \frac{2t^2 + 1}{t^2 + 7t} = 2$, so $y = 2$ is the only horizontal asymptote.

$R(t) = \dfrac{2t^2 + 1}{t^2 + 7t} = \dfrac{2t^2 + 1}{t(t + 7)}$, so $t = 0$ and $t = -7$ are not in the domain of $R$. We determine the one-sided limits at these numbers. As $t \to 0^-$, the numerator of $R$ approaches $0 + 1 = 1$. In the denominator, the factor $t$ approaches 0 through negative values while $(t + 7) \to 7$. Thus the denominator of $R$ approaches 0 through negative values and $\displaystyle\lim_{t \to 0^-} R(t) = -\infty$.

As $t \to 0^+$, the denominator approaches 0 through positive values, so $\displaystyle\lim_{t \to 0^+} R(t) = \infty$. $\displaystyle\lim_{t \to -7^-} R(t) = \infty$ because the numerator of $R$ approaches 99, and the denominator approaches 0 through positive values. Similarly, $\displaystyle\lim_{t \to -7^+} R(t) = -\infty$ because the numerator approaches 99 and the denominator approaches 0 through negative values. Thus, $t = 0$ and $t = -7$ are vertical asymptotes.

**6.** Since $f'(2) = 0$, 2 is a critical number of $f$. By the Second Derivative Test, since $f''(2) < 0$, $f$ has a local maximum value at $x = 2$.

**7.** (a) $g'(1) = 0$, so 1 is a critical number of $g$. Since $g'$ changes from negative to positive at $x = 1$, $g$ has a local minimum value at $x = 1$ by the First Derivative Test.

(b) Because $g''$ changes from negative to positive at $x = 1$, $g$ has an inflection point at $x = 1$.

**8.** (a) The graph of $f$ has a vertical asymptote of $x = 5$.

(b) The graph has a horizontal asymptote of $y = 5$.

---

### Exercises

*Abbreviations*: inc, increasing; dec, decreasing; loc, local: max, maximum; min, minimum; CU, concave upward; CD, concave downward; IP, inflection point; HA, horizontal asymptote; VA, vertical asymptote

**1.** (a) $y = f(x) = 2x^2 - 8x + 3 \Rightarrow f'(x) = 4x - 8 = 4(x - 2)$. $f'(x) = 0$ when $x = 2$, so 2 is the only critical number. $f'(x) < 0$ for $x < 2$ and $f'(x) > 0$ for $x > 2$, so $f$ is inc on $(2, \infty)$ and dec on $(-\infty, 2)$.

(b) $f'$ changes from negative to positive at $x = 2$, so $f(2) = -5$ is a loc min by the First Derivative Test.

(c) $f''(x) = 4$, so $f''$ is positive everywhere and $f$ is CU on $(-\infty, \infty)$. There are no inflection points.

(d) $\lim\limits_{x \to \infty} (2x^2 - 8x + 3) = \lim\limits_{x \to \infty} [2x(x - 4) + 3] = \infty$ because both $2x$ and $(x - 4)$ become arbitrarily large as $x \to \infty$.

Similarly, $\lim\limits_{x \to -\infty} [2x(x - 4) + 3] = \infty$.

(e) Notice that the $y$-intercept is 3, so the graph passes through the point $(0, 3)$.

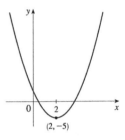

**3.** (a) $y = f(x) = x^3 + x \Rightarrow f'(x) = 3x^2 + 1$. $f'(x) > 0$ for all $x$, so $f$ is inc on $(-\infty, \infty)$.

(b) There are no critical numbers, so there are no local maximum or minimum values.

(c) $f''(x) = 6x$, so $f''(x) = 0$ for $x = 0$, $f''(x) > 0$ for $x > 0$ and $f''(x) < 0$ for $x < 0$. Thus $f$ is CU on $(0, \infty)$ and CD on $(-\infty, 0)$. $f''(x)$ changes sign at $x = 0$, so $(0, 0)$ is an IP.

(d) $\lim\limits_{x \to \infty} (x^3 + x) = \infty$ because both $x^3$ and $x$ become arbitrarily large as $x \to \infty$. $\lim\limits_{x \to -\infty} (x^3 + x) = -\infty$ because both $x^3$ and $x$ become arbitrarily large negative as $x \to -\infty$.

(e) The curve passes through the origin because $f(0) = 0$.

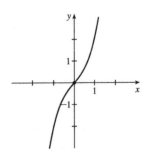

**5.** (a) $y = f(x) = 2 - 15x + 9x^2 - x^3 \Rightarrow f'(x) = -15 + 18x - 3x^2 = -3(x^2 - 6x + 5) = -3(x - 1)(x - 5)$, so the critical numbers, where $f'(x) = 0$, are $x = 1$ and $x = 5$. $f'(x) > 0$ when $1 < x < 5$, and $f'(x) < 0$ when $x < 1$ or $x > 5$, so $f$ is increasing on $(1, 5)$ and decreasing on $(-\infty, 1)$ and $(5, \infty)$.

(b) Loc max $f(5) = 27$ because $f'$ changes from positive to negative there, and loc min $f(1) = -5$ because $f'$ changes from negative to positive there.

(c) $f''(x) = 18 - 6x = -6(x - 3)$, so $f''(x) = 0$ when $x = 3$, $f''(x) > 0$ when $x < 3$, and $f''(x) < 0$ when $x > 3$. Thus $f$ is CU on $(-\infty, 3)$ and CD on $(3, \infty)$. IP at $(3, 11)$ since $f''$ changes sign at $x = 3$.

(d) We can write $f(x) = 2 - x(15 - 9x + x^2) = 2 - x[x(x - 9) + 15]$, so $\lim\limits_{x \to \infty} f(x) = -\infty$ because both $x$ and $(x - 9)$, and hence $[x(x - 9) + 15]$, become arbitrarily large as $x \to \infty$, so $2 - x[x(x - 9) + 15]$ becomes arbitrarily large negative. $\lim\limits_{x \to -\infty} f(x) = \infty$ because $[x(x - 9) + 15]$ becomes arbitrarily large positive as $x \to -\infty$, so $2 - x[x(x - 9) + 15]$ becomes arbitrarily large positive.

(e) Notice that $f(0) = 2$, so the graph passes through $(0, 2)$.

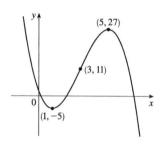

7. (a) $y = f(x) = x^4 + 4x^3 \Rightarrow f'(x) = 4x^3 + 12x^2 = 4x^2(x + 3)$. We have critical numbers where $f'(x) = 0$, that is, at $x = 0$ and $x = -3$. $4x^2 \geq 0$, so $f'(x) > 0$ when $-3 < x < 0$ or $x > 0$, and $f'(x) < 0$ when $x < -3$. Thus we can say that $f$ is inc on $(-3, \infty)$, and $f$ is dec on $(-\infty, -3)$.

(b) Loc min $f(-3) = -27$, no loc max. [The critical number 0 does not correspond to a loc max or min because $f'$ does not change sign there.]

(c) $f''(x) = 12x^2 + 24x = 12x(x + 2)$, and $f''(x) = 0$ for $x = 0$ or $x = -2$, $f''(x) > 0$ when $x < -2$ or $x > 0$, $f''(x) < 0$ when $-2 < x < 0$. Thus $f$ is CD on $(-2, 0)$ and CU on $(-\infty, -2)$ and $(0, \infty)$. IP at $(0, 0)$ and $(-2, -16)$.

(d) $\lim\limits_{x \to \infty} (x^4 + 4x^3) = \lim\limits_{x \to \infty} x^3(x + 4) = \infty$ because both $x^3$ and $(x + 4)$ become arbitrarily large as $x \to \infty$.

$\lim\limits_{x \to -\infty} x^3(x + 4) = \infty$ because both $x^3$ and $(x + 4)$ become arbitrarily large negative as $x \to -\infty$. So $x^3(x + 4)$ becomes arbitrarily large positive.

(e) Notice that the curve has a horizontal tangent at $(0, 0)$.

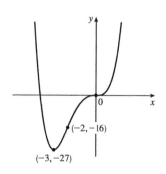

9. $y = f(x) = x/(x - 1)$. The domain is $\{x \mid x \neq 1\} = (-\infty, 1) \cup (1, \infty)$. $f'(x) = \dfrac{(x - 1) \cdot 1 - x \cdot 1}{(x - 1)^2} = \dfrac{-1}{(x - 1)^2}$, and $f'(x) < 0$ for $x \neq 1$, so $f$ is decreasing on $(-\infty, 1)$ and $(1, \infty)$. There are no local extreme points.

$$f''(x) = \frac{0 - (-1) \cdot 2(x-1)(1)}{(x-1)^4} = \frac{2}{(x-1)^3}, \text{ and } f''(x) > 0 \text{ when } x > 1, \; f''(x) < 0 \text{ when } x < 1. \text{ Thus } f \text{ is CU on}$$

$(1, \infty)$ and CD on $(-\infty, 1)$. No IP. ($f''$ changes sign at $x = 1$, but 1 is not in the domain of $f$.)

$$\lim_{x \to \pm\infty} \frac{x}{x-1} = \lim_{x \to \pm\infty} \frac{x/x}{x/x - 1/x} = \lim_{x \to \pm\infty} \frac{1}{1 - 1/x} = 1, \text{ so } y = 1 \text{ is a}$$

HA. $\displaystyle\lim_{x \to 1^-} \frac{x}{x-1} = -\infty, \; \lim_{x \to 1^+} \frac{x}{x-1} = \infty$, so $x = 1$ is a VA.   Also we

have an $x$-intercept of 0 and $y$-intercept $= f(0) = 0$.

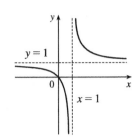

11. $y = f(x) = \dfrac{1}{x^2 - 9} = \dfrac{1}{(x+3)(x-3)}$, and the domain is $\{x \mid x \ne \pm 3\} = (-\infty, -3) \cup (-3, 3) \cup (3, \infty)$.

$f(x) = (x^2 - 9)^{-1} \;\Rightarrow\; f'(x) = -(x^2 - 9)^{-2}(2x) = -\dfrac{2x}{(x^2 - 9)^2}.$   $f'(x) > 0$ when $x < 0 \; (x \ne -3), \; f'(x) < 0$ when

$x > 0 \; (x \ne 3)$, so $f$ is increasing on $(-\infty, -3)$ and $(-3, 0)$ and decreasing on $(0, 3)$ and $(3, \infty)$.   $f(0) = -\frac{1}{9}$ is a loc max.

$$f''(x) = -\frac{(x^2 - 9)^2 \cdot 2 - 2x \cdot 2(x^2 - 9)(2x)}{[(x^2 - 9)^2]^2} = -\frac{2(x^2 - 9)\left[(x^2 - 9) - 4x^2\right]}{(x^2 - 9)^4}$$

$$= -\frac{2(-3x^2 - 9)}{(x^2 - 9)^3} = \frac{6(x^2 + 3)}{(x^2 - 9)^3}$$

$f''(x) > 0$ when $x^2 > 9 \;\Rightarrow\; x > 3$ or $x < -3$, $f''(x) < 0$ when $-3 < x < 3$.

Thus $f$ is CU on $(-\infty, -3)$ and $(3, \infty)$ and CD on $(-3, 3)$. No IP.

$$\lim_{x \to \pm\infty} \frac{1}{x^2 - 9} = \lim_{x \to \pm\infty} \frac{1/x^2}{1 - 9/x^2} = 0, \text{ so } y = 0 \text{ is a HA.}$$

$$\lim_{x \to 3^-} \frac{1}{x^2 - 9} = -\infty, \; \lim_{x \to 3^+} \frac{1}{x^2 - 9} = \infty, \; \lim_{x \to -3^-} \frac{1}{x^2 - 9} = \infty,$$

$$\lim_{x \to -3^+} \frac{1}{x^2 - 9} = -\infty, \text{ so } x = 3 \text{ and } x = -3 \text{ are VA.}$$

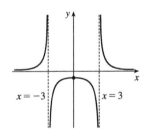

13. $y = f(x) = \dfrac{x - 1}{x^2}$. The domain is $\{x \mid x \ne 0\} = (-\infty, 0) \cup (0, \infty)$.

$$f'(x) = \frac{x^2 \cdot 1 - (x-1) \cdot 2x}{(x^2)^2} = \frac{-x^2 + 2x}{x^4} = \frac{-(x-2)}{x^3}, \text{ so } f'(x) > 0 \text{ when } 0 < x < 2 \text{ and } f'(x) < 0 \text{ when } x < 0$$

or $x > 2$. Thus $f$ is increasing on $(0, 2)$ and decreasing on $(-\infty, 0)$ and $(2, \infty)$. No loc min, loc max $f(2) = \frac{1}{4}$.

$$f''(x) = \frac{x^3 \cdot (-1) - [-(x-2)] \cdot 3x^2}{(x^3)^2} = \frac{2x^3 - 6x^2}{x^6} = \frac{2(x-3)}{x^4}.$$

$f''(x)$ is negative on $(-\infty, 0)$ and $(0, 3)$ and positive on $(3, \infty)$, so $f$ is CD

on $(-\infty, 0)$ and $(0, 3)$ and CU on $(3, \infty)$. IP at $\left(3, \frac{2}{9}\right)$.

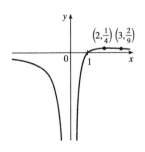

$$\lim_{x \to \pm\infty} \frac{x - 1}{x^2} = \lim_{x \to \pm\infty} \frac{1/x - 1/x^2}{1} = 0, \text{ so } y = 0 \text{ is a HA.}$$

$$\lim_{x \to 0} \frac{x - 1}{x^2} = -\infty, \text{ so } x = 0 \text{ is a VA.}$$

**15.** $h(a) = \sqrt{a}(a^2 - 4a)$ is defined only when $a \geq 0$, so the domain is $[0, \infty)$.

$$h'(a) = \sqrt{a} \cdot (2a - 4) + (a^2 - 4a) \cdot \tfrac{1}{2} a^{-1/2} = \sqrt{a}(2a - 4) + \frac{a^2 - 4a}{2\sqrt{a}}$$

$$= \frac{1}{2\sqrt{a}} \left[ 2a(2a - 4) + a^2 - 4a \right] = \frac{5a^2 - 12a}{2\sqrt{a}} = \frac{a(5a - 12)}{2\sqrt{a}}$$

The domain of $h'$ is $(0, \infty)$ and $h'(a) = 0$ when $a = \frac{12}{5}$, so the only critical number is $\frac{12}{5} = 2.4$. $h'(a) > 0$ for $a > \frac{12}{5}$,

$h'(a) < 0$ for $0 < a < \frac{12}{5}$, so $h$ is inc on $\left( \frac{12}{5}, \infty \right)$ and dec on $\left( 0, \frac{12}{5} \right)$. Loc min $h \left( \frac{12}{5} \right) = -\frac{192\sqrt{3}}{25\sqrt{5}} \approx -5.95$.

$$h''(a) = \frac{2\sqrt{a} \cdot (10a - 12) - (5a^2 - 12a) \cdot 2 \left( \tfrac{1}{2} a^{-1/2} \right)}{(2\sqrt{a})^2}$$

$$= \frac{a^{-1/2} \left[ 2a(10a - 12) - (5a^2 - 12a) \right]}{4a} = \frac{15a^2 - 12a}{4a^{3/2}} = \frac{a(15a - 12)}{4a^{3/2}} = \frac{15a - 12}{4\sqrt{a}}$$

$h''(a) = 0$ for $a = \frac{4}{5}$, $h''(a) > 0$ for $a > \frac{4}{5}$ and $h''(a) < 0$ for $0 < a < \frac{4}{5}$, so $h$ is

CU on $\left( \frac{4}{5}, \infty \right)$ and CD on $\left( 0, \frac{4}{5} \right)$. IP $\left( \frac{4}{5}, -\frac{128}{25\sqrt{5}} \right) \approx (0.8, -2.29)$.

$\lim\limits_{x \to \infty} h(a) = \lim\limits_{x \to \infty} \sqrt{a}(a)(a - 4) = \infty$ because $\sqrt{a}$, $a$, and $(a - 4)$ all become

arbitrarily large as $a \to \infty$. The graph begins at the origin because $h(0) = 0$.

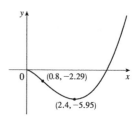

**17.** $y = f(x) = 1/(1 + e^{-x}) = (1 + e^{-x})^{-1}$. The domain is $\mathbb{R}$. $f'(x) = -(1 + e^{-x})^{-2} \cdot e^{-x}(-1) = e^{-x}/(1 + e^{-x})^2$. This is

positive for all $x$, so $f$ is increasing on $\mathbb{R}$ and there are no local extreme points.

$$f''(x) = \frac{(1 + e^{-x})^2 (-e^{-x}) - e^{-x} \cdot 2(1 + e^{-x})(-e^{-x})}{(1 + e^{-x})^4} = \frac{e^{-x}(1 + e^{-x}) \left[ -(1 + e^{-x}) + 2e^{-x} \right]}{(1 + e^{-x})^4} = \frac{e^{-x}(e^{-x} - 1)}{(1 + e^{-x})^3},$$

$f''(x) = 0$ when $e^{-x} - 1 = 0 \Rightarrow e^{-x} = 1 \Rightarrow x = 0$, $f''(x) > 0$ when $x < 0$, $f''(x) < 0$ when $x > 0$, so $f$ is CU on

$(-\infty, 0)$ and CD on $(0, \infty)$. IP at $\left( 0, \frac{1}{2} \right)$. $\lim\limits_{x \to \infty} 1/(1 + e^{-x}) = \frac{1}{1 + 0} = 1$, and

$\lim\limits_{x \to -\infty} 1/(1 + e^{-x}) = 0$ since $e^{-x}$, and hence $1 + e^{-x}$, becomes arbitrarily large

as $x \to -\infty$. Thus $f$ has horizontal asymptotes $y = 0$ and $y = 1$.

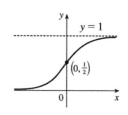

**19.** $y = f(x) = xe^{-x^2}$

| $x$ | $f(x)$ | | $x$ | $f(x)$ |
|---|---|---|---|---|
| 1 | 0.37 | | $-1$ | $-0.37$ |
| 2 | 0.037 | | $-2$ | $-0.037$ |
| 5 | $6.9 \times 10^{-11}$ | | $-5$ | $-6.9 \times 10^{-11}$ |
| 10 | $3.7 \times 10^{-43}$ | | $-10$ | $-3.7 \times 10^{-43}$ |

From the tables, it appears that $\lim\limits_{x \to \infty} f(x) = \lim\limits_{x \to -\infty} f(x) = 0$, so $y = 0$ is a HA. The domain of $f$ is $\mathbb{R}$.

$f'(x) = x \cdot e^{-x^2}(-2x) + e^{-x^2} \cdot 1 = e^{-x^2}(1 - 2x^2)$, and $f'(x) = 0$ when $1 - 2x^2 = 0 \Rightarrow x^2 = \frac{1}{2} \Rightarrow x = \pm\frac{1}{\sqrt{2}}$, so

these are the two critical numbers. $f'(x) < 0$ when $x < -\frac{1}{\sqrt{2}}$ or $x > \frac{1}{\sqrt{2}}$, and $f'(x) > 0$ when

$-\frac{1}{\sqrt{2}} < x < \frac{1}{\sqrt{2}}$. Thus $f$ is increasing on $\left(-\frac{1}{\sqrt{2}}, \frac{1}{\sqrt{2}}\right)$ and decreasing on $\left(-\infty, -\frac{1}{\sqrt{2}}\right)$

and $\left(\frac{1}{\sqrt{2}}, \infty\right)$. $f\left(\frac{1}{\sqrt{2}}\right) = 1/\sqrt{2e}$ is a local max, and $f\left(-\frac{1}{\sqrt{2}}\right) = -1/\sqrt{2e}$ is a local min.

$f''(x) = e^{-x^2} \cdot (-4x) + (1 - 2x^2) \cdot e^{-x^2}(-2x) = 2xe^{-x^2}\left[-2 - (1 - 2x^2)\right] = 2x(2x^2 - 3)e^{-x^2}$. $f''(x) = 0$ when

$x = 0$ or $2x^2 - 3 = 0 \Rightarrow x^2 = \frac{3}{2} \Rightarrow x = \pm\sqrt{\frac{3}{2}}$, $f''(x) > 0$ when $x > \sqrt{\frac{3}{2}}$ or $-\sqrt{\frac{3}{2}} < x < 0$, $f''(x) < 0$ when

$0 < x < \sqrt{\frac{3}{2}}$ or $x < -\sqrt{\frac{3}{2}}$, so $f$ is CU on $\left(-\sqrt{\frac{3}{2}}, 0\right)$ and $\left(\sqrt{\frac{3}{2}}, \infty\right)$ and

CD on $\left(-\infty, -\sqrt{\frac{3}{2}}\right)$ and $\left(0, \sqrt{\frac{3}{2}}\right)$. IP are $(0, 0)$, $\left(\sqrt{\frac{3}{2}}, \sqrt{\frac{3}{2}}\,e^{-3/2}\right)$,

and $\left(-\sqrt{\frac{3}{2}}, -\sqrt{\frac{3}{2}}\,e^{-3/2}\right)$.

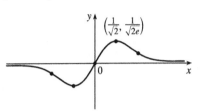

**21.** $f'(0) = f'(4) = 0$ means that we have horizontal tangents at $x = 0$ and $x = 4$ (and 0, 4 are critical numbers).

$f'(x) > 0$ if $x < 0 \Rightarrow f$ is increasing on $(-\infty, 0)$.

$f'(x) < 0$ if $0 < x < 4$ or if $x > 4 \Rightarrow f$ is decreasing on $(0, 4)$ and $(4, \infty)$.

$f''(x) > 0$ if $2 < x < 4 \Rightarrow f$ is concave upward on $(2, 4)$.

$f''(x) < 0$ if $x < 2$ or $x > 4 \Rightarrow f$ is concave downward on $(-\infty, 2)$ and $(4, \infty)$.

There are inflection points when $x = 2$ and 4.

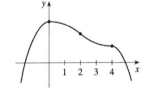

**23.** $f'(0) = f'(2) = f'(4) = 0 \Rightarrow$ horizontal tangents at $x = 0, 2, 4$ (and these are critical numbers).

$f'(x) > 0$ if $x < 0$ or $2 < x < 4 \Rightarrow f$ is increasing on $(-\infty, 0)$ and $(2, 4)$.

$f'(x) < 0$ if $0 < x < 2$ or $x > 4 \Rightarrow f$ is decreasing on $(0, 2)$ and $(4, \infty)$.

$f''(x) > 0$ if $1 < x < 3 \Rightarrow f$ is concave upward on $(1, 3)$.

$f''(x) < 0$ if $x < 1$ or $x > 3 \Rightarrow f$ is concave downward on $(-\infty, 1)$ and $(3, \infty)$.

There are inflection points when $x = 1$ and 3.

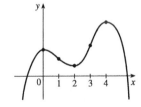

**25.** $f'(x) > 0$ if $x \neq 2 \Rightarrow f$ is increasing on $(-\infty, 2)$ and $(2, \infty)$. $f$ is concave

upward on $(-\infty, 2)$ and concave downward on $(2, \infty)$. There is an inflection point

at $(2, 5)$. $\lim\limits_{x \to \infty} f(x) = 8 \Rightarrow y = 8$ is a horizontal asymptote (at the right), and

$\lim\limits_{x \to -\infty} f(x) = 0 \Rightarrow y = 0$ is a horizontal asymptote (at the left).

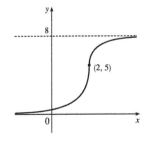

**27.** $f'(5) = 0 \Rightarrow$ horizontal tangent at $x = 5$.

$f'(x) < 0$ when $x < 5 \Rightarrow f$ is decreasing on $(-\infty, 5)$.

$f'(x) > 0$ when $x > 5 \Rightarrow f$ is increasing on $(5, \infty)$.

$f''(2) = 0$, $f''(8) = 0$, $f''(x) < 0$ when $x < 2$ or $x > 8$,

$f''(x) > 0$ for $2 < x < 8 \Rightarrow f$ is concave upward on $(2, 8)$ and concave downward on $(-\infty, 2)$ and $(8, \infty)$.

There are inflection points at $x = 2$ and $x = 8$.

$\lim\limits_{x \to \infty} f(x) = 3$, $\lim\limits_{x \to -\infty} f(x) = 3 \Rightarrow y = 3$ is a horizontal asymptote.

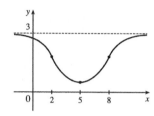

**29.** $f(x) = 4x^4 - 32x^3 + 89x^2 - 95x + 29 \Rightarrow f'(x) = 16x^3 - 96x^2 + 178x - 95 \Rightarrow f''(x) = 48x^2 - 192x + 178$.

$f(x) = 0$ when $x \approx 0.5$, $1.60$; $f'(x) = 0$ when $x \approx 0.92$, $2.5$, $2.58$ and $f''(x) = 0$ when $x \approx 1.46$, $2.54$.

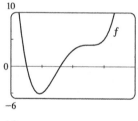

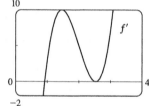

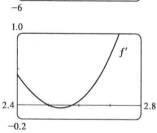

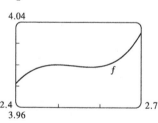

From the graphs of $f'$, we estimate that $f' < 0$ and that $f$ is decreasing on $(-\infty, 0.92)$ and $(2.5, 2.58)$, and that $f' > 0$ and $f$ is increasing on $(0.92, 2.5)$ and $(2.58, \infty)$ with local minimum values $f(0.92) \approx -5.12$ and $f(2.58) \approx 3.998$ and local maximum value $f(2.5) = 4$. The graphs of $f'$ make it clear that $f$ has a maximum and a minimum near $x = 2.5$, shown more clearly in the fourth graph.

From the graph of $f''$, we estimate that $f'' > 0$ and that $f$ is CU on $(-\infty, 1.46)$ and $(2.54, \infty)$, and that $f'' < 0$ and $f$ is CD on $(1.46, 2.54)$. There are inflection points at about $(1.46, -1.40)$ and $(2.54, 3.999)$.

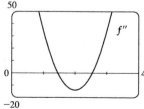

**31.** $f(x) = x^6 - 10x^5 - 400x^4 + 2500x^3 \Rightarrow f'(x) = 6x^5 - 50x^4 - 1600x^3 + 7500x^2 \Rightarrow$

$f''(x) = 30x^4 - 200x^3 - 4800x^2 + 1500x.$

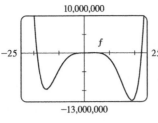

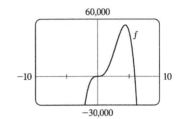

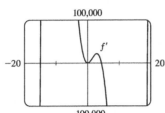

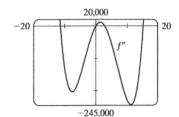

From the graph of $f'$, we estimate that $f$ is decreasing on $(-\infty, -15)$, increasing on $(-15, 4.40)$, decreasing on

$(4.40, 18.93)$, and increasing on $(18.93, \infty)$, with local minimum values of $f(-15) \approx -9{,}700{,}000$ and

$f(18.93) \approx -12{,}700{,}000$ and local maximum value $f(4.40) \approx 53{,}800$. From the graph of $f''$, we estimate that $f$ is CU on

$(-\infty, -11.34)$, CD on $(-11.34, 0)$, CU on $(0, 2.92)$, CD on $(2.92, 15.08)$, and CU on $(15.08, \infty)$. There are inflection

points at $(0, 0)$ and at about $(-11.34, -6{,}250{,}000)$, $(2.92, 31{,}800)$, and $(15.08, -8{,}150{,}000)$.

**33.** $f(x) = x^4 + cx^2 = x^2(x^2 + c)$. For $c \geq 0$, the only $x$-intercept is the point $(0, 0)$. We calculate

$f'(x) = 4x^3 + 2cx = 4x\left(x^2 + \frac{1}{2}c\right) \Rightarrow f''(x) = 12x^2 + 2c$. If $c \geq 0$, $x = 0$ is the only critical number and there is no

inflection point. As we can see from the examples, there is no change in the basic shape of the graph for $c \geq 0$; it merely

becomes steeper as $c$ increases. For $c = 0$, the graph is the simple curve $y = x^4$. For $c < 0$, there are $x$-intercepts at $0$ and

at $\pm\sqrt{-c}$. Also, there is a local maximum at $(0, 0)$, and there are

local minimum values at $\left(\pm\sqrt{-\frac{1}{2}c}, -\frac{1}{4}c^2\right)$. As $c \to -\infty$, the

$x$-coordinates of these minimum values get larger in absolute value,

and the minimum points move downward. There are inflection points

at $\left(\pm\sqrt{-\frac{1}{6}c}, -\frac{5}{36}c^2\right)$, which also move away from the origin

as $c \to -\infty$.

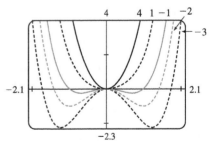

**35.** Note that $c = 0$ is a transitional value at which the graph consists of the line $y = 0$. Also, we can see that if we substitute $-c$

for $c$, the function $f(x) = \dfrac{cx}{1 + c^2x^2}$ becomes opposite and the graph will be reflected about the $x$-axis, so we investigate only

positive values of $c$ (except $c = -1$, as a demonstration of this reflective property). $\lim\limits_{x \to \pm\infty} f(x) = 0$, so $y = 0$ is a horizontal

asymptote for all $c$. We calculate $f'(x) = \dfrac{(1 + c^2x^2)c - cx(2c^2x)}{(1 + c^2x^2)^2} = -\dfrac{c(c^2x^2 - 1)}{(1 + c^2x^2)^2}$. $f'(x) = 0$ when $c^2x^2 - 1 = 0 \Rightarrow$

$x = \pm 1/c$. So there is a local (and absolute) maximum value of $f(1/c) = \frac{1}{2}$ and local (and absolute) minimum value of $f(-1/c) = -\frac{1}{2}$. These extreme values have the same value regardless of $c$, but the maximum points move closer to the $y$-axis as $c$ increases.

$$f''(x) = \frac{(1 + c^2x^2)^2(-2c^3x) - (-c^3x^2 + c)[2(1 + c^2x^2)(2c^2x)]}{(1 + c^2x^2)^4}$$

$$= \frac{(1 + c^2x^2)(-2c^3x) + (c^3x^2 - c)(4c^2x)}{(1 + c^2x^2)^3} = \frac{2c^3x(c^2x^2 - 3)}{(1 + c^2x^2)^3}$$

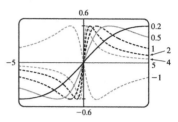

$f''(x) = 0$ when $x = 0$ or $\pm\sqrt{3}/c$, so there are inflection points at $(0, 0)$ and

at $(\sqrt{3}/c, \sqrt{3}/4)$, $(-\sqrt{3}/c, -\sqrt{3}/4)$. Again, the $y$-coordinate of the inflection points does not depend on $c$, but as $c$ increases, both inflection points approach the $y$-axis.

**37.** $f(x) = xe^{-cx} \Rightarrow f'(x) = x \cdot e^{-cx}(-c) + e^{-cx} \cdot 1 = (1 - cx)e^{-cx} \Rightarrow$

$f''(x) = (1 - cx) \cdot e^{-cx}(-c) + e^{-cx} \cdot (-c) = c(cx - 2)e^{-cx}$.

The origin is the only $x$- and $y$-intercept for all values of $c$. For $c > 0$, there is a local maximum value $f(1/c) = 1/(ce)$ and inflection point $(2/c, 2/(ce^2))$ to the right of the $y$-axis that move closer to the origin as $c$ increases. For $c < 0$,

$f(1/c) = 1/(ce)$ is a local minimum value that moves closer to the origin as $c$ decreases, as does the inflection point $(2/c, 2/(ce^2))$. From the graphed examples it appears that $y = 0$ is a horizontal asymptote, at the right for $c > 0$ and at the left for $c < 0$. $c = 0$ is a transitional value: when $c$ is replaced by $-c$, the curve is reflected about the origin.

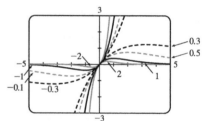

**39.** $y = f(x) = x^3 - 3a^2x + 2a^3$, $a > 0$. The $y$-intercept is $f(0) = 2a^3$.

$f'(x) = 3x^2 - 3a^2 = 3(x^2 - a^2) = 3(x + a)(x - a)$. The critical numbers are $-a$ and $a$. $f'(x) < 0$ when $-a < x < a$, and $f'(x) > 0$ when $x < -a$ or $x > a$, so $f$ is decreasing on $(-a, a)$ and $f$ is increasing on $(-\infty, -a)$ and $(a, \infty)$.

$f(-a) = 4a^3$ is a local maximum value and $f(a) = 0$ is a local minimum value. Since $f(a) = 0$, $a$ is an $x$-intercept, and $x - a$ is a factor of $f$. Dividing $y = x^3 - 3a^2x + 2a^3$ by $x - a$ gives us the following result:

$y = x^3 - 3a^2x + 2a^3 = (x - a)(x^2 + ax - 2a^2) = (x - a)(x - a)(x + 2a) = (x - a)^2(x + 2a)$, which tells

us that the only $x$-intercepts are $-2a$ and $a$. $f'(x) = 3x^2 - 3a^2 \Rightarrow f''(x) = 6x$,

so $f''(x) > 0$ on $(0, \infty)$ and $f''(x) < 0$ on $(-\infty, 0)$. This tells us that $f$ is CU on

$(0, \infty)$ and CD on $(-\infty, 0)$. There is an inflection point at $(0, 2a^3)$. The graph illustrates these features.

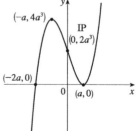

What the curves in the family have in common is that they are all CD on $(-\infty, 0)$,

CU on $(0, \infty)$, and have the same basic shape. But as $a$ increases, the four key points shown in the figure move further away from the origin.

**41.** (a) $p(t) = \dfrac{1}{2}$ $\Rightarrow$ $\dfrac{1}{2} = \dfrac{1}{1 + ae^{-kt}}$ $\Rightarrow$ $1 + ae^{-kt} = 2$ $\Rightarrow$ $ae^{-kt} = 1$ $\Rightarrow$ $e^{-kt} = \dfrac{1}{a}$ $\Rightarrow$

$\ln e^{-kt} = \ln \dfrac{1}{a}$ $\Rightarrow$ $-kt = \ln a^{-1}$ $\Rightarrow$ $-kt = -\ln a$ $\Rightarrow$ $t = \dfrac{\ln a}{k}$, which is when half the population will have

heard the rumor.

(b) The rate of spread is given by $p'(t) = \dfrac{0 - 1(ae^{-kt}(-k))}{(1 + ae^{-kt})^2} = \dfrac{ake^{-kt}}{(1 + ae^{-kt})^2}$. To find the greatest rate of spread, we'll

apply the First Derivative Test to $p'(t)$ [not $p(t)$].

$$[p'(t)]' = p''(t) = \frac{(1 + ae^{-kt})^2(-ak^2e^{-kt}) - ake^{-kt} \cdot 2(1 + ae^{-kt})(-ake^{-kt})}{[(1 + ae^{-kt})^2]^2}$$

$$= \frac{(1 + ae^{-kt})(-ake^{-kt})[k(1 + ae^{-kt}) - 2ake^{-kt}]}{(1 + ae^{-kt})^4}$$

$$= \frac{-ake^{-kt}(k)(1 - ae^{-kt})}{(1 + ae^{-kt})^3} = \frac{ak^2e^{-kt}(ae^{-kt} - 1)}{(1 + ae^{-kt})^3}$$

$p''(t) > 0$ when $ae^{-kt} > 1$ $\Rightarrow$ $-kt > \ln a^{-1}$ $\Rightarrow$ $t < \dfrac{\ln a}{k}$, so $p'(t)$ is increasing for $t < \dfrac{\ln a}{k}$ and $p'(t)$ is

decreasing for $t > \dfrac{\ln a}{k}$. Thus, $p'(t)$, the rate of spread of the rumor, is greatest at the same time, $\dfrac{\ln a}{k}$, as when half the

population [by part (a)] has heard it. Note that this is where $p'$ changes from increasing to decreasing, and hence an

inflection point of $p$.

(c) $p(0) = \dfrac{1}{1 + a}$ and $\lim\limits_{t \to \infty} p(t) = 1$. The graph is shown

with $a = 4$ and $k = \frac{1}{2}$.

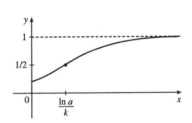

## 4.6 Optimization

**Prepare Yourself**

1. The Pythagorean Theorem states that if $a$ and $b$ are the lengths of the legs of a right triangle and $c$ is the length of the

hypotenuse, then $a^2 + b^2 = c^2$. Thus here $x^2 + (20 - x)^2 = c^2$, so the length of the hypotenuse is $c = \sqrt{x^2 + (20 - x)^2}$.

2. As in the previous exercise, $a^2 + b^2 = c^2$ $\Rightarrow$ $(x - 2)^2 + b^2 = (2x + 5)^2$. Then the length of the other leg is

$b = \sqrt{(2x + 5)^2 - (x - 2)^2}$.

3. The volume of the box is the area of the base multiplied by the height $h$. Thus $x(x + 4)h = 840$ $\Rightarrow$ $h = \dfrac{840}{x(x + 4)}$ inches.

4. The line passes through the points $(0, 0)$ and $(a, b)$, so the slope is $\dfrac{\Delta y}{\Delta x} = \dfrac{y_2 - y_1}{x_2 - x_1} = \dfrac{b - 0}{a - 0} = \dfrac{b}{a}$.

**5.** $3x - \dfrac{4}{x^2} = 0 \;\Rightarrow\; 3x = \dfrac{4}{x^2} \;\Rightarrow\; 3x \cdot x^2 = 4 \;\Rightarrow\; x^3 = 4/3 \;\Rightarrow\; x = \sqrt[3]{4/3}$

**6.** $A = \dfrac{p+m}{cm^2 - bm} \;\Rightarrow\;$

$$\frac{dA}{dm} = \frac{(cm^2 - bm) \cdot 1 - (p+m) \cdot (2cm - b)}{(cm^2 - bm)^2} = \frac{(cm^2 - bm) - (2pcm - bp + 2cm^2 - bm)}{(cm^2 - bm)^2}$$

$$= \frac{cm^2 - bm - 2pcm + bp - 2cm^2 + bm}{(cm^2 - bm)^2} = \frac{-cm^2 - 2pcm + bp}{(cm^2 - bm)^2}$$

**7.** rate × time = distance, so time $= \dfrac{\text{distance}}{\text{rate}}$. The time spent jogging is $\dfrac{x \text{ mi}}{3 \text{ mi/h}} = \dfrac{x}{3}$ h and the time spent walking is

$\dfrac{x - 2 \text{ mi}}{1.5 \text{ mi/h}} = \dfrac{x-2}{1.5}$ h, so the total time is $\dfrac{x}{3} + \dfrac{x-2}{1.5}$ h.

---

**Exercises**

**1.** (a)

| First Number | Second Number | Product |
|:---:|:---:|:---:|
| 1 | 22 | 22 |
| 2 | 21 | 42 |
| 3 | 20 | 60 |
| 4 | 19 | 76 |
| 5 | 18 | 90 |
| 6 | 17 | 102 |
| 7 | 16 | 112 |
| 8 | 15 | 120 |
| 9 | 14 | 126 |
| 10 | 13 | 130 |
| 11 | 12 | 132 |

We needn't consider pairs for which the first number is larger than the second, since we can just interchange the numbers in such cases. The answer appears to be 11 and 12, but we have considered only integers in the table.

(b) Call the two numbers $x$ and $y$. Then $x + y = 23$, so $y = 23 - x$. Call the product $P$. Then

$P = xy = x(23 - x) = 23x - x^2$, so we wish to maximize the function $P(x) = 23x - x^2$. Since $P'(x) = 23 - 2x$,

we see that $P'(x) = 0$ when $x = \frac{23}{2} = 11.5$, so this is the only critical number. $P'(x) > 0$ for $x < 11.5$ and $P'(x) < 0$

for $x > 11.5$, so by the First Derivative Test for Absolute Extreme Values (see page 258), the absolute maximum value of

$P$ is $P(11.5) = (11.5)^2 = 132.25$ and it occurs when $x = y = 11.5$.

*Or:* Note that $P''(x) = -2 < 0$ for all $x$, so $P$ is everywhere concave downward, and the local maximum at $x = 11.5$

must be an absolute maximum.

**3.** Call the two numbers $x$ and $y$. Then $xy = 100 \;\Rightarrow\; y = \dfrac{100}{x}$, where $x > 0$. We wish to minimize $f(x) = x + \dfrac{100}{x}$ for

$x > 0$. $f'(x) = 1 + 100(-x^{-2}) = 1 - \dfrac{100}{x^2} = \dfrac{x^2 - 100}{x^2}$, and $f'(x) = 0$ when $x^2 - 100 = 0 \;\Rightarrow\; x^2 = 100 \;\Rightarrow$

$x = 10$ (since $x > 0$). Thus 10 is the only critical number. Since $f'(x) < 0$ for $0 < x < 10$ and $f'(x) > 0$ for $x > 10$, there is an absolute minimum at $x = 10$ (by the First Derivative Test for Absolute Extreme Values). The numbers are 10 and 10.

**5.** If the rectangle has dimensions $x$ and $y$, then its perimeter is $2x + 2y = 100$ m, so $2y = 100 - 2x \Rightarrow y = 50 - x$. Thus, the area is $A = xy = x(50 - x) = 50x - x^2$. We wish to maximize the function $A(x) = 50x - x^2$, where $0 < x < 50$ (so that both $x$ and $y = 50 - x$ are positive). $A'(x) = 50 - 2x$, so $A'(x) = 0$ when $x = 25$, $A'(x) > 0$ for $0 < x < 25$ and $A'(x) < 0$ for $25 < x < 50$. Thus, $A$ has an absolute maximum at $x = 25$. The dimensions of the rectangle with maximum area are $x = y = 25$ m. (The rectangle is a square.)

**7.** (a)

 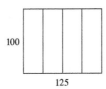 

The areas of the three figures are 12,500, 12,500, and 9000 ft$^2$. There appears to be a maximum area of at least 12,500 ft$^2$.

(b) Let $x$ denote the length of each of two sides and three dividers.

Let $y$ denote the length of the other two sides.

(c) Area $A = \text{length} \times \text{width} = y \cdot x = xy$

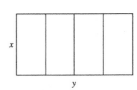

(d) Length of fencing $= 750 \Rightarrow 5x + 2y = 750$

(e) $5x + 2y = 750 \Rightarrow 2y = 750 - 5x \Rightarrow y = 375 - \frac{5}{2}x$, so the area is given by

$$A(x) = x\left(375 - \tfrac{5}{2}x\right) = 375x - \tfrac{5}{2}x^2.$$

(f) Note that $x \geq 0$ and $x \leq 150$ (otherwise we do not have enough fencing). Thus we wish to maximize

$A(x) = 375x - \frac{5}{2}x^2$ for $0 \leq x \leq 150$. $A'(x) = 375 - 5x$ and $A'(x) = 0$ when $x = 75$, so this is the only critical number. We have $A(0) = 0$, $A(75) = 14{,}062.5$, and $A(150) = 0$. By the Closed Interval Method, the absolute maximum occurs when $x = 75$ and $y = 375 - \frac{5}{2}(75) = \frac{375}{2} = 187.5$. The largest possible total area is $A(75) = 14{,}062.5$ ft$^2$.

These values of $x$ and $y$ are between the values in the first and second figures in part (a). Our original estimate was low.

**9.** Let $b$ be the length of the base of the box and $h$ the height. The area of the base is $b^2$ and the four sides all have area $bh$, so the total surface area is $b^2 + 4bh$. The surface area must be 1200 cm$^2$, so we have $b^2 + 4bh = 1200 \Rightarrow 4bh = 1200 - b^2 \Rightarrow$

$h = \dfrac{1200 - b^2}{4b}$. The volume is $V = \text{length} \times \text{width} \times \text{height} = b \cdot b \cdot \dfrac{1200 - b^2}{4b} = \frac{1}{4}b(1200 - b^2) = 300b - \frac{1}{4}b^3$. Then

$V'(b) = 300 - \frac{3}{4}b^2$ and $V'(b) = 0$ when $300 = \frac{3}{4}b^2 \Rightarrow b^2 = 400 \Rightarrow b = \sqrt{400} = 20$ (since $b > 0$). Since $V'(b) > 0$

for $0 < b < 20$ and $V'(b) < 0$ for $b > 20$, there is an absolute maximum when $b = 20$ by the First Derivative Test for

Absolute Extreme Values (see page 258). The largest possible volume is $V(20) = 300(20) - \frac{1}{4}(20)^3 = 4000$ cm$^3$.

**11.** (a) Let the rectangle have sides $x$ and $y$ and area $A$, so $A = xy$ or $y = A/x$. The perimeter is $P = 2x + 2y = 2x + 2A/x$,

so we want to find the absolute minimum of $P(x) = 2x + 2A/x$. Now $P'(x) = 2 - 2A/x^2$ and $P'(x) = 0$ when

$2 = 2A/x^2 \Rightarrow x^2 = A \Rightarrow x = \sqrt{A}$, so this is the only critical number. Since $P'(x) < 0$ for $0 < x < \sqrt{A}$ and

$P'(x) > 0$ for $x > \sqrt{A}$, there is an absolute minimum at $x = \sqrt{A}$ (since $x > 0$). The sides of the rectangle are $\sqrt{A}$ and

$A/\sqrt{A} = \sqrt{A}$, so the rectangle is a square.

(b) Let $p$ be the perimeter and $x$ and $y$ the lengths of the sides, so $p = 2x + 2y \Rightarrow 2y = p - 2x \Rightarrow y = \frac{1}{2}p - x$.

The area is $A(x) = x\left(\frac{1}{2}p - x\right) = \frac{1}{2}px - x^2$. Now $A'(x) = 0 \Rightarrow \frac{1}{2}p - 2x = 0 \Rightarrow 2x = \frac{1}{2}p \Rightarrow x = \frac{1}{4}p$. Since

$A''(x) = -2 < 0$ for all $x$, there is an absolute maximum for $A$ when $x = \frac{1}{4}p$. The sides of the rectangle are $\frac{1}{4}p$ and

$\frac{1}{2}p - \frac{1}{4}p = \frac{1}{4}p$, so the rectangle is a square.

**13.**

The volume is $V = \pi r^2 h$ and the surface area is $A = \pi r^2 + 2\pi rh$.

We are given that $V = 231$ in$^3$, so $\pi r^2 h = 231 \Rightarrow h = \dfrac{231}{\pi r^2}$ and then

$A(r) = \pi r^2 + 2\pi r\left(\dfrac{231}{\pi r^2}\right) = \pi r^2 + \dfrac{462}{r}.$

To minimize the cost of metal, we minimize the surface area. The derivative is $A'(r) = 2\pi r - \dfrac{462}{r^2}$ and $A'(r) = 0$ when

$2\pi r = \dfrac{462}{r^2} \Rightarrow r^3 = \dfrac{462}{2\pi} \Rightarrow r = \sqrt[3]{\dfrac{231}{\pi}}$ in. This gives an absolute minimum since $A'(r) < 0$ for $0 < r < \sqrt[3]{\dfrac{231}{\pi}}$

and $A'(r) > 0$ for $r > \sqrt[3]{\dfrac{231}{\pi}}$. When $r = \sqrt[3]{\dfrac{231}{\pi}} \approx 4.19$ in, $h = \dfrac{231}{\pi r^2} = \dfrac{231}{\pi(231/\pi)^{2/3}} = \dfrac{231^{1/3}}{\pi^{1/3}} = \sqrt[3]{\dfrac{231}{\pi}} \approx 4.19$ in.

**15.** We need to maximize $Y$ for $N \geq 0$. $Y = \dfrac{kN}{1 + N^2} \Rightarrow$

$\dfrac{dY}{dN} = \dfrac{(1 + N^2)k - kN(2N)}{(1 + N^2)^2} = \dfrac{k(1 - N^2)}{(1 + N^2)^2} = \dfrac{k(1 + N)(1 - N)}{(1 + N^2)^2}$. Then $\dfrac{dY}{dN} = 0$ when $N = 1$ (this is the only critical

number for $N \geq 0$). We have $\dfrac{dY}{dN} > 0$ for $0 < N < 1$ and $\dfrac{dY}{dN} < 0$ for $N > 1$. Thus, $Y$ has an absolute maximum of

$Y = \frac{1}{2}k$ at $N = 1$.

**17.** $P(R) = \dfrac{E^2 R}{(R + r)^2} \Rightarrow$

$P'(R) = \dfrac{(R + r)^2 \cdot E^2 - E^2 R \cdot 2(R + r)(1)}{[(R + r)^2]^2} = \dfrac{(R^2 + 2Rr + r^2)E^2 - 2E^2 R^2 - 2E^2 Rr}{(R + r)^4}$

$= \dfrac{E^2 r^2 - E^2 R^2}{(R + r)^4} = \dfrac{E^2(r^2 - R^2)}{(R + r)^4} = \dfrac{E^2(r + R)(r - R)}{(R + r)^4} = \dfrac{E^2(r - R)}{(R + r)^3}$

$P'(R) = 0$ when $R = r \Rightarrow P(r) = \dfrac{E^2 r}{(r + r)^2} = \dfrac{E^2 r}{4r^2} = \dfrac{E^2}{4r}.$

The expression for $P'(R)$ shows that $P'(R) > 0$ for $R < r$ and $P'(R) < 0$ for $R > r$. Thus, the maximum value of the

power is $E^2/(4r)$, and this occurs when $R = r$.

**19.** $y = \ln x + e^x$ $\Rightarrow$ $\dfrac{dy}{dx} = \dfrac{1}{x} + e^x$, so the slope of the tangent line to the curve at $x = a$ is $m(a) = \dfrac{1}{a} + e^a$. We want to

minimize $m$, so we graph $m'(a) = -\dfrac{1}{a^2} + e^a$ for $a > 0$ and estimate that the $a$-intercept is approximately $0.703$.

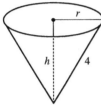

We see from the graph that $m'(a) < 0$ for $a < 0.703$ and $m'(a) > 0$ for

$a > 0.703$, so $a \approx 0.703$ corresponds to an absolute minimum, and the point

on the curve where the tangent line has the smallest slope is about

$\left(0.703, \ln 0.703 + e^{0.703}\right) \approx (0.703, 1.667)$.

**21.**

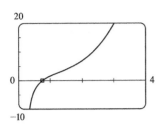

Let $r$ be the radius of the top of the cup and $h$ the height. Then from the Pythagorean

Theorem we have $h^2 + r^2 = 4^2$ $\Rightarrow$ $r^2 = 16 - h^2$. The volume of the cup is

$V(h) = \frac{1}{3}\pi r^2 h = \frac{1}{3}\pi(16 - h^2)h = \frac{\pi}{3}(16h - h^3)$. $V'(h) = \frac{\pi}{3}(16 - 3h^2)$, and

$V'(h) = 0$ when $16 = 3h^2$ $\Rightarrow$ $h = \sqrt{\dfrac{16}{3}} = \dfrac{4}{\sqrt{3}}$.

This gives an absolute maximum, since $V'(h) > 0$ for $0 < h < \dfrac{4}{\sqrt{3}}$ and $V'(h) < 0$ for $h > \dfrac{4}{\sqrt{3}}$. The maximum volume is

$$V\left(\dfrac{4}{\sqrt{3}}\right) = \dfrac{\pi}{3}\left[16 \cdot \dfrac{4}{\sqrt{3}} - \left(\dfrac{4}{\sqrt{3}}\right)^3\right] = \dfrac{\pi}{3}\left(\dfrac{64}{\sqrt{3}} - \dfrac{64}{3\sqrt{3}}\right) = \dfrac{\pi}{3}\left(\dfrac{128}{3\sqrt{3}}\right) = \dfrac{128}{9\sqrt{3}}\pi \approx 25.80 \text{ in}^3.$$

**23.**

Let $x$ be the length of the wire used for the square. Then $10 - x$ m is used for the

equilateral triangle, $\dfrac{10 - x}{3}$ for each side. If $h$ is the height of the triangle, then

$$\left(\dfrac{10 - x}{6}\right)^2 + h^2 = \left(\dfrac{10 - x}{3}\right)^2 \text{ by the Pythagorean Theorem, and}$$

$h^2 = \dfrac{(10 - x)^2}{9} - \dfrac{(10 - x)^2}{36} = \dfrac{(10 - x)^2}{12}$ $\Rightarrow$ $h = \dfrac{(10 - x)}{2\sqrt{3}} \cdot \dfrac{\sqrt{3}}{\sqrt{3}} = \dfrac{\sqrt{3}}{6}(10 - x)$. The total area is

$$A(x) = \left(\dfrac{x}{4}\right)^2 + \dfrac{1}{2}\left(\dfrac{10 - x}{3}\right)\dfrac{\sqrt{3}}{6}(10 - x)$$

$$= \tfrac{1}{16}x^2 + \dfrac{\sqrt{3}}{36}(10 - x)^2, \ 0 \leq x \leq 10$$

$A'(x) = \frac{1}{8}x + \dfrac{\sqrt{3}}{36} \cdot 2(10 - x)(-1) = \frac{1}{8}x - \dfrac{\sqrt{3}}{18}(10 - x)$, so $A'(x) = 0$ when $\frac{1}{8}x - \dfrac{10\sqrt{3}}{18} + \dfrac{\sqrt{3}}{18}x = 0$ $\Rightarrow$

$9x - 40\sqrt{3} + 4\sqrt{3}\,x$ (multiplying both sides by 72) $\Rightarrow$ $(9 + 4\sqrt{3})x = 40\sqrt{3}$ $\Rightarrow$ $x = \dfrac{40\sqrt{3}}{9 + 4\sqrt{3}} \approx 4.35$, so this is the

only critical number.

Now $A(0) = \left(\dfrac{\sqrt{3}}{36}\right)100 \approx 4.81$, $A(10) = \frac{1}{16}(100) = 6.25$ and $A\left(\dfrac{40\sqrt{3}}{9 + 4\sqrt{3}}\right) \approx 2.72$, so by the Closed Interval Method:

(a) The maximum area occurs when $x = 10$ m, and all the wire is used for the square.

(b) The minimum area occurs when $x = \dfrac{40\sqrt{3}}{9 + 4\sqrt{3}} \approx 4.35$ m.

**25.** There are $(6 - x)$ km over land and $\sqrt{x^2 + 4}$ km under the river. (See the diagram.)

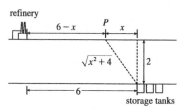

We need to minimize the cost $C$ (measured in \$100,000) of the pipeline for $0 \le x \le 6$.

$C(x) = (6 - x)(4) + \left(\sqrt{x^2 + 4}\right)(8) = 24 - 4x + 8\sqrt{x^2 + 4} \quad \Rightarrow$

$C'(x) = -4 + 8 \cdot \frac{1}{2}(x^2 + 4)^{-1/2}(2x) = -4 + \dfrac{8x}{\sqrt{x^2 + 4}}$. Then $C'(x) = 0$ when $4 = \dfrac{8x}{\sqrt{x^2 + 4}} \quad \Rightarrow \quad \sqrt{x^2 + 4} = 2x \quad \Rightarrow$

$x^2 + 4 = 4x^2 \quad \Rightarrow \quad 4 = 3x^2 \quad \Rightarrow \quad x^2 = \frac{4}{3} \quad \Rightarrow \quad x = 2/\sqrt{3}$ (since $x \ge 0$). Using the Closed Interval Method,

$C(0) = 24 + 16 = 40$, $C(2/\sqrt{3}) = 24 - 8/\sqrt{3} + 32/\sqrt{3} = 24 + 24/\sqrt{3} \approx 37.9$, and $C(6) = 0 + 8\sqrt{40} \approx 50.6$. So the

minimum cost is about $37.9 \times \$100,000 = \$3.79$ million when $P$ is $6 - 2/\sqrt{3} \approx 4.85$ km east of the refinery.

**27.**

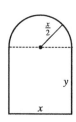

The perimeter is $\frac{1}{2} \cdot 2\pi r + x + 2y = \pi\left(\dfrac{x}{2}\right) + x + 2y$, and we are given that the

perimeter is 30 ft, so $\pi\left(\dfrac{x}{2}\right) + x + 2y = 30 \quad \Rightarrow \quad 2y = 30 - x - \frac{1}{2}\pi x \quad \Rightarrow$

$y = \frac{1}{2}\left(30 - x - \frac{1}{2}\pi x\right) = 15 - \frac{1}{2}x - \frac{1}{4}\pi x.$

We want to maximize the total area of the window, which is the area of the rectangle plus the area of the semicircle,

or $xy + \frac{1}{2}\pi\left(\dfrac{x}{2}\right)^2$, so $A(x) = x\left(15 - \frac{1}{2}x - \frac{1}{4}\pi x\right) + \frac{1}{8}\pi x^2 = 15x - \frac{1}{2}x^2 - \frac{\pi}{8}x^2.$

$A'(x) = 15 - x - \frac{\pi}{4}x = 15 - \left(1 + \frac{\pi}{4}\right)x$, and $A'(x) = 0$ when $\left(1 + \frac{\pi}{4}\right)x = 15 \quad \Rightarrow \quad x = \dfrac{15}{1 + \pi/4} = \dfrac{60}{4 + \pi}.$

$A''(x) = -\left(1 + \dfrac{\pi}{4}\right) < 0$ for all $x$, so this gives a maximum. The dimensions are $x = \dfrac{60}{4 + \pi} \approx 8.40$ ft and

$y = 15 - \dfrac{30}{4 + \pi} - \dfrac{15\pi}{4 + \pi} = \dfrac{60 + 15\pi - 30 - 15\pi}{4 + \pi} = \dfrac{30}{4 + \pi} \approx 4.20$ ft, so the height of the rectangle is half the base.

**29.**

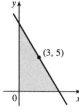

The line with slope $m$ (where $m < 0$) through $(3, 5)$ has equation $y - 5 = m(x - 3)$

or $y = mx + (5 - 3m)$. The $y$-intercept is $5 - 3m$ and the $x$-intercept is $-5/m + 3$.

So the triangle has area $A(m) = \frac{1}{2}(5 - 3m)(-5/m + 3) = 15 - 25/(2m) - \frac{9}{2}m.$

Now $A'(m) = \dfrac{25}{2m^2} - \dfrac{9}{2}$, and $A'(m) = 0$ when $\dfrac{25}{2m^2} = \dfrac{9}{2} \quad \Rightarrow \quad m^2 = \frac{25}{9} \quad \Rightarrow \quad m = -\frac{5}{3}$ (since $m < 0$).

$A''(m) = -\dfrac{25}{m^3} > 0$ for all $m < 0$, so there is an absolute minimum when $m = -\frac{5}{3}$. Thus, an equation of the line is

$y - 5 = -\frac{5}{3}(x - 3)$ or $y = -\frac{5}{3}x + 10.$

**31. (a)**

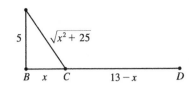

If $k$ = energy/km over land, then energy/km over water = $1.4k$.

So the total energy is $E(x) = 1.4k\sqrt{x^2 + 25} + k(13 - x)$, $0 \le x \le 13$,

and so $E'(x) = \dfrac{1.4kx}{(x^2 + 25)^{1/2}} - k$.

Then $E'(x) = 0$ when $1.4kx = k(x^2 + 25)^{1/2} \Rightarrow 1.96x^2 = x^2 + 25 \Rightarrow 0.96x^2 = 25 \Rightarrow x = \dfrac{5}{\sqrt{0.96}} \approx 5.1$.

Testing against the value of $E$ at the endpoints: $E(0) = 1.4k(5) + 13k = 20k$, $E(5.1) \approx 17.9k$, $E(13) \approx 19.5k$.

Thus, to minimize energy, the bird should fly to a point about 5.1 km from $B$.

**(b)** If $W/L$ is large, the bird would fly to a point $C$ that is closer to $B$ than to $D$ to minimize the energy used flying over water.

If $W/L$ is small, the bird would fly to a point $C$ that is closer to $D$ than to $B$ to minimize the distance of the flight.

$E(x) = W\sqrt{x^2 + 25} + L(13 - x) \Rightarrow E'(x) = \dfrac{Wx}{\sqrt{x^2 + 25}} - L = 0$ when $\dfrac{W}{L} = \dfrac{\sqrt{x^2 + 25}}{x}$. By the same sort of

argument as in part (a), this ratio will give the minimal expenditure of energy if the bird heads for the point $x$ km from $B$.

**(c)** For flight direct to $D$, $x = 13$, so from part (b), $W/L = \dfrac{\sqrt{13 + 25^2}}{13} \approx 1.07$. There is no value of $W/L$ for which the bird

should fly directly to $B$. But note that $\lim\limits_{x \to 0^+} (W/L) = \infty$, so if the point at which $E$ is a minimum is close to $B$, then

$W/L$ is large.

**(d)** Assuming that the birds instinctively choose the path that minimizes the energy expenditure, we can use the equation for

$E'(x) = 0$ from part (a) with $1.4k = c$, $x = 4$, and $k = 1$: $c \cdot 4 = 1 \cdot (4^2 + 25)^{1/2} \Rightarrow c = \sqrt{41}/4 \approx 1.6$.

**33.**

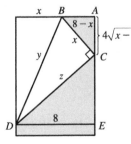

$y^2 = x^2 + z^2$, but triangles $CDE$ and $BCA$ are similar, so

$z/8 = x/\left(4\sqrt{x - 4}\right) \Rightarrow z = 2x/\sqrt{x - 4}$. Thus, we minimize

$f(x) = y^2 = x^2 + \dfrac{4x^2}{x - 4} = \dfrac{x^2(x - 4) + 4x^2}{x - 4} = \dfrac{x^3}{x - 4}$, $4 < x \le 8$.

(The minimum of $y^2$ occurs for the same value of $x$ as the minimum of $y$, and

the computations are easier.)

$f'(x) = \dfrac{(x - 4)(3x^2) - x^3}{(x - 4)^2} = \dfrac{x^2[3(x - 4) - x]}{(x - 4)^2} = \dfrac{2x^2(x - 6)}{(x - 4)^2}$, and $f'(x) = 0$ when $x = 6$. $f'(x) < 0$ when $x < 6$, and

$f'(x) > 0$ when $x > 6$, so the minimum occurs when $x = 6$ in.

## 4.7  Optimization in Business and Economics

**Prepare Yourself**

**1.** $f(x) = 3250 + 4.2x - 0.3x^2 + 0.002x^3 \Rightarrow$

$f'(x) = 0 + 4.2 - 0.3(2x) + 0.002(3x^2) = 4.2 - 0.6x + 0.006x^2$, $f''(x) = 0 - 0.6 + 0.006(2x) = -0.6 + 0.012x$

**2.** $P(t) = 12e^{-0.4t} \Rightarrow P'(t) = 12e^{-0.4t}(-0.4) = -4.8e^{-0.4t}$, $P''(t) = -4.8e^{-0.4t}(-0.4) = 1.92e^{-0.4t}$

**3.** $g(x) = 60 \ln(4x) \;\Rightarrow\; g'(x) = 60 \cdot \dfrac{1}{4x}(4) = \dfrac{60}{x} = 60x^{-1} \;\Rightarrow\; g''(x) = 60(-x^{-2}) = -\dfrac{60}{x^2}$

**4.** $h(a) = 5 + \sqrt{3a} = 5 + (3a)^{1/2} \;\Rightarrow\; h'(a) = 0 + \frac{1}{2}(3a)^{-1/2}(3) = \frac{3}{2}(3a)^{-1/2} = \dfrac{3}{2\sqrt{3a}}$ or $\dfrac{\sqrt{3}}{2\sqrt{a}} \;\Rightarrow$

$h''(a) = \frac{3}{2}\left[-\frac{1}{2}(3a)^{-3/2}(3)\right] = -\frac{9}{4}(3a)^{-3/2} = -\dfrac{9}{4(3a)^{3/2}}$ or $-\dfrac{\sqrt{3}}{4a^{3/2}}$

**5.** We have $g(6000) = 16$ and $g(7200) = 12$, so the line passes through the points $(6000, 16)$ and $(7200, 12)$, and the slope is

$m = \dfrac{y_2 - y_1}{x_2 - x_1} = \dfrac{12 - 16}{7200 - 6000} = \dfrac{-4}{1200} = -\dfrac{1}{300}$. An equation for the line is $y - y_1 = m(x - x_1) \;\Rightarrow$

$y - 16 = -\frac{1}{300}(x - 6000) \;\Rightarrow\; y = -\frac{1}{300}x + 36$, and the linear function is $g(x) = -\frac{1}{300}x + 36$.

**6.** (a) $x^{0.7} = 18 \;\Rightarrow\; \left(x^{0.7}\right)^{1/0.7} = (18)^{1/0.7} \;\Rightarrow\; x = 18^{1/0.7} \approx 62.12$

(b) $15e^{0.48x} = 113 \;\Rightarrow\; e^{0.48x} = \frac{113}{15} \;\Rightarrow\; \ln e^{0.48x} = \ln \frac{113}{15} \;\Rightarrow\; 0.48x = \ln \frac{113}{15} \;\Rightarrow$

$x = \frac{1}{0.48} \ln \frac{113}{15} \approx 4.21$

---

**Exercises**

**1.** (a) $C(0)$ represents the fixed costs of production, such as rent, utilities, machinery etc., which are incurred even when nothing is produced.

(b) The inflection point is the point at which $C''(q)$ changes from negative to positive; that is, the marginal cost $C'(q)$ changes from decreasing to increasing. Thus, the marginal cost is minimized at the inflection point.

(c) The marginal cost function is $C'(q)$.

We graph it as in Example 1 in Section 2.4.

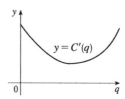

**3.** $c(q) = 21.4 - 0.002q$ and $c(q) = C(q)/q \;\Rightarrow\; C(q) = q \cdot c(q) = 21.4q - 0.002q^2$. $C'(q) = 21.4 - 0.004q$ and $C'(1000) = \$17.4/\text{unit}$. This means that the cost of producing the 1001st unit is about $17.40.

**5.** (a) $C(q) = 16{,}000 + 200q + 4q^{3/2}$, so $C(1000) = 16{,}000 + 200(1000) + 4(1000)^{3/2} \approx \$342{,}491$. The average cost at a production level of 1000 units is $c(1000) = C(1000)/1000 \approx \$342{,}491/1000 \approx \$342.49/\text{unit}$. The marginal cost is $C'(q) = 200 + 4 \cdot \frac{3}{2}q^{1/2} = 200 + 6\sqrt{q}$, so $C'(1000) = 200 + 6\sqrt{1000} \approx \$389.74/\text{unit}$.

(b) The average cost is $c(q) = \dfrac{C(q)}{q} = \dfrac{16{,}000 + 200q + 4q^{3/2}}{q} = \dfrac{16{,}000}{q} + 200 + 4q^{1/2}$, so

$c'(q) = -16{,}000q^{-2} + 0 + 4 \cdot \frac{1}{2}q^{-1/2} = -\dfrac{16{,}000}{q^2} + \dfrac{2}{\sqrt{q}}$. $c'(q) = 0$ when $\dfrac{2}{\sqrt{q}} = \dfrac{16{,}000}{q^2} \;\Rightarrow$

$2q^2 = 16{,}000\sqrt{q} \;\Rightarrow\; q^{3/2} = 8000 \;\Rightarrow\; q = (8000)^{2/3} = 400$, so 400 is the only critical number. (Alternatively, we could determine when average cost and marginal cost are equal.) $c'(q) < 0$ for $q < 400$ and $c'(q) > 0$ for $q > 400$, so $c(q)$ has an absolute minimum at $q = 400$.

(c) The minimum average cost is $c(400) = 40 + 200 + 80 = \$320/\text{unit}$.

**7.** (a) $C(q) = 3700 + 5q - 0.04q^2 + 0.0003q^3 \implies C'(q) = 5 - 0.08q + 0.0009q^2$ (marginal cost).

$$c(q) = \frac{C(q)}{q} = \frac{3700}{q} + 5 - 0.04q + 0.0003q^2 \text{ (average cost).}$$

(b)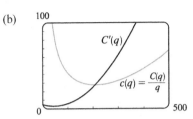

Marginal cost and average cost are equal where the graphs intersect, at approximately $(208.51, 27.45)$, so the production level that minimizes average cost is between 208 and 209 units.

(c) $c'(q) = -\dfrac{3700}{q^2} - 0.04 + 0.0006q$, and $c'(q) = 0$ when $-3700 - 0.04q^2 + 0.0006q^3 = 0 \implies$

$3700 + 0.04q^2 - 0.0006q^3 = 0 \implies q \approx 208.51$ (solve graphically).  $c'(q) < 0$ for $0 < q < 208.51$ (approximately) and $c'(q) > 0$ for $q > 208.51$, so the minimum average cost is about $c(209) \approx \$27.45/\text{unit}$.

(d) The marginal cost is $M(q) = C'(q) = 5 - 0.08q + 0.0009q^2$, so $M'(q) = -0.08 + 0.0018q$, and $M'(q) = 0$ when $0.0018q = 0.08 \implies q = \frac{0.08}{0.0018} \approx 44.44$.  $M''(q) = 0.0018$ is positive for all $q$, so $M$ is concave upward and $q \approx 44.44$ is a minimum. The minimum marginal cost is about $C'(44.44) \approx \$3.22/\text{unit}$. (This corresponds to an inflection point of $C$.)

**9.** The profit function is $P(q) = R(q) - C(q) = (12.2q - 0.002q^2) - (2250 + 3.5q + 0.004q^2) = 8.7q - 0.006q^2 - 2250$.

$P'(q) = 8.7 - 0.012q$, so $P'(q) = 0$ when $8.7 = 0.012q \implies q = \frac{8.7}{0.012} = 725$ and 725 is the only critical number of $P$. Because $P''(q) = -0.012 < 0$, $P$ is concave downward everywhere and the maximum profit is achieved for $q = 725$ units. Alternatively, we can solve $R'(q) = C'(q)$ and verify that $R''(q) < C''(q)$ at that value.

**11.** The marginal cost function, $C'(q)$, starts to increase when its derivative, $C''(q)$, becomes positive. In other words, $C$ changes from concave downward to concave upward, and so $C$ has an inflection point.

$C(q) = 0.0008q^3 - 0.72q^2 + 325.3q + 78{,}000 \implies C'(q) = 0.0024q^2 - 1.44q + 325.3 \implies C''(q) = 0.0048q - 1.44$.

$C''(q) = 0$ when $q = \frac{1.44}{0.0048} = 300$, and $C''(q) > 0$ when $q > 300$. Thus, the marginal cost function starts to increase when the production level is 300 units.

**13.** (a) $C(x) = 1200 + 12x - 0.1x^2 + 0.0005x^3$.

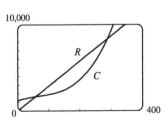

$R(x) = \text{units} \times \text{price per unit} = x \cdot p(x) = 29x - 0.00021x^2$.

Since the profit is maximized when $R'(x) = C'(x)$ $[R''(x) < C''(x)]$, we examine the curves $R$ and $C$ in the figure, looking for $x$-values at which the slopes of the tangent lines are equal. It appears that $x = 200$ is a good estimate. (This is also where the vertical distance between the curves is maximal.)

(b) $P(x) = R(x) - C(x) = (29x - 0.00021x^2) - (1200 + 12x - 0.1x^2 + 0.0005x^3)$

$= 17x + 0.09979x^2 - 1200 - 0.0005x^3$

so $P'(x) = 17 + 0.19958x - 0.0015x^2$ and $P'(x) = 0$ when $-0.0015x^2 + 0.19958x + 17 = 0 \Rightarrow$

$x = \dfrac{-0.19958 \pm \sqrt{(0.19958)^2 - 4(-0.0015)(17)}}{2(-0.0015)} \approx 192.06$ (using the Quadratic Formula with $x > 0$). $P'(x) > 0$ for

$0 < x < 192.06$ (approximately) and $P'(x) < 0$ for $x > 192.06$, so the maximum profit occurs at a production level of

about 192 yards of fabric.

[We could also have solved $R'(x) = C'(x)$ for $x$ and verified that $R''(x) < C''(x)$ there.]

**15.** (a) We are given that the demand function $D$ is linear and $D(27{,}000) = 10$, $D(33{,}000) = 8$, so the slope is

$\frac{10-8}{27{,}000-33{,}000} = -\frac{1}{3000}$ and an equation of the line is $p - 10 = \left(-\frac{1}{3000}\right)(q - 27{,}000) \Rightarrow p = D(x) = -\frac{1}{3000}q + 19$.

(b) The revenue is $R(q) = q \cdot D(q) = -\frac{1}{3000}q^2 + 19q \Rightarrow R'(q) = -\frac{1}{1500}q + 19$, and $R'(q) = 0$ when

$q = 19(1500) = 28{,}500$. Since $R''(q) = -\frac{1}{1500} < 0$, the maximum revenue occurs when $q = 28{,}500 \Rightarrow$ the price is

$D(28{,}500) = \$9.50$.

**17.** (a) The demand function $D$ is linear and we are given that $D(1000) = 450$. Since a \$10 reduction in price increases sales by

100 per week, we have $D(1100) = 440$. The slope for $D$ is $\frac{440-450}{1100-1000} = -\frac{1}{10}$, so an equation is

$p - 450 = -\frac{1}{10}(q - 1000)$ or $p = D(q) = -\frac{1}{10}q + 550$.

(b) $R(q) = q \cdot D(q) = -\frac{1}{10}q^2 + 550q$. $R'(q) = -\frac{1}{5}q + 550$, and $R'(q) = 0$ when $q = 5(550) = 2750$.

$R''(q) = -\frac{1}{5} < 0$, so this is a maximum. $D(2750) = 275$, so the rebate should be $450 - 275 = \$175$.

(c) $C(q) = 68{,}000 + 150q \Rightarrow$

$P(q) = R(q) - C(q) = \left(-\frac{1}{10}q^2 + 550q\right) - (68{,}000 + 150q) = -\frac{1}{10}q^2 + 400q - 68{,}000 \Rightarrow P'(q) = -\frac{1}{5}q + 400$,

and $P'(q) = 0$ when $q = 5(400) = 2000$. Since $P''(q) = -\frac{1}{5} < 0$, this is an absolute maximum. We have

$D(2000) = 350$, so the rebate to maximize profits should be $450 - 350 = \$100$.

**19.** (a) $p = 155 - 0.035q$, so when $p = 65$ we have $65 = 155 - 0.035q \Rightarrow -90 = -0.035q \Rightarrow q = \frac{-90}{-0.035} \approx 2571$.

Thus about 2571 pairs of sunglasses can be sold at a price of \$65.

(b) $dp/dq = -0.035$, so by Equation 3 the elasticity of demand is $E(2571) = -\left.\dfrac{p/q}{dp/dq}\right|_{q=2571} = -\dfrac{65/2571}{-0.035} \approx 0.72$. This

means that when the sunglasses are priced at \$65, changes in demand are proportionally about 72% of changes in price.

Because $E(2571) < 1$, demand is inelastic.

**21.** Tablet computers are highly desirable to consumers but many have not purchased one yet, so it is likely that a decrease in price

would cause a proportionally greater increase in demand. Thus demand is elastic, that is, $E(q) > 1$.

**23.** $1.25p + q = 860$, so when $p = 225$ we have $1.25(225) + q = 860 \Rightarrow q = 860 - 281.25 = 578.75$. Also

$1.25p = 860 - q \Rightarrow p = \frac{1}{1.25}(860 - q) = 688 - 0.8q$, so $dp/dq = -0.8$. Using Equation 3, when $p = 225$ the elasticity

of demand is $E(578.75) = -\left.\dfrac{p/q}{dp/dq}\right|_{q=578.75} = -\dfrac{225/578.75}{-0.8} \approx 0.49$. Because $E(578.75) < 1$, demand is inelastic.

**25.** $q + 80\ln(2p) = 600 \Rightarrow q = 600 - 80\ln(2p) \Rightarrow dq/dp = -80 \cdot \dfrac{1}{2p}(2) = -\dfrac{80}{p}$. When $p = 400$ we have

$q = 600 - 80\ln(800) \approx 65.23$. Using the alternate formula $E(p) = -\dfrac{dq/dp}{q/p}$, the elasticity of demand is

$E(400) = -\dfrac{dq/dp}{q/p}\bigg|_{p=400} = -\dfrac{-80/400}{65.23/400} \approx 1.23$. Because demand is larger than 1, it is elastic.

**27.** $D(q) = 45e^{-0.4q} \Rightarrow D'(q) = 45e^{-0.4q}(-0.4) = -18e^{-0.4q}$, so $E(q) = -\dfrac{D(q)}{qD'(q)} = -\dfrac{45e^{-0.4q}}{q(-18e^{-0.4q})} = \dfrac{5}{2q}$ and

$E(q) = 1 \Rightarrow \dfrac{5}{2q} = 1 \Rightarrow q = 2.5$. Thus the elasticity of demand is equal to 1 when the price is

$D(2.5) = 45e^{-0.4(2.5)} \approx \$16.55$. If $q = 2$, then the price is $D(2) = 45e^{-0.4(2)} \approx \$20.22$ and $E(2) = \frac{5}{2(2)} > 1$, so demand

is elastic. If $q = 3$, then $D(3) = 45e^{-0.4(3)} \approx \$13.55$ and $E(3) = \frac{5}{2(3)} < 1$, inelastic. Here the demand is elastic for

$p > 16.55$ and inelastic for $p < 16.55$.

**29.** If $p = D(q) = M/q$, then $D'(q) = -M/q^2$ and $E(q) = -\dfrac{D(q)}{qD'(q)} = -\dfrac{M/q}{q(-M/q^2)} = -\dfrac{M/q}{-M/q} = 1$.

**31.** (a) $p = D(q) = 58{,}000 - 32q \Rightarrow D'(q) = -32$ and $E(q) = -\dfrac{D(q)}{qD'(q)} = -\dfrac{58{,}000 - 32q}{-32q} = \dfrac{58{,}000 - 32q}{32q}$. When

$p = 28{,}000$ we have $28{,}000 = 58{,}000 - 32q \Rightarrow q = \frac{30{,}000}{32} = 937.5$ and the elasticity of demand is

$E(937.5) = \dfrac{58{,}000 - 32(937.5)}{32(937.5)} \approx 0.93$.

(b) Because $E(937.5) < 1$, the demand is inelastic and revenue will increase if prices are raised from $28{,}000.

(c) Revenue is maximized when $E(q) = 1 \Rightarrow \dfrac{58{,}000 - 32q}{32q} = 1 \Rightarrow 58{,}000 - 32q = 32q \Rightarrow$

$58{,}000 = 64q \Rightarrow q = \frac{58{,}000}{64} = 906.25$, corresponding to a price of $D(906.25) = 58{,}000 - 32(906.25) = \$29{,}000$.

**33.** $p = D(q) = 150 - 4\sqrt{q} \Rightarrow D'(q) = -4 \cdot \frac{1}{2}q^{-1/2} = -2/\sqrt{q} \Rightarrow$

$E(q) = -\dfrac{D(q)}{qD'(q)} = -\dfrac{150 - 4\sqrt{q}}{q(-2/\sqrt{q})} = \dfrac{150 - 4\sqrt{q}}{2\sqrt{q}}$. Revenue is maximized when $E(q) = 1 \Rightarrow \dfrac{150 - 4\sqrt{q}}{2\sqrt{q}} = 1 \Rightarrow$

$150 - 4\sqrt{q} = 2\sqrt{q} \Rightarrow 150 = 6\sqrt{q} \Rightarrow$

$\sqrt{q} = 25 \Rightarrow q = 625$, and the corresponding price is $D(625) = 150 - 4\sqrt{625} = \$50$.

**35.** (a) A graphing calculator gives an exponential model for the data as approximately $D(q) = (418.6)(0.9768^q)$, where $q$ is

measured in thousands of units.

(b) $D(q) = 189 \Rightarrow (418.6)(0.9768^q) = 189 \Rightarrow 0.9768^q = \frac{189}{418.6} \Rightarrow \ln 0.9768^q = \ln \frac{189}{418.6} \Rightarrow$

$q \ln 0.9768 = \ln \frac{189}{418.6} \Rightarrow q = \dfrac{\ln(189/418.6)}{\ln 0.9768} \approx 33.9$ thousand units.

(c) $D'(q) = (418.6)(0.9768^q)\ln 0.9768 \Rightarrow$

$E(q) = -\dfrac{D(q)}{qD'(q)} = -\dfrac{(418.6)(0.9768^q)}{q(418.6)(0.9768^q)\ln 0.9768} = -\dfrac{1}{q \ln 0.9768}$ and $E(33.9) = -\dfrac{1}{33.9 \ln 0.9768} \approx 1.26$.

Because $E(33.9) > 1$, revenue will increase if the price is lowered.

(d) Revenue = price per unit × number of units:

| Price per unit | Units sold (thousands) | Revenue (thousands) |
|---|---|---|
| $125 | 51.6 | $6450 |
| $150 | 43.4 | $6510 |
| $175 | 37.6 | $6580 |
| $200 | 31.6 | $6320 |
| $225 | 26.3 | $5917.5 |

It appears that revenue is maximized when the price is about $175.

(e) Revenue is maximized when $E(q) = -\dfrac{1}{q \ln 0.9768} = 1 \ \Rightarrow \ q \ln 0.9768 = -1 \ \Rightarrow \ q = -1/\ln 0.9768 \approx 42.6$.

The corresponding price is $D(42.6) \approx \$154$. This is somewhat lower than the estimated price in part (d).

**37.** If the reorder quantity is $x$, then the manager places $800/x$ orders per year. The average number of cases stored during the

year is $\frac{1}{2}x$, so storage costs for the year are $\frac{1}{2}x \cdot 4 = 2x$ dollars. Handling costs are $100 per delivery, for a total of

$\dfrac{800}{x} \cdot 100 = \dfrac{80{,}000}{x}$ dollars. The total costs for the year are $C(x) = \dfrac{80{,}000}{x} + 2x$. To minimize $C(x)$, we calculate

$C'(x) = -\dfrac{80{,}000}{x^2} + 2$, and there is a critical number when $C'(x) = 0 \ \Rightarrow \ \dfrac{80{,}000}{x^2} = 2 \ \Rightarrow \ x^2 = 40{,}000 \ \Rightarrow$

$x = \sqrt{40{,}000} = 200$ (since $x > 0$). $C'(x) < 0$ when $0 < x < 200$ and $C'(x) > 0$ when $x > 200$, so cost is minimized when

$x = 200$ cases are ordered each time. The manager will place 4 orders per year for a total cost of $C(200) = \$800$.

**39.** Let $x$ be the number of boxes of envelopes in each order, so the number of orders per year is $740/x$. The average number of

boxes stored during the year is $\frac{1}{2}x$, and the total costs are $C(x) = \$225(740/x) + \$4(x/2) = 166{,}500/x + 2x$.

$C'(x) = -166{,}500/x^2 + 2$ and we have a critical number when $C'(x) = 0 \ \Rightarrow \ 166{,}500/x^2 = 2 \ \Rightarrow \ x^2 = 83{,}250 \ \Rightarrow$

$x = \sqrt{83{,}250} \approx 289$ (since $x > 0$). $C'(x) < 0$ for $0 < x < 289$ and $C'(x) > 0$ for $x > 289$, so cost is minimized when

$x \approx 289$ boxes. Thus the company should place an order $740/289 \approx 2.56$ times per year, or about every 4.7 months

(4 months 21 days).

**41.** (a) $E(q) = -\dfrac{D(q)}{q D'(q)}$ and $R(q) = q \cdot D(q) \ \Rightarrow$

$R'(q) = q \cdot D'(q) + D(q) \cdot 1 = q \cdot D'(q) \left[ 1 + \dfrac{D(q)}{q D'(q)} \right] = q \cdot D'(q) \left[ 1 - \left( -\dfrac{D(q)}{q D'(q)} \right) \right] = q \cdot D'(q) \left[ 1 - E(q) \right]$.

(b) When $0 < E(q) < 1$ we have $[1 - E(q)] > 0$. $q$ is positive and $D'(q)$ is negative (since demand curves are decreasing),

so $R'(q) = q \cdot D'(q) \left[ 1 - E(q) \right] < 0$. When $E(q) > 1$, $[1 - E(q)] < 0$ and $R'(q) > 0$.

(c) The revenue function has a local (and absolute) maximum, which must occur at a critical number. Since

$R'(q) = q \cdot D'(q) \left[ 1 - E(q) \right]$ and $q > 0$, $D'(q) < 0$, the only critical number occurs when $1 - E(q) = 0 \ \Rightarrow$

$E(q) = 1$.

## 4 Review

**1.** $f(x) = x^3 - 6x^2 + 9x + 1$, $[2, 4]$. $f'(x) = 3x^2 - 12x + 9 = 3(x^2 - 4x + 3) = 3(x - 1)(x - 3)$. $f'(x) = 0 \Rightarrow$
$x = 1$ or $x = 3$, but 1 is not in the interval. $f'(x) > 0$ for $3 < x < 4$ and $f'(x) < 0$ for $2 < x < 3$, so $f(3) = 1$ is a local
minimum value. Checking the endpoints, we find $f(2) = 3$ and $f(4) = 5$. By the Closed Interval Method, $f(3) = 1$ is the
absolute minimum value and $f(4) = 5$ is the absolute maximum value.

**3.** $g(t) = 4t^3 - 3t + 2$, $[-1, 1]$. $g'(t) = 12t^2 - 3 = 3(4t^2 - 1) = 3(2t + 1)(2t - 1)$. $g'(t) = 0 \Rightarrow t = -\frac{1}{2}$ or $t = \frac{1}{2}$.
$g'(t) > 0$ for $-1 < t < -\frac{1}{2}$ and $\frac{1}{2} < t < 1$, and $g'(t) < 0$ for $-\frac{1}{2} < t < \frac{1}{2}$, so $g(-\frac{1}{2}) = 3$ is a local maximum value and
$g(\frac{1}{2}) = 1$ is a local minimum value. Checking the endpoints, we find $g(-1) = 1$ and $g(1) = 3$. By the Closed Interval
Method, $g(-1) = g(\frac{1}{2}) = 1$ is the absolute minimum value and $g(-\frac{1}{2}) = g(1) = 3$ is the absolute maximum value.

**5.** $f(x) = \dfrac{3x - 4}{x^2 + 1}$, $[-2, 2]$. $f'(x) = \dfrac{(x^2 + 1) \cdot 3 - (3x - 4) \cdot 2x}{(x^2 + 1)^2} = \dfrac{-(3x^2 - 8x - 3)}{(x^2 + 1)^2} = \dfrac{-(3x + 1)(x - 3)}{(x^2 + 1)^2}$.
$f'(x) = 0 \Rightarrow x = -\frac{1}{3}$ or $x = 3$, but 3 is not in the interval. $f'(x) > 0$ for $-\frac{1}{3} < x < 2$ and $f'(x) < 0$ for
$-2 < x < -\frac{1}{3}$, so $f(-\frac{1}{3}) = \frac{-5}{10/9} = -\frac{9}{2}$ is a local minimum value. Checking the endpoints, we find $f(-2) = -2$ and
$f(2) = \frac{2}{5}$. Thus, $f(-\frac{1}{3}) = -\frac{9}{2}$ is the absolute minimum value and $f(2) = \frac{2}{5}$ is the absolute maximum value.

**7.** $f(x) = \dfrac{x^2 + 1}{e^{\sqrt{x^2 + 1}}}$, $0 \le x \le 4$

Using a graphing calculator, the absolute maximum is approximately
$f(1.73) \approx 0.54$, and the absolute minimum occurs at the right
endpoint: $f(4) \approx 0.28$.

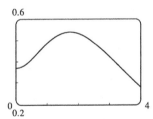

**9.** (a) $N(t) = t^3 - 3t^2 - 24t + 5 \Rightarrow N'(t) = 3t^2 - 6t - 24 = 3(t^2 - 2t - 8) = 3(t + 2)(t - 4)$. We have critical numbers
where $N'(t) = 0$, at $t = -2$ and $t = 4$.

| Interval | $t + 2$ | $t - 4$ | $N'(t)$ | $N$ |
|---|---|---|---|---|
| $t < -2$ | $-$ | $-$ | $+$ | increasing on $(-\infty, -2)$ |
| $-2 < t < 4$ | $+$ | $-$ | $-$ | decreasing on $(-2, 4)$ |
| $t > 4$ | $+$ | $+$ | $+$ | increasing on $(4, \infty)$ |

(b) $N'$ changes from positive to negative at $t = -2$ and from negative to positive at $t = 4$. Thus, $N(-2) = 33$ is a local
maximum value and $N(4) = -75$ is a local minimum value.

(c) $N''(t) = 6t - 6$. Then $N''(t) = 0$ when $t = 1$, $N''(t) < 0$ for $t < 1$, and $N''(t) > 0$ for $t > 1$. Thus $N$ is concave
upward on $(1, \infty)$, concave downward on $(-\infty, 1)$, and since concavity changes direction at 1, $(1, -21)$ is an inflection
point.

**11.** (a) $f(x) = x + \sqrt{1-x}$ which is defined when $1 - x \geq 0 \Rightarrow x \leq 1$.

$f'(x) = 1 + \frac{1}{2}(1-x)^{-1/2}(-1) = 1 - \frac{1}{2\sqrt{1-x}}$, and $f'(x) = 0$ when $1 = \frac{1}{2\sqrt{1-x}} \Rightarrow 2\sqrt{1-x} = 1 \Rightarrow$

$\sqrt{1-x} = \frac{1}{2} \Rightarrow 1 - x = \frac{1}{4} \Rightarrow x = \frac{3}{4}$. $f'(x)$ is undefined when $x = 1$. Thus we have critical numbers 1 (an

endpoint of the domain of $f$) and $\frac{3}{4}$. $f'(x) > 0$ when $x < \frac{3}{4}$ and $f'(x) < 0$ when $\frac{3}{4} < x < 1$, so $f$ is increasing on

$\left(-\infty, \frac{3}{4}\right)$ and decreasing on $\left(\frac{3}{4}, 1\right)$.

(b) $f\left(\frac{3}{4}\right) = \frac{3}{4} + \sqrt{1 - \frac{3}{4}} = \frac{3}{4} + \sqrt{\frac{1}{4}} = \frac{3}{4} + \frac{1}{2} = \frac{5}{4}$ is a local maximum value because $f'$ changes from positive to negative

there.

(c) $f'(x) = 1 - \frac{1}{2}(1-x)^{-1/2} \Rightarrow f''(x) = -\frac{1}{2}\left(-\frac{1}{2}\right)(1-x)^{-3/2}(-1) = -\frac{1}{4(1-x)^{3/2}}$ which is negative for $x < 1$,

so $f$ is concave downward on $(-\infty, 1)$. (There is no inflection point.)

**13.** (a) $y = f(x) = xe^{4x} \Rightarrow f'(x) = x \cdot e^{4x}(4) + e^{4x} \cdot 1 = (4x+1)e^{4x}$, and $e^{4x}$ is never zero, so $f'(x) = 0$ when $x = -\frac{1}{4}$.

$f'(x) < 0$ when $x < -\frac{1}{4}$ and $f'(x) > 0$ when $x > -\frac{1}{4}$, so $f$ is increasing on $\left(-\frac{1}{4}, \infty\right)$ and decreasing on $\left(-\infty, -\frac{1}{4}\right)$.

(b) $f\left(-\frac{1}{4}\right) = -\frac{1}{4}e^{-1} = -1/(4e) \approx 0.0920$ is a local minimum value because $f'$ changes from negative to positive there.

(c) $f''(x) = (4x+1) \cdot e^{4x}(4) + e^{4x} \cdot 4 = 8(2x+1)e^{4x}$; $f''(x) = 0$ when $x = -\frac{1}{2}$, $f''(x) < 0$ when $x < -\frac{1}{2}$, and

$f''(x) > 0$ when $x > -\frac{1}{2}$. Thus $f$ is concave upward on $\left(-\frac{1}{2}, \infty\right)$, concave downward on $\left(-\infty, -\frac{1}{2}\right)$, and $\left(-\frac{1}{2}, -\frac{1}{2e^2}\right)$

is an inflection point.

**15.** (a) $f$ is increasing on the intervals where $f'(x) > 0$, namely, $(-\infty, -1)$ and $(1, 4)$. $f$ is decreasing on the intervals where

$f'(x) < 0$: $(-1, 1)$ and $(4, \infty)$.

(b) $f$ has a local maximum where $f'$ changes from positive to negative, at $x = -1$ and $x = 4$. Where $f'$ changes from

negative to positive, $f$ has a local minimum, at $x = 1$.

(c) When $f'$ is increasing, its derivative $f''$ is positive and hence, $f$ is concave upward. This happens on approximately

$(0, 2.5)$. $f$ is concave downward when $f'$ is decreasing—that is, on $(-\infty, 0)$ and $(2.5, \infty)$.

(d) $f$ has inflection points at $x = 0$ and $x = 2.5$, since the direction of concavity changes at each of these values.

**17.** As $x$ approaches 2 while $x < 2$, $x - 2$ is negative and approaches 0. The numerator approaches 10, so $\frac{2x+6}{x-2}$ becomes larger

and larger negative, and $\lim\limits_{x \to 2^-} \frac{2x+6}{x-2} = -\infty$.

**19.** Divide both the numerator and denominator by $x^2$ (the highest power of $x$ that occurs in the denominator).

$$\lim_{x \to \infty} \frac{8x^2 - 5x}{2x^2 + x + 1} = \lim_{x \to \infty} \frac{\dfrac{8x^2}{x^2} - \dfrac{5x}{x^2}}{\dfrac{2x^2}{x^2} + \dfrac{x}{x^2} + \dfrac{1}{x^2}} = \lim_{x \to \infty} \frac{8 - \dfrac{5}{x}}{2 + \dfrac{1}{x} + \dfrac{1}{x^2}} = \frac{\lim\limits_{x \to \infty} 8 - \lim\limits_{x \to \infty} \dfrac{5}{x}}{\lim\limits_{x \to \infty} 2 + \lim\limits_{x \to \infty} \dfrac{1}{x} + \lim\limits_{x \to \infty} \dfrac{1}{x^2}} = \frac{8 - 0}{2 + 0 + 0} = 4$$

**21.** Both $2^u$ and $u$ grow arbitrarily large as $u \to \infty$, and the values of $\ln t$ grow arbitrarily large as $t$ does, so

$$\lim_{u \to \infty} \ln(2^u + u) = \infty.$$

**23.** (a) $f(x) = 2 - 2x - x^3$ is defined for all $x$, so there are no vertical asymptotes. $\displaystyle\lim_{x \to \infty} f(x) = \lim_{x \to \infty} \left[2 - x(2 + x^2)\right] = -\infty$

and $\displaystyle\lim_{x \to -\infty} f(x) = \infty$, so there are no horizontal asymptotes (but we see the end behavior).

(b) $f'(x) = -2 - 3x^2 = -(3x^2 + 2)$, so $f'(x) < 0$ for all $x$ and $f$ is decreasing on $(-\infty, \infty)$.

(c) $f$ is all decreasing so there are no local extreme values.

(d) $f''(x) = -6x \;\Rightarrow\; f''(x) = 0$ when $x = 0$, $f''(x) < 0$ on

$(0, \infty)$, and $f''(x) > 0$ on $(-\infty, 0)$. Thus $f$ is concave

downward on $(0, \infty)$ and concave upward on $(-\infty, 0)$.

There is an inflection point at $(0, 2)$.

(e)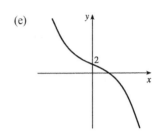

**25.** (a) $H(w) = \dfrac{3w}{2w - 4}$ is undefined at $w = 2$, and $\displaystyle\lim_{w \to 2^-} H(w) = -\infty$, $\displaystyle\lim_{w \to 2^+} H(w) = \infty$, so $w = 2$ is a vertical asymptote.

$\displaystyle\lim_{w \to \pm\infty} \dfrac{3w}{2w - 4} = \lim_{w \to \pm\infty} \dfrac{3}{2 - 4/w} = \dfrac{3}{2 - 0} = \dfrac{3}{2}$, so $y = \dfrac{3}{2}$ is a horizontal asymptote.

(b) $H'(w) = \dfrac{(2w - 4) \cdot 3 - 3w \cdot 2}{(2w - 4)^2} = \dfrac{-12}{(2w - 4)^2}$ and $H'(w) < 0$ for all $w \neq 2$, so $H$ is decreasing on $(-\infty, 2)$

and $(2, \infty)$.

(c) There are no critical numbers, so $H$ has no local extreme values.

(d) $H''(w) = \dfrac{(2w - 4)^2 \cdot 0 - (-12) \cdot 2(2w - 4)(2)}{[(2w - 4)^2]^2}$

$= \dfrac{48(2w - 4)}{(2w - 4)^4} = \dfrac{48}{(2w - 4)^3}$

and $H''(w) > 0$ when $w > 2$, $H''(w) < 0$ when $w < 2$, so $H$ is concave

upward on $(2, \infty)$ and concave downward on $(-\infty, 2)$. There are no

inflection points.

(e)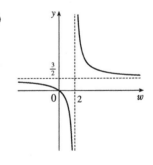

**27.** (a) $y = f(x) = \ln(x^2 + 4)$ is defined for all $x$, so there are no vertical asymptotes. $\displaystyle\lim_{x \to \pm\infty} f(x) = \infty$ because $x^2 + 4 \to \infty$

as $x \to \infty$, and $\ln t$ grows arbitrarily large as $t \to \infty$. Thus there are no horizontal asymptotes.

(b) $f'(x) = \dfrac{2x}{x^2 + 4} \;\Rightarrow\; f'(x) = 0$ when $x = 0$, $f'(x) < 0$ when $x < 0$, and $f'(x) > 0$ when $x > 0$, so $f$ is decreasing on

$(-\infty, 0)$ and increasing on $(0, \infty)$.

(c) $f(0) = \ln 4$ is a local (and absolute) minimum; no local maximum

(d) $f''(x) = \dfrac{(x^2 + 4) \cdot 2 - 2x \cdot 2x}{(x^2 + 4)^2} = \dfrac{8 - 2x^2}{(x^2 + 4)^2} = \dfrac{-2(x+2)(x-2)}{(x^2+4)^2} \quad \Rightarrow$ (e)

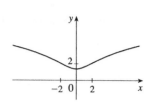

$f''(x) = 0$ when $x = \pm 2$, $f''(x) > 0$ when $-2 < x < 2$, $f''(x) < 0$ when

$x < -2$ or $x > 2$. Thus $f$ is concave upward on $(-2, 2)$ and concave

downward on $(-\infty, -2)$ and $(2, \infty)$. There are inflection points at

$(-2, \ln 8)$ and $(2, \ln 8)$.

**29.** $f'(3) = 0 \quad \Rightarrow$ horizontal tangent at $x = 3$. $f'(x) > 0$ when $x < 3 \quad \Rightarrow \quad f$ is increasing on $(-\infty, 3)$.

$f'(x) < 0$ when $x > 3 \quad \Rightarrow \quad f$ is decreasing on $(3, \infty)$. Thus we have a local maximum at $x = 3$.

$f''(-1) = 0$, $f''(x) > 0$ when $x < -1$, $f''(x) < 0$ when $x > -1 \quad \Rightarrow \quad f$ is concave upward on $(-\infty, -1)$ and concave

downward on $(-1, \infty)$. There is an inflection point at $x = -1$.

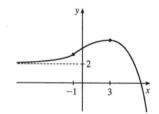

$\lim\limits_{x \to -\infty} f(x) = 2 \quad \Rightarrow \quad y = 2$ is a horizontal asymptote (at the left).

$\lim\limits_{x \to \infty} f(x) = -\infty \quad \Rightarrow$ there is no horizontal asymptote at the right,

but the curve heads downward.

**31.** $f(x) = \dfrac{x^2 - 1}{x^3} \quad \Rightarrow \quad f'(x) = \dfrac{x^3 \cdot 2x - (x^2 - 1) \cdot 3x^2}{x^6} = \dfrac{3 - x^2}{x^4} \quad \Rightarrow$

$f''(x) = \dfrac{x^4(-2x) - (3 - x^2)4x^3}{x^8} = \dfrac{2x^2 - 12}{x^5}$

From the graphs of $f'$ and $f''$, it appears that $f$ is increasing on $(-1.73, 0)$ and $(0, 1.73)$ and decreasing on $(-\infty, -1.73)$

and $(1.73, \infty)$;

$f$ has a local maximum of about $f(1.73) = 0.38$ and a local minimum of about $f(-1.73) = -0.38$; $f$ is concave upward on

$(-2.45, 0)$ and $(2.45, \infty)$, and concave downward on $(-\infty, -2.45)$ and $(0, 2.45)$; and $f$ has inflection points at about

$(-2.45, -0.34)$ and $(2.45, 0.34)$.

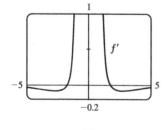

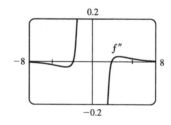

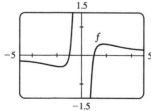

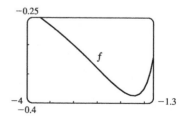

**33.** $f(x) = 3x^6 - 5x^5 + x^4 - 5x^3 - 2x^2 + 2 \quad \Rightarrow \quad f'(x) = 18x^5 - 25x^4 + 4x^3 - 15x^2 - 4x \quad \Rightarrow$

$f''(x) = 90x^4 - 100x^3 + 12x^2 - 30x - 4$

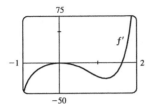 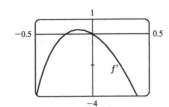 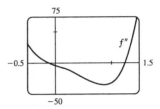

From the graphs of $f'$ and $f''$, it appears that $f$ is increasing on $(-0.23, 0)$ and $(1.62, \infty)$ and decreasing on $(-\infty, -0.23)$ and $(0, 1.62)$; $f$ has a local maximum of about $f(0) = 2$ and local minimum values of about $f(-0.23) = 1.96$ and

$f(1.62) = -19.2$.

$f$ is concave upward on $(-\infty, -0.12)$ and $(1.24, \infty)$ and concave downward on $(-0.12, 1.24)$; and $f$ has inflection points at

about $(-0.12, 1.98)$ and $(1.24, -12.1)$.

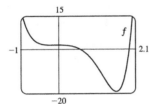 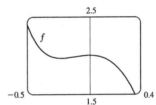

**35.**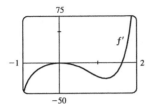

From the graph, we estimate the points of inflection to be about $(\pm 0.82, 0.22)$.

$f(x) = e^{-1/x^2} \quad \Rightarrow \quad f'(x) = e^{-x^{-2}}(2x^{-3}) = 2x^{-3}e^{-1/x^2} \quad \Rightarrow$

$f''(x) = 2[x^{-3} \cdot e^{-1/x^2}(2x^{-3}) + e^{-1/x^2} \cdot (-3x^{-4})] = 2x^{-6}e^{-1/x^2}(2 - 3x^2).$

$f''(x) = 0$ when $2 - 3x^2 = 0 \quad \Rightarrow \quad x^2 = \frac{2}{3} \quad \Rightarrow \quad x = \pm\sqrt{\frac{2}{3}}$, so the inflection

points are $\left( \pm\sqrt{\frac{2}{3}}, e^{-3/2} \right)$.

**37.** $f(x) = x^4 + x^3 + cx^2 \quad \Rightarrow \quad f'(x) = 4x^3 + 3x^2 + 2cx$. This is 0 when $x(4x^2 + 3x + 2c) = 0 \quad \Rightarrow \quad x = 0$ or

$4x^2 + 3x + 2c = 0$. Using the quadratic formula, we find that the solutions of this last equation are $x = \dfrac{-3 \pm \sqrt{9 - 32c}}{8}$.

Now if $9 - 32c < 0 \quad \Leftrightarrow \quad c > \frac{9}{32}$, then $(0, 0)$ is the only critical point, a minimum. If $c = \frac{9}{32}$, then there are two critical

points (a minimum at $x = 0$, and a horizontal tangent with no maximum or minimum at $x = -\frac{3}{8}$) and if $c < \frac{9}{32}$, then there are

three critical points except when $c = 0$, in which case the solution with the $+$ sign coincides with the critical point at $x = 0$.

For $0 < c < \frac{9}{32}$, there is a minimum at $x = -\dfrac{3}{8} - \dfrac{\sqrt{9 - 32c}}{8}$, a maximum at $x = -\dfrac{3}{8} + \dfrac{\sqrt{9 - 32c}}{8}$, and a minimum at

$x = 0$. For $c = 0$, there is a minimum at $x = -\frac{3}{4}$ and a horizontal tangent with no extreme value at $x = 0$, and for $c < 0$,

there is a maximum at $x = 0$, and there are minimum values at $x = -\dfrac{3}{8} \pm \dfrac{\sqrt{9 - 32c}}{8}$. Now we calculate

$f''(x) = 12x^2 + 6x + 2c.$ $f''(x) = 0$ when $x = \dfrac{-6 \pm \sqrt{36 - 4 \cdot 12 \cdot 2c}}{24}$. So if $36 - 96c \le 0 \iff c \ge \frac{3}{8}$, then there is no

inflection point. If $c < \frac{3}{8}$, then there are two inflection points at $x = -\dfrac{1}{4} \pm \dfrac{\sqrt{9 - 24c}}{12}$.

| Value of $c$ | Number of critical points | Number of inflection points |
|:---:|:---:|:---:|
| $c < 0$ | 3 | 2 |
| $c = 0$ | 2 | 2 |
| $0 < c < \frac{9}{32}$ | 3 | 2 |
| $c = \frac{9}{32}$ | 2 | 2 |
| $\frac{9}{32} < c < \frac{3}{8}$ | 1 | 2 |
| $c \ge \frac{3}{8}$ | 1 | 0 |

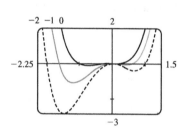
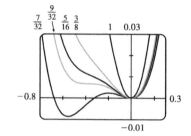

**39.** $P(t) = \dfrac{825}{1 + 2.97e^{-0.224t}} \quad \Rightarrow$

$P'(t) = \dfrac{(1 + 2.97e^{-0.224t}) \cdot 0 - 825 \cdot (2.97e^{-0.224t})(-0.224)}{(1 + 2.97e^{-0.224t})^2} = \dfrac{548.856e^{-0.224t}}{(1 + 2.97e^{-0.224t})^2}.$

The population was increasing most rapidly when $P'$ is maximal. From the

graph, this occurs at $t \approx 4.86$, so about 4.86 years after December 31, 1990

(November 1995).

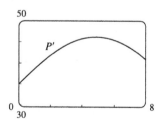

**41.** $v = K\sqrt{\dfrac{L}{C} + \dfrac{C}{L}} \quad \Rightarrow \quad \dfrac{dv}{dL} = K \cdot \dfrac{1}{2}\left(\dfrac{L}{C} + \dfrac{C}{L}\right)^{-1/2}\left(\dfrac{1}{C} - CL^{-2}\right) = \dfrac{K}{2\sqrt{(L/C) + (C/L)}}\left(\dfrac{1}{C} - \dfrac{C}{L^2}\right)$, and $\dfrac{dv}{dL} = 0$

when $\dfrac{1}{C} = \dfrac{C}{L^2} \quad \Rightarrow \quad L^2 = C^2 \quad \Rightarrow \quad L = C$ (since $L$ and $C$ are positive). This gives the minimum velocity since

$dv/dL < 0$ for $0 < L < C$ and $dv/dL > 0$ for $L > C$.

**43.** $p = 1.3e^{-0.04q}$, and $p$ and $q$ are changing with time $t$. Differentiating with respect to $t$ gives

$\dfrac{dp}{dt} = 1.3e^{-0.04q}\left(-0.04\,\dfrac{dq}{dt}\right) = -0.52e^{-0.04q}\,\dfrac{dq}{dt}.$ We are given that $\dfrac{dp}{dt} = 0.05$, and when $q = 30$ ($q$ is measured in

thousands), $0.05 = -0.52e^{-0.04(30)}\,\dfrac{dq}{dt} \quad \Rightarrow \quad \dfrac{dq}{dt} = -\dfrac{0.05}{0.52}e^{1.2} \approx -0.319.$ Thus demand will decrease at a rate of about

3.19 thousand discs per month.

**45.** Let $h$ be the height of the balloon, $z$ the distance from the boy to the balloon, and $x$ the horizontal distance. (See the diagram.) We are given that $dh/dt = 5$ and $dx/dt = 15$, and we want to find $dz/dt$. From the Pythagorean Theorem, $z^2 = x^2 + h^2$, and differentiating with respect to $t$ gives $2z \dfrac{dz}{dt} = 2x \dfrac{dx}{dt} + 2h \dfrac{dh}{dt}$. After 3 seconds ($t = 3$), $h = 45 + 3(5) = 60$ and

$x = 15(3) = 45$, and $z^2 = 45^2 + 60^2 \Rightarrow z = \sqrt{45^2 + 60^2} = 75$.

Then $2(75) \dfrac{dz}{dt} = 2(45)(15) + 2(60)(5) \Rightarrow \dfrac{dz}{dt} = \dfrac{1950}{150} = 13$ ft/s.

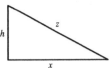

**47.** Call the two integers $x$ and $y$. Then $x + 4y = 1000$, so $x = 1000 - 4y$. Their product is $P = xy = (1000 - 4y)y$, so our problem is to maximize the function $P(y) = 1000y - 4y^2$, where $0 < y < 250$ and $y$ is an integer. $P'(y) = 1000 - 8y$, so $P'(y) = 0$ when $y = 125$. $P''(y) = -8 < 0$, so $P(125) = 62{,}500$ is an absolute maximum. Since the optimal $y$ turned out to be an integer, we have found the desired pair of numbers, namely $x = 1000 - 4(125) = 500$ and $y = 125$.

**49.**

| | | $y$ | | |
|---|---|---|---|---|
| $x$ | $x$ | | $x$ | |

building

Let $x$ denote the length of each of three fence portions perpendicular to the building, and let $y$ denote the length of the side parallel to the building. The total length of fence is 2100 ft, so $3x + y = 2100 \Rightarrow y = 2100 - 3x$.

We wish to maximize the area $A(x) = xy = x(2100 - 3x) = 2100x - 3x^2$ for $0 \le x \le 700$. (If $x$ is larger than 700, there won't be enough fencing material.) $A'(x) = 2100 - 6x$, and $A'(x) = 0$ when $x = \frac{2100}{6} = 350$, so this is the only critical number. Comparing the values of $A$ at the endpoints, we have $A(0) = 0$, $A(350) = 367{,}500$, and $A(700) = 0$, so by the Closed Interval Method, the largest area occurs when $x = 350$ ft and $y = 2100 - 3(350) = 1050$ ft.

**51.** (a) $C(q) = 6200 + 7.3q + 0.002q^2 \Rightarrow C'(q) = 7.3 + 0.004q$, and $R(q) = 31q - 0.003q^2$. The cost to produce 2000 helmets is $C(2000) = \$28{,}800$, the marginal cost is $C'(2000) = \$15.30$/helmet, the average cost is $C(2000)/2000 = \$28{,}800/2000 = \$14.40$/helmet, and the profit is

$P(2000) = R(2000) - C(2000) = \$50{,}000 - \$28{,}800 = \$21{,}200$.

(b) The average cost is $c(q) = \dfrac{C(q)}{q} = \dfrac{6200 + 7.3q + 0.002q^2}{q} = \dfrac{6200}{q} + 7.3 + 0.002q$, so $c'(q) = -6200q^{-2} + 0.002$.

$c'(q) = 0$ when $\dfrac{6200}{q^2} = 0.002 \Rightarrow q^2 = \dfrac{6200}{0.002} = 3{,}100{,}000 \Rightarrow q = \sqrt{3{,}100{,}000} \approx 1761$. (Alternatively, we could determine when average cost and marginal cost are equal.) $c'(q) < 0$ for $q < 1761$ (approximately) and $c'(q) > 0$ for $q > 1761$, so $c(q)$ has an absolute minimum when the production level is $q \approx 1761$ helmets.

(c) The profit function is $P(q) = R(q) - C(q) = (31q - 0.003q^2) - (6200 + 7.3q + 0.002q^2) = 23.7q - 0.005q^2 - 6200$.

$P'(q) = 23.7 - 0.01q$, so $P'(q) = 0$ when $23.7 = 0.01q \Rightarrow q = \frac{23.7}{0.01} = 2370$. Because $P''(q) = -0.01 < 0$, $P$ is concave downward everywhere and the maximum profit is achieved for $q = 2370$ helmets.

Alternatively, we can solve $R'(q) = C'(q)$ and verify that $R''(q) < C''(q)$ at that value.

**53.** The demand function $D$ is linear and $D(11{,}000) = 12$, $D(12{,}000) = 11$, so the slope is $\frac{11-12}{12{,}000-12{,}000} = -\frac{1}{1000}$ and an

equation of the line is $p - 12 = \left(-\frac{1}{1000}\right)(q - 11{,}000)$ $\Rightarrow$ $p = D(q) = -\frac{1}{1000}q + 23$. The revenue is

$R(q) = q \cdot D(q) = -\frac{1}{1000}q^2 + 23q$, which we wish to maximize for $0 \leq q \leq 15{,}000$. $R'(q) = -\frac{1}{500}q + 23$, and $R'(q) = 0$

when $q = 23(500) = 11{,}500$. $R(11{,}500) = 132{,}250$, $R(0) = 0$, and $R(15{,}000) = 120{,}000$, so by the Closed Interval

Method, the maximum revenue occurs when $q = 11{,}500$ and the price is $D(11{,}500) = \$11.50$.

**55.** (a) $D(q) = 308 - \frac{1}{24}q$ $\Rightarrow$ $D'(q) = -\frac{1}{24}$ and $E(q) = -\dfrac{D(q)}{qD'(q)} = -\dfrac{308 - q/24}{-q/24} = \dfrac{7392 - q}{q}$.

(b) $E(q) < 1$ when $7392 - q < q$ $\Rightarrow$ $7392 < 2q$ $\Rightarrow$ $q > 3696$, and $E(q) > 1$ when $q < 3696$. $D(3696) = 154$, and

$D$ is a decreasing function, so $E(q) < 1$ for prices below $\$154$ and $E(q) > 1$ for prices above $\$154$.

(c) Revenue is maximized when $E(q) = 1$ $\Rightarrow$ $\dfrac{7392 - q}{q} = 1$ $\Rightarrow$ $q = 3696$. The corresponding price is

$D(3696) = \$154$.

(d) $R(q) = q \cdot D(q) = 308q - \frac{1}{24}q^2$ $\Rightarrow$ $R'(q) = 308 - \frac{1}{12}q$, and $R'(q) = 0$ when $308 = \frac{1}{12}q$ $\Rightarrow$

$q = 12(308) = 3696$. Since $R''(q) = -\frac{1}{12}$ is negative for all $q$, this is a maximum. This agrees with our previous result.

The maximum revenue is $R(3696) = \$569{,}184$.

# 5 INTEGRALS

## 5.1 Cost, Area, and the Definite Integral

1. (a) We take batches of 2000 units and use the initial marginal cost as the marginal cost for each unit in the batch. Thus, the estimated cost is

$$(\$3.22 \times 2000) + (\$2.76 \times 2000) + (\$2.55 \times 2000) + (\$2.26 \times 2000) + (\$2.13 \times 2000) + (\$2.19 \times 2000)$$
$$= \$6440 + \$5520 + \$5100 + \$4520 + \$4260 + \$4380$$
$$= \$30{,}220$$

(b) We take batches of 1000 units and use the ending marginal cost as the marginal cost for each unit in the batch. Thus, the estimated cost is

$$(\$2.40 \times 1000) + (\$2.26 \times 1000) + (\$2.14 \times 1000) + (\$2.13 \times 1000) + (\$2.15 \times 1000) + (\$2.19 \times 1000) + (\$2.33 \times 1000)$$
$$= \$2400 + \$2260 + \$2140 + \$2130 + \$2150 + \$2190 + \$2330$$
$$= \$15{,}600$$

3. For each time interval, the distance traveled is velocity $\times$ time, so we estimate the total distance during the first three seconds to be

$$(6.2\ \text{ft/s})(0.5\ \text{s}) + (10.8)(0.5) + (14.9)(0.5) + (18.1)(0.5) + (19.4)(0.5) + (20.2)(0.5) = 44.8\ \text{ft}$$

5. (a) First, convert the velocities from mi/h to ft/s by multiplying each velocity by (5280/3600):

| Time (s) | Velocity (mi/h) | Velocity (ft/s) |
|---|---|---|
| 0 | 182.9 | 268.3 |
| 10 | 168.0 | 246.4 |
| 20 | 106.6 | 156.3 |
| 30 | 99.8 | 146.4 |
| 40 | 124.5 | 182.6 |
| 50 | 176.1 | 258.3 |
| 60 | 175.6 | 257.5 |

Using the ending velocity for each time interval, the distance traveled after one minute is approximately

$$(246.4\ \text{ft/s})(10\ \text{s}) + (156.3\ \text{ft/s})(10\ \text{s}) + (146.4\ \text{ft/s})(10\ \text{s}) + (182.6\ \text{ft/s})(10\ \text{s})$$
$$+ (258.3\ \text{ft/s})(10\ \text{s}) + (257.5\ \text{ft/s})(10\ \text{s}) = 12{,}475\ \text{ft} \approx 2.36\ \text{mi}$$

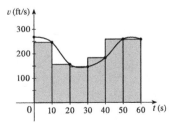

(b) For three subintervals, each has length 20 and the midpoints of the subintervals are $t = 10$, $t = 30$, and $t = 50$. Then we estimate the distance to be

$$(246.4 \text{ ft/s})(20 \text{ s}) + (146.4)(20) + (258.3)(20) = 13{,}022 \text{ ft}$$
$$\approx 2.47 \text{ mi}$$

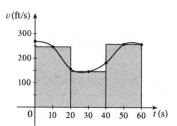

**7.** Using subintervals of length $\Delta t = 2$ hours and left endpoints, the total amount of oil leaked is approximately

$$(8.7 \text{ L/h})(2 \text{ h}) + 7.6(2) + 6.8(2) + 6.2(2) + 5.7(2) = 70.0 \text{ L}$$

Using right endpoints, we estimate that

$$7.6(2) + 6.8(2) + 6.2(2) + 5.7(2) + 5.3(2) = 63.2 \text{ L of oil leaked out.}$$

**9.** (a) Using 4 subintervals, $\Delta x = \frac{8-0}{4} = 2$. Using right endpoints, we estimate the area to be

$$R_4 = f(x_1)\,\Delta x + f(x_2)\,\Delta x + f(x_3)\,\Delta x + f(x_4)\,\Delta x$$
$$= f(2) \cdot 2 + f(4) \cdot 2 + f(6) \cdot 2 + f(8) \cdot 2$$
$$\approx 3.75(2) + 5(2) + 5.75(2) + 6(2) = 41$$

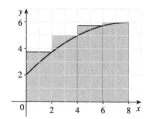

This estimate is higher than the actual area.

(b) Using left endpoints,

$$L_4 = f(x_0)\,\Delta x + f(x_1)\,\Delta x + f(x_2)\,\Delta x + f(x_3)\,\Delta x$$
$$= f(0) \cdot 2 + f(2) \cdot 2 + f(4) \cdot 2 + f(6) \cdot 2$$
$$\approx 2(2) + 3.75(2) + 5(2) + 5.75(2) = 33$$

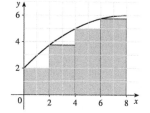

This estimate is lower than the actual area.

(c) With 8 subintervals, $\Delta x = 1$. Using right endpoints,

$$R_8 = f(x_1)\,\Delta x + f(x_2)\,\Delta x + \cdots + f(x_8)\,\Delta x$$
$$= f(1) \cdot 1 + f(2) \cdot 1 + \cdots + f(8) \cdot 1$$
$$\approx 2.9(1) + 3.75(1) + 4.4(1) + 5(1) + 5.4(1) + 5.75(1) + 5.9(1) + 6(1)$$
$$= 39.1$$

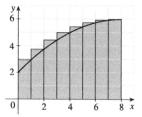

**11.** (a) Here, $\Delta x - \frac{5-1}{4} = 1$ and

$$R_4 = f(x_1) \cdot 1 + f(x_2) \cdot 1 + f(x_3) \cdot 1 + f(x_4) \cdot 1$$
$$= f(2) + f(3) + f(4) + f(5)$$
$$= \tfrac{1}{2} + \tfrac{1}{3} + \tfrac{1}{4} + \tfrac{1}{5} = \tfrac{77}{60} \approx 1.28$$

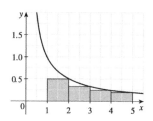

Since $f$ is decreasing on $[1, 5]$, this is an underestimate.

(b) $L_4 = f(x_0) \cdot 1 + f(x_1) \cdot 1 + f(x_2) \cdot 1 + f(x_3) \cdot 1$

$\quad = f(1) + f(2) + f(3) + f(4)$

$\quad = 1 + \frac{1}{2} + \frac{1}{3} + \frac{1}{4} = \frac{25}{12} \approx 2.08$

This is an overestimate.

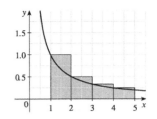

**13.** (a) $f(x) = 1 + x^2$ and with $n = 3$ rectangles, we have $\Delta x = \dfrac{2 - (-1)}{3} = 1 \quad \Rightarrow$

$\quad R_3 = f(0) \cdot 1 + f(1) \cdot 1 + f(2) \cdot 1 = 1 \cdot 1 + 2 \cdot 1 + 5 \cdot 1 = 8.$

With $n = 6$ rectangles, $\Delta x = \dfrac{2 - (-1)}{6} = 0.5 \quad \Rightarrow$

$\quad R_6 = 0.5[f(-0.5) + f(0) + f(0.5) + f(1) + f(1.5) + f(2)]$

$\quad = 0.5(1.25 + 1 + 1.25 + 2 + 3.25 + 5)$

$\quad = 0.5(13.75) = 6.875$

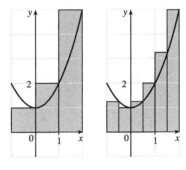

(b) $L_3 = f(-1) \cdot 1 + f(0) \cdot 1 + f(1) \cdot 1 = 2 \cdot 1 + 1 \cdot 1 + 2 \cdot 1 = 5.$

$\quad L_6 = 0.5[f(-1) + f(-0.5) + f(0) + f(0.5) + f(1) + f(1.5)]$

$\quad = 0.5(2 + 1.25 + 1 + 1.25 + 2 + 3.25)$

$\quad = 0.5(10.75) = 5.375$

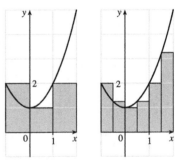

(c) $M_3 = f(-0.5) \cdot 1 + f(0.5) \cdot 1 + f(1.5) \cdot 1$

$\quad = 1.25 \cdot 1 + 1.25 \cdot 1 + 3.25 \cdot 1 = 5.75$

$\quad M_6 = 0.5[f(-0.75) + f(-0.25) + f(0.25) + f(0.75) + f(1.25) + f(1.75)]$

$\quad = 0.5(1.5625 + 1.0625 + 1.0625 + 1.5625 + 2.5626 + 4.0625)$

$\quad = 0.5(11.875) = 5.9375$

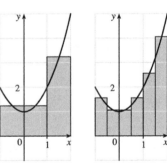

(d) $M_6$ appears to be the best estimate.

**15.** (a) Since $g(t) > 0$ on the interval $[0, 4]$, the net area $\int_0^4 g(t)\, dt$ is positive.

(b) It appears that $g(t) \leq 0$ on the interval $[6, 8]$, so the net area $\int_6^8 g(t)\, dt$ is negative.

(c) It appears that $g(t) \geq 0$ on the interval $[4, 6]$ and that $g(t) \leq 0$ on the interval $[6, 10]$. Because the graph encloses more area below the $t$-axis than above the $t$-axis for $4 \leq t \leq 10$, the net area is negative and $\int_4^{10} g(t)\, dt$ is negative.

(d) It appears that $g(t) \geq 0$ on the interval $[0, 6]$ and that $g(t) \leq 0$ on the interval $[6, 8]$. Because the graph encloses more area above the $t$-axis than below the $t$-axis for $0 \leq t \leq 8$, the net area is positive and $\int_0^8 g(t)\, dt$ is positive.

**17.** (a) $\int_{-4}^{0} f(x)\,dx = $ (area of region above $x$-axis) $-$ (area of region below $x$-axis)

$\quad\quad = $ (area of region $B$) $-$ (area of region $A$) $= 3 - 3 = 0$

(b) $\int_{-4}^{2} f(x)\,dx = $ (area of region above $x$-axis) $-$ (area of region below $x$-axis)

$\quad\quad = $ (area of region $B$) $-$ (area of region $A$ + area of region $C$) $= 3 - (3 + 3) = -3$

**19.** (a) For $0 \leq x \leq 2$, the area below the graph and above the $x$-axis consists of a rectangle with base 2 and height 1 surmounted

by a triangle with base 2 and height 2. Thus, $\int_{0}^{2} f(x)\,dx = (2)(1) + \frac{1}{2}(2)(2) = 4$.

(b) $\int_{0}^{5} f(x)\,dx = \int_{0}^{2} f(x)\,dx + \int_{2}^{3} f(x)\,dx + \int_{3}^{5} f(x)\,dx$. From part (a), $\int_{0}^{2} f(x)\,dx = 4$.

$\int_{2}^{3} f(x)\,dx = $ area of a rectangle with base 1 and height $3 = 1 \cdot 3 = 3$, and

$\int_{3}^{5} f(x)\,dx = $ area of a triangle with base 2 and height $3 = \frac{1}{2} \cdot 2 \cdot 3 = 3$.

Thus, $\int_{0}^{5} f(x)\,dx = 4 + 3 + 3 = 10$.

(c) $\int_{5}^{7} f(x)\,dx$ is the negative of the area of a triangle with base 2 and height 3, so $\int_{5}^{7} f(x)\,dx = -\frac{1}{2} \cdot 2 \cdot 3 = -3$.

(d) $\int_{7}^{9} f(x)\,dx$ is the negative of the area of a 2-by-2 square and a triangle with base 2 and height 1, thus

$\int_{7}^{9} f(x)\,dx = -\left(2 \cdot 2 + \frac{1}{2} \cdot 2 \cdot 1\right) = -5$, thus

$\int_{0}^{9} f(x)\,dx = \int_{0}^{5} f(x)\,dx + \int_{5}^{7} f(x)\,dx + \int_{7}^{9} f(x)\,dx = 10 + (-3) + (-5) = 2$.

**21.** $\int_{-1}^{4} (4 - 2x)\,dx$ can be interpreted as the area of the triangle above the $x$-axis

minus the area of the triangle below the $x$-axis; that is

$\frac{1}{2}(3)(6) - \frac{1}{2}(2)(4) = 9 - 4 = 5$.

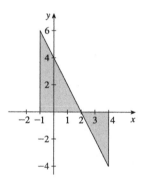

**23.** $\int_{-1}^{2} |x|\,dx$ can be interpreted as the sum of the areas of the two shaded

triangles; that is, $\frac{1}{2}(1)(1) + \frac{1}{2}(2)(2) = \frac{1}{2} + \frac{4}{2} = \frac{5}{2} = 2.5$.

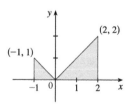

**25.** $\Delta x = (10 - 2)/4 = 2$, so the endpoints are 2, 4, 6, 8, and 10, and the midpoints are 3, 5, 7, and 9. The Midpoint Rule gives

$$\int_{2}^{10} \sqrt{x^3 + 1}\,dx \approx f(\overline{x}_1)\,\Delta x + f(\overline{x}_2)\,\Delta x + f(\overline{x}_3)\,\Delta x + f(\overline{x}_4)\,\Delta x$$

$$= \Delta x[f(3) + f(5) + f(7) + f(9)]$$

$$= 2\left(\sqrt{3^3 + 1} + \sqrt{5^3 + 1} + \sqrt{7^3 + 1} + \sqrt{9^3 + 1}\right) \approx 124.1644$$

**27.** $\Delta x = (2-1)/5 = 0.2$, so the endpoints are 1, 1.2, 1.4, 1.6, 1.8, and 2, and the midpoints are 1.1, 1.3, 1.5, 1.7, and 1.9.

The Midpoint Rule gives

$$\int_1^2 4e^{-0.6t}\, dt \approx f(\overline{x}_1)\,\Delta x + f(\overline{x}_2)\,\Delta x + f(\overline{x}_3)\,\Delta x + f(\overline{x}_4)\,\Delta x + f(\overline{x}_5)\,\Delta x$$

$$= \Delta x\left[f(1.1) + f(1.3) + f(1.5) + f(1.7) + f(1.9)\right]$$

$$= 0.2[4e^{-0.6(1.1)} + 4e^{-0.6(1.3)} + 4e^{-0.6(1.5)} + 4e^{-0.6(1.7)} + 4e^{-0.6(1.9)}] \approx 1.6498$$

**29.** Since $f$ is increasing, $L_5 \le \int_{10}^{30} f(x)\, dx \le R_5$.

$$\text{lower estimate} = L_5 = f(x_0)\,\Delta x + f(x_1)\,\Delta x + f(x_2)\,\Delta x + f(x_3)\,\Delta x + f(x_4)\,\Delta x$$

$$= \Delta x\left[f(10) + f(14) + f(18) + f(22) + f(26)\right]$$

$$= 4[-12 + (-6) + (-2) + 1 + 3] = 4(-16) = -64$$

$$\text{upper estimate} = R_5 = f(x_1)\,\Delta x + f(x_2)\,\Delta x + f(x_3)\,\Delta x + f(x_4)\,\Delta x + f(x_5)\,\Delta x$$

$$= \Delta x\left[f(14) + f(18) + f(22) + f(26) + f(30)\right]$$

$$= 4[-6 + (-2) + 1 + 3 + 8] = 4(4) = 16$$

**31.** $\Delta x = (8-0)/4 = 2$

(a) $\int_0^8 f(x)\, dx \approx R_4 = f(x_1)\,\Delta x + f(x_2)\,\Delta x + f(x_3)\,\Delta x + f(x_4)\,\Delta x$

$$= \Delta x[f(2) + f(4) + f(6) + f(8)] \approx 2[1 + 2 + (-2) + 1] = 4$$

(b) $\int_0^8 f(x)\, dx \approx L_4 = f(x_0)\,\Delta x + f(x_1)\,\Delta x + f(x_2)\,\Delta x + f(x_3)\,\Delta x$

$$= \Delta x[f(0) + f(2) + f(4) + f(6)] \approx 2[2 + 1 + 2 + (-2)] = 6$$

(c) $\int_0^8 f(x)\, dx \approx M_4 = f(\overline{x}_1)\,\Delta x + f(\overline{x}_2)\,\Delta x + f(\overline{x}_3)\,\Delta x + f(\overline{x}_4)\,\Delta x$

$$= \Delta x[f(1) + f(3) + f(5) + f(7)] \approx 2[3 + 2 + 1 + (-1)] = 10$$

**33.** (a) Yes. Since $f(x) > 0$ for $2 \le x \le 6$, its graph lies above the $x$-axis, and $\int_2^6 f(x)\, dx$ can be interpreted as the area of the region below the graph and above the $x$-axis, which must be positive.

(b) No. $\int_2^6 f(x)\, dx > 0$ tells us only that the *net* area is positive. The graph of $f$ may go below the $x$-axis on portions of the interval $[2, 6]$. For example, if $f$ is the function graphed in Exercise 19, then $f(x) < 0$ for $5 \le x \le 6$, but $\int_2^6 f(x)\, dx > 0$.

**35.** $\int_0^3 f(x)\, dx$ is clearly less than $-1$ and has the smallest value. The slope of the tangent line of $f$ at $x = 1$, $f'(1)$, has a value between $-1$ and $0$, so it has the next smallest value. The largest value is $\int_3^8 f(x)\, dx$ because this integral represents the largest area above the $x$-axis, followed by $\int_4^8 f(x)\, dx$, which has a value about 1 unit less than $\int_3^8 f(x)\, dx$. Still positive, but with a smaller value than $\int_4^8 f(x)\, dx$, is $\int_0^8 f(x)\, dx$. Ordering these quantities from smallest to largest gives us

$$\int_0^3 f(x)\, dx < f'(1) < \int_0^8 f(x)\, dx < \int_4^8 f(x)\, dx < \int_3^8 f(x)\, dx \ \text{ or } \ B < E < A < D < C$$

**37.** (a) We divide the interval $[0, 4]$ into $n$ subintervals, each of equal width $\Delta x = \dfrac{4-0}{n} = \dfrac{4}{n}$. Then the endpoints of the

subintervals are $x_0 = 0$, $x_1 = \dfrac{4}{n}$, $x_2 = 2 \cdot \dfrac{4}{n}$, $x_3 = 3 \cdot \dfrac{4}{n}$, ..., $x_i = i \cdot \dfrac{4}{n}$, ..., $x_n = n \cdot \dfrac{4}{n} = 4$. By Definition 5 with

$f(x) = x$,

$$\int_0^4 f(x)\,dx = \lim_{n \to \infty} \left[ f(x_1)\,\Delta x + f(x_2)\,\Delta x + \cdots + f(x_n)\,\Delta x \right]$$

$$= \lim_{n \to \infty} \Delta x \left[ f\left(\frac{4}{n}\right) + f\left(2 \cdot \frac{4}{n}\right) + f\left(3 \cdot \frac{4}{n}\right) + \cdots + f\left(n \cdot \frac{4}{n}\right) \right]$$

$$= \lim_{n \to \infty} \frac{4}{n} \left[ \frac{4}{n} + 2 \cdot \frac{4}{n} + 3 \cdot \frac{4}{n} + \cdots + n \cdot \frac{4}{n} \right]$$

$$= \lim_{n \to \infty} \frac{4}{n} \cdot \frac{4}{n} [1 + 2 + 3 + \cdots + n]$$

$$= \lim_{n \to \infty} \frac{16}{n^2} \left[ \tfrac{1}{2} n(n+1) \right] = \lim_{n \to \infty} \frac{16}{n^2} \cdot \frac{n^2 + n}{2}$$

$$= \lim_{n \to \infty} 8 \cdot \frac{n^2 + n}{n^2} = 8 \lim_{n \to \infty} \left( \frac{n^2}{n^2} + \frac{n}{n^2} \right)$$

$$= 8 \lim_{n \to \infty} \left( 1 + \frac{1}{n} \right) = 8(1 + 0) = 8$$

(b) $\int_0^4 x\,dx$ is the area of the region under the line $y = x$ and above the $x$-axis for $0 \le x \le 4$. The region is a triangle with

base 4 and height 4 whose area is $\tfrac{1}{2}(4)(4) = 8$.

## 5.2   The Fundamental Theorem of Calculus

**Prepare Yourself**

**1.** $(4t + 3)^2 = (4t + 3)(4t + 3) = 16t^2 + 12t + 12t + 9 = 16t^2 + 24t + 9$

**2.** $\sqrt{x}(2x^2 - 3) = x^{1/2}(2x^2 - 3) = 2x^{2+1/2} - 3x^{1/2} = 2x^{5/2} - 3x^{1/2}$

**3.** (a) $\sqrt{x} = x^{1/2}$ 
(b) $\left( \sqrt[3]{x} \right)^2 = (x^{1/3})^2 = x^{(1/3)\cdot 2} = x^{2/3}$

(c) $1/x^2 = x^{-2}$ 
(d) $\dfrac{1}{x\sqrt{x}} = \dfrac{1}{x \cdot x^{1/2}} = \dfrac{1}{x^{3/2}} = x^{-3/2}$

**4.** $\dfrac{2x^4 + 5x^3 + 2}{x^2} = \dfrac{2x^4}{x^2} + \dfrac{5x^3}{x^2} + \dfrac{2}{x^2} = 2x^2 + 5x + 2x^{-2}$

**5.** (a) $y = \tfrac{1}{4} x^4 \ \Rightarrow \ \dfrac{dy}{dx} = \tfrac{1}{4}(4x^{4-1}) = x^3$ 
(b) $B(t) = \tfrac{2}{3} t^{3/2} \ \Rightarrow \ B'(t) = \tfrac{2}{3} \left( \tfrac{3}{2} t^{3/2-1} \right) = t^{1/2} = \sqrt{t}$

(c) $L(u) = \ln|u| \ \Rightarrow \ L'(u) = \dfrac{1}{u}$ 
(d) $P = 7.3e^t \ \Rightarrow \ \dfrac{dP}{dt} = 7.3e^t$

(e) $g(t) = -\dfrac{1}{0.2} e^{-0.2t} \ \Rightarrow \ g'(t) = -\dfrac{1}{0.2} e^{-0.2t}(-0.2) = e^{-0.2t}$

(f) $f(v) = 2\sqrt{v} = 2v^{1/2} \Rightarrow f'(v) = 2 \cdot \frac{1}{2}v^{-1/2} = v^{-1/2} = 1/\sqrt{v}$

(g) $h(x) = 1/x^3 = x^{-3} \Rightarrow h'(x) = -3x^{-3-1} = -3x^{-4} = -\dfrac{3}{x^4}$

(h) $A = 5^t \Rightarrow dA/dt = 5^t \ln 5$

## Exercises

**1.** $f(x) = 6x^2 - 8x + 3 \Rightarrow F(x) = 6 \cdot \dfrac{x^3}{3} - 8 \cdot \dfrac{x^2}{2} + 3x + C = 2x^3 - 4x^2 + 3x + C$

Check: $F'(x) = 2 \cdot 3x^2 - 4 \cdot 2x + 3 + 0 = 6x^2 - 8x + 3 = f(x)$

**3.** $f(x) = 5x^{1/4} - 7x^{3/4} \Rightarrow F(x) = 5 \cdot \dfrac{x^{1/4+1}}{\frac{1}{4}+1} - 7 \cdot \dfrac{x^{3/4+1}}{\frac{3}{4}+1} + C = 5 \cdot \dfrac{x^{5/4}}{5/4} - 7 \cdot \dfrac{x^{7/4}}{7/4} + C = 4x^{5/4} - 4x^{7/4} + C$

**5.** $f(x) = 3\sqrt{x} + \dfrac{5}{x^6} = 3x^{1/2} + 5x^{-6} \Rightarrow$

$F(x) = 3 \cdot \dfrac{x^{1/2+1}}{1/2+1} + 5 \cdot \dfrac{x^{-6+1}}{-6+1} + C = 3 \cdot \dfrac{x^{3/2}}{3/2} + 5 \cdot \dfrac{x^{-5}}{-5} + C = 2x^{3/2} - x^{-5} + C$ (on any interval that doesn't

include 0)

**7.** $q(s) = 5e^s + 3.7 \Rightarrow Q(s) = 5e^s + 3.7s + C$

**9.** $f(q) = 1 + 2e^{0.8q} \Rightarrow F(q) = q + 2\left(\dfrac{1}{0.8}e^{0.8q}\right) + C = q + 2.5e^{0.8q} + C$

**11.** $g(x) = \dfrac{5 - 4x^3 + 2x^6}{x^6} = \dfrac{5}{x^6} - \dfrac{4x^3}{x^6} + \dfrac{2x^6}{x^6} = 5x^{-6} - 4x^{-3} + 2 \Rightarrow$

$G(x) = 5\left(\dfrac{x^{-5}}{-5}\right) - 4\left(\dfrac{x^{-2}}{-2}\right) + 2x + C = -x^{-5} + 2x^{-2} + 2x + C = -\dfrac{1}{x^5} + \dfrac{2}{x^2} + 2x + C$ (on any interval that doesn't

include 0)

**13.** $f'(x) = 8x^2 - 3/x = 8x^2 - 3 \cdot \dfrac{1}{x} \Rightarrow f(x) = 8 \cdot \dfrac{x^3}{3} - 3\ln|x| + C = \frac{8}{3}x^3 - 3\ln|x| + C$

**15.** $f'(x) = \sqrt{x}(6 + 5x) = x^{1/2}(6 + 5x) = 6x^{1/2} + 5x^{3/2} \Rightarrow f(x) = 6 \cdot \dfrac{x^{3/2}}{3/2} + 5 \cdot \dfrac{x^{5/2}}{5/2} + C = 4x^{3/2} + 2x^{5/2} + C.$

$f(1) = 6 + C$ and $f(1) = 10 \Rightarrow C = 4$, so $f(x) = 4x^{3/2} + 2x^{5/2} + 4.$

**17.** (a) $f''(x) = 24x^2 + 2x + 10 \Rightarrow f'(x) = 24 \cdot \dfrac{x^3}{3} + 2 \cdot \dfrac{x^2}{2} + 10x + C = 8x^3 + x^2 + 10x + C.$ Then $f'(1) = 19 + C$

and we were given that $f'(1) = -3$, so $C = -22$ and $f'(x) = 8x^3 + x^2 + 10x - 22.$

(b) $f'(x) = 8x^3 + x^2 + 10x - 22 \Rightarrow f(x) = 8 \cdot \dfrac{x^4}{4} + \dfrac{x^3}{3} + 10 \cdot \dfrac{x^2}{2} - 22x + C = 2x^4 + \frac{1}{3}x^3 + 5x^2 - 22x + C.$ Then

$f(1) = 2 + \frac{1}{3} + 5 - 22 + C = -\frac{44}{3} + C$ and we were given that $f(1) = 5$, so $C = 5 + \frac{44}{3} = \frac{59}{3}$ and

$f(x) = 2x^4 + \frac{1}{3}x^3 + 5x^2 - 22x + \frac{59}{3}.$

**19.** (a)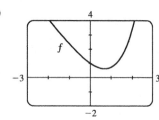

(b)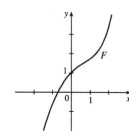

Since $F(0) = 1$, the graph passes through the point $(0, 1)$. Since $f$ is always positive, $F$ is always increasing.

(c) $f(x) = e^x - 2x \quad \Rightarrow \quad F(x) = e^x - 2 \cdot \dfrac{x^2}{2} + C = e^x - x^2 + C. \ F(0) = 1 \quad \Rightarrow \quad e^0 - 0^2 + C = 1 \quad \Rightarrow$

$1 - 0 + C = 1 \quad \Rightarrow \quad C = 0$, so $F(x) = e^x - x^2$.

(d) If we use a graphing device to graph $F(x) = e^x - x^2$, we see that the graph looks similar to the one in part (b).

**21.** $\int_0^4 (6x - 5)\, dx = \left[6 \cdot \frac{1}{2}x^2 - 5x\right]_0^4 = \left[3x^2 - 5x\right]_0^4 = (3 \cdot 4^2 - 5 \cdot 4) - (3 \cdot 0^2 - 5 \cdot 0) = 28 - 0 = 28$

**23.** $\displaystyle\int_{-1}^3 x^5\, dx = \left[\dfrac{x^6}{6}\right]_{-1}^3 = \dfrac{3^6}{6} - \dfrac{(-1)^6}{6} = \dfrac{729 - 1}{6} = \dfrac{364}{3}$

**25.** $\int_0^2 (6x^2 - 4x + 5)\, dx = \left[6 \cdot \frac{1}{3}x^3 - 4 \cdot \frac{1}{2}x^2 + 5x\right]_0^2 = \left[2x^3 - 2x^2 + 5x\right]_0^2 = (16 - 8 + 10) - 0 = 18$

**27.** $\displaystyle\int_1^9 4\sqrt{z}\, dz = \int_1^9 4z^{1/2}\, dz = \left[4 \cdot \dfrac{z^{3/2}}{3/2}\right]_1^9 = \left[\tfrac{8}{3}z^{3/2}\right]_1^9 = \tfrac{8}{3}(9)^{3/2} - \tfrac{8}{3}(1)^{3/2} = \tfrac{8}{3}(27) - \tfrac{8}{3} = \tfrac{208}{3}$

**29.** $\int_{-1}^0 (2x - e^x)\, dx = \left[x^2 - e^x\right]_{-1}^0 = (0 - 1) - (1 - e^{-1}) = -2 + 1/e$

**31.** $\displaystyle\int_0^4 (1 + 2e^{-0.6q})\, dq = \left[q + 2\left(\dfrac{1}{-0.6}e^{-0.6q}\right)\right]_0^4 = \left[q - \tfrac{10}{3}e^{-0.6q}\right]_0^4 = \left(4 - \tfrac{10}{3}e^{-0.6(4)}\right) - \left(0 - \tfrac{10}{3}e^0\right)$

$= 4 - \tfrac{10}{3}e^{-2.4} + \tfrac{10}{3} = \tfrac{22}{3} - \tfrac{10}{3}e^{-2.4}$

**33.** $\int_0^4 (2v + 5)(3v - 1)\, dv = \int_0^4 (6v^2 + 13v - 5)\, dv = \left[6 \cdot \tfrac{1}{3}v^3 + 13 \cdot \tfrac{1}{2}v^2 - 5v\right]_0^4 = \left[2v^3 + \tfrac{13}{2}v^2 - 5v\right]_0^4$

$= (128 + 104 - 20) - 0 = 212$

**35.** $\displaystyle\int_1^2 \dfrac{v^3 + 3v^6}{v^4}\, dv = \int_1^2 \left(\dfrac{v^3}{v^4} + \dfrac{3v^6}{v^4}\right)\, dv = \int_1^2 \left(\dfrac{1}{v} + 3v^2\right)\, dv = \left[\ln|v| + v^3\right]_1^2 = (\ln 2 + 8) - (\ln 1 + 1) = \ln 2 + 7$

**37.** $\displaystyle\int_1^9 \dfrac{1}{2x}\, dx = \dfrac{1}{2}\int_1^9 \dfrac{1}{x}\, dx = \tfrac{1}{2}\left[\ln|x|\right]_1^9 = \tfrac{1}{2}(\ln 9 - \ln 1) = \tfrac{1}{2}\ln 9 - 0 = \tfrac{1}{2}\ln 9$ or equivalently $\ln 9^{1/2} = \ln 3$

**39.** $\displaystyle\int_0^1 5(2^z)\, dz = 5\left[\dfrac{1}{\ln 2}2^z\right]_0^1 = \dfrac{5}{\ln 2}(2 - 1) = \dfrac{5}{\ln 2}$

**41.** $4x + 0.3e^x > 0$ for $1 \le x \le 4$, so the area under the graph is

$\int_1^4 (4x + 0.3e^x)\, dx = 2x^2 + 0.3e^x\Big]_1^4 = (2 \cdot 4^2 + 0.3e^4) - (2 \cdot 1^2 + 0.3e^1) = 32 + 0.3e^4 - 2 - 0.3e$

$= 30 + 0.3e^4 - 0.3e \approx 45.56$

**43.** $5\sqrt{t} \geq 0$, so the area under the graph is

$$\int_0^4 5\sqrt{t}\,dt = \int_0^4 5t^{1/2}\,dt = \left[5\left(\frac{t^{3/2}}{3/2}\right)\right]_0^4 = \left[\frac{10}{3}t^{3/2}\right]_0^4 = \frac{10}{3}(4^{3/2} - 0) = \frac{10}{3}\cdot 8 = \frac{80}{3} \approx 26.67$$

**45.** The graph shows that $y = x + x^2 - x^4$ has $x$-intercepts at $x = 0$ and at
$x \approx 1.32$. So the area of the region that lies under the curve and above the
$x$-axis is approximately

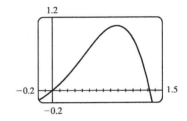

$$\int_0^{1.32}(x + x^2 - x^4)\,dx = \left[\frac{1}{2}x^2 + \frac{1}{3}x^3 - \frac{1}{5}x^5\right]_0^{1.32}$$
$$= \left[\frac{1}{2}(1.32)^2 + \frac{1}{3}(1.32)^3 - \frac{1}{5}(1.32)^5\right] - 0$$
$$\approx 0.84$$

**47.** $\int_{-1}^2 x^3\,dx = \left[\frac{1}{4}x^4\right]_{-1}^2 = 4 - \frac{1}{4} = \frac{15}{4} = 3.75.$

The value of the integral is the area labeled with a plus sign minus
the area labeled with a minus sign.

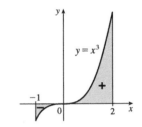

**49.** $\int_0^3 (t^2 + 2t - 3)\,dt = \frac{t^3}{3} + 2\cdot\frac{t^2}{2} - 3t\bigg]_0^3 = \frac{1}{3}t^3 + t^2 - 3t\bigg]_0^3$

$$= \left[\frac{1}{3}(3)^3 + 3^2 - 3(3)\right] - 0 = 9 + 9 - 9 = 9$$

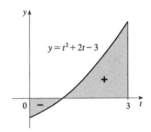

**51.** $\int 12x^3\,dx = 12\cdot\frac{x^4}{4} + C = 3x^4 + C$

**53.** $\int (t^2 + 3t + 4)\,dt = \frac{t^3}{3} + 3\cdot\frac{t^2}{2} + 4t + C = \frac{1}{3}t^3 + \frac{3}{2}t^2 + 4t + C$

**55.** $\int (0.01q^3 + 0.6q^2 + 3.5q + 14.9)\,dq = 0.01\left(\frac{1}{4}q^4\right) + 0.6\left(\frac{1}{3}q^3\right) + 3.5\left(\frac{1}{2}q^2\right) + 14.9q + C$
$$= 0.0025q^4 + 0.2q^3 + 1.75q^2 + 14.9q + C$$

**57.** $\int 5e^{2t}\,dt = 5\left(\frac{1}{2}e^{2t}\right) + C = \frac{5}{2}e^{2t} + C$

**59.** $\int (6u^2 - 3\sqrt{u})\,du = \int (6u^2 - 3u^{1/2})\,du = 6\left(\frac{1}{3}u^3\right) - 3\left(\frac{2}{3}u^{3/2}\right) + C = 2u^3 - 2u^{3/2} + C$

**61.** $\int (1 - t)(2 + t^2)\,dt = \int (2 - 2t + t^2 - t^3)\,dt = 2t - 2\cdot\frac{t^2}{2} + \frac{t^3}{3} - \frac{t^4}{4} + C$
$$= 2t - t^2 + \frac{1}{3}t^3 - \frac{1}{4}t^4 + C$$

**63.** $\int x\sqrt{x}\,dx = \int x\cdot x^{1/2}\,dx = \int x^{3/2}\,dx = \dfrac{x^{5/2}}{5/2}+C$

$\qquad = \frac{2}{5}x^{5/2}+C$

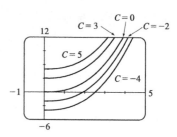

**65.** (a) $\int f(x)\,dx = \int\left(\frac{1}{2}x^3+4x-1\right)dx = \frac{1}{2}\left(\frac{1}{4}x^4\right)+4\left(\frac{1}{2}x^2\right)-x+C = \frac{1}{8}x^4+2x^2-x+C$

(b) $\int_0^2 f(x)\,dx = \int_0^2\left(\frac{1}{2}x^3+4x-1\right)dx = \left[\frac{1}{8}x^4+2x^2-x\right]_0^2 = \left[\frac{1}{8}(2)^4+2(2)^2-2\right]-\left[\frac{1}{8}(0)^4+2(0)^2-0\right] = 8$

**67.** $\dfrac{d}{dx}\left[\sqrt{x^2+1}+C\right] = \dfrac{d}{dx}\left[(x^2+1)^{1/2}+C\right] = \frac{1}{2}(x^2+1)^{-1/2}(2x)+0 = \dfrac{x}{\sqrt{x^2+1}}$

**69.** $\int_0^9 [2f(x)+3g(x)]\,dx = 2\int_0^9 f(x)\,dx + 3\int_0^9 g(x)\,dx = 2(37)+3(16) = 122$

**71.** $\int_{-2}^2 f(x)\,dx + \int_2^5 f(x)\,dx - \int_{-2}^{-1} f(x)\,dx = \int_{-2}^5 f(x)\,dx - \int_{-2}^{-1} f(x)\,dx$  [by Property 8]

$\qquad\qquad = \int_{-2}^5 f(x)\,dx + \int_{-1}^{-2} f(x)\,dx$  [by Property 7]

$\qquad\qquad = \int_{-1}^5 f(x)\,dx$  [by Property 8]

**73.** The area is $\int_0^2 \left(2y-y^2\right)dy = \left[y^2-\frac{1}{3}y^3\right]_0^2 = \left(4-\frac{8}{3}\right)-0 = \frac{4}{3}$

**75.** $\int_0^8 g(x)\,dx = \int_0^5 g(x)\,dx + \int_5^8 g(x)\,dx = \int_0^5 (0.2x^2+4)\,dx + \int_5^8 (3x-6)\,dx$

$\qquad = \left[\dfrac{0.2}{3}x^3+4x\right]_0^5 + \left[\frac{3}{2}x^2-6x\right]_5^8 = \left[\dfrac{0.2}{3}(125)+20-0\right] + \left[\frac{3}{2}(64)-48-\frac{3}{2}(25)+30\right]$

$\qquad = \frac{85}{3}+\frac{81}{2} = \frac{413}{6} \approx 68.83$

**77.** The second derivative is the derivative of the first derivative, so $h'$ is an antiderivative of $h''$. Thus

$\int_1^2 h''(u)\,du = h'(u)\Big]_1^2 = h'(2)-h'(1) = 5-2 = 3$. The other information is unnecessary.

## 5.3  The Net Change Theorem and Average Value

**1.** By the Net Change Theorem, $\int_{300}^{500} C'(q)\,dq = C(500)-C(300)$, so it represents the increase in cost, in thousands of dollars, when production is increased from 300 guitars to 500 guitars.

**3.** If $w'(t)$ is the rate of growth in pounds per year, then $w(t)$ represents the weight in pounds of the child at age $t$. We know from the Net Change Theorem that $\int_5^{10} w'(t)\,dt = w(10)-w(5)$, so the integral represents the increase in the child's weight (in pounds) between the ages of 5 and 10.

**5.** Since $r(t)$ is the rate at which oil leaks, $\int_0^{120} r(t)\,dt$ represents the net change in the amount of leaked oil from $t=0$ to $t=120$. In other words, it is the number of gallons of oil that leaked from the tank in the first two hours (120 minutes).

**7.** The slope of the trail is the rate of change of the elevation $E$, so $f(x) = E'(x)$. By the Net Change Theorem, $\int_3^5 f(x)\,dx = \int_3^5 E'(x)\,dx = E(5) - E(3)$ is the change in the elevation $E$ between $x = 3$ (horizontal) miles and $x = 5$ miles from the start of the trail.

**9.** $f(t)$ is the rate of data flow, in mB/s, after $t$ seconds, so the units for $\int_a^b f(t)\,dt$ are $(\text{mB/s}) \cdot \text{s} = \text{mB}$, and the integral represents the net change in the amount of data that flowed through the router, in mB, between $a$ seconds and $b$ seconds after midnight. Thus $\int_{21,600}^{28,800} f(t)\,dt = 18{,}350$ means that the total amount of data that flowed through the router between 21,600 seconds (6 hours) and 28,800 seconds (8 hours) after midnight was 18,350 mB. In other words, between 6:00 AM and 8:00 AM, 18.35 GB of data flowed through the router.

**11.** From the Net Change Theorem, the increase in cost if the production level is raised from 2000 yards to 4000 yards is

$$\int_{2000}^{4000} C'(y)\,dy = \int_{2000}^{4000}(3 - 0.01y + 0.000006y^2)\,dy = \Big[3y - 0.005y^2 + 0.000002y^3\Big]_{2000}^{4000}$$

$$= 60{,}000 - 2{,}000 = 58{,}000$$

When the production is raised from 2000 yards to 4000 yards, the cost increases by $58,000.

**13.** According to the Net Change Theorem, the net change in the number of subscribers in 2010 is

$$\int_0^{12} s(t)\,dt = \int_0^{12}(-15{,}000e^{-0.04t})\,dt = -15{,}000\left(\frac{1}{-0.04}\,e^{-0.04t}\right)\Bigg]_0^{12} = 375{,}000(e^{-0.04(12)} - e^0)$$

$$= 375{,}000(e^{-0.48} - 1) \approx -142{,}956$$

Thus the newspaper lost 142,956 subscribers during 2010.

**15.** By the Net Change Theorem, the change (in thousands) in the bacteria population is

$$\int_0^2 r(t)\,dt = \int_0^2 6e^{0.3t}\,dt = 6\left(\frac{1}{0.3}\,e^{0.3t}\right)\Bigg]_0^2 = 20e^{0.3t}\Big]_0^2 = 20e^{0.6} - 20e^0$$

$$= 20(e^{0.6} - 1) \approx 16.442$$

After two hours, the population increased by 16,442 bacteria.

**17.** $r(t)$ is the rate at which the oil leaks, so the amount of oil that leaks during the first 60 minutes is

$$\int_0^{60} r(t)\,dt = \int_0^{60} 100e^{-0.01t}\,dt = 100\left(\frac{1}{-0.01}\,e^{-0.01t}\right)\Bigg]_0^{60} = -10{,}000(e^{-0.6} - e^0)$$

$$= -10{,}000(e^{-0.6} - 1) \approx 4512 \text{ liters}$$

**19.** The units of $P(t)$ are thousands of people per year and the units of $t$ are years, so the units of the integral $\int_a^b P(t)\,dt$ are (thousands of people/year)(years) = thousands of people.

**21.** As in Example 3, power is the rate of change of energy. By the Net Change Theorem, the total amount of energy consumed during that day is

$$\int_0^{24} p(t)\,dt = \int_0^{24}(-0.016t^3 + 0.44t^2 - 1.4t + 12.1)\,dt$$

$$= \Big[-0.016\big(\tfrac{1}{4}t^4\big) + 0.44\big(\tfrac{1}{3}t^3\big) - 1.4\big(\tfrac{1}{2}t^2\big) + 12.1t\Big]_0^{24} \approx 587.6 \text{ kilowatt-hours}$$

**23.** Acceleration is the rate of change of velocity, so the net change in velocity from $t = 5$ to $t = 8$ is

$$\int_5^8 a(t)\, dt = \int_5^8 (0.4 + 0.12t)\, dt = \left[0.4t + 0.06t^2\right]_5^8 = 7.04 - 3.5 = 3.54$$

The units are $(\text{ft/s}^2) \times \text{s} = \text{ft/s}$, so the hawk's velocity (speed) increased 3.54 ft/s.

**25.** (a) By Equation 1, the displacement is $\int_0^3 v(t)\, dt = \int_0^3 (3t - 5)\, dt = \left[\frac{3}{5}t^2 - 5t\right]_0^3 = \frac{27}{2} - 15 = -\frac{3}{2}$ m

(b) By Equation 2, the distance traveled is $\int_0^3 |v(t)|\, dt = \int_0^3 |3t - 5|\, dt$. Now $3t - 5 \geq 0$ for $t \geq \frac{5}{3}$, and $3t - 5 < 0$ for

$t < \frac{5}{3}$. Thus

$$\int_0^3 |v(t)|\, dt = \int_0^{5/3} [-v(t)]\, dt + \int_{5/3}^3 v(t)\, dt = \int_0^{5/3}(5 - 3t)\, dt + \int_{5/3}^3 (3t - 5)\, dt$$

$$= \left[5t - \frac{3}{2}t^2\right]_0^{5/3} + \left[\frac{3}{2}t^2 - 5t\right]_{5/3}^3 = \frac{25}{3} - \frac{3}{2}\cdot\frac{25}{9} + \frac{27}{2} - 15 - \left(\frac{3}{2}\cdot\frac{25}{9} - \frac{25}{3}\right) = \frac{41}{6}$$ m

**27.** (a) $v'(t) = a(t) = t + 4 \ \Rightarrow\ v(t) = \int (t + 4)\, dt = \frac{1}{2}t^2 + 4t + C$. Then $v(0) = C$, but we were given that $v(0) = 5$, so

$C = 5$ and $v(t) = \frac{1}{2}t^2 + 4t + 5$ m/s.

(b) Note that $v(t) > 0$ for $0 \leq t \leq 10$, so $|v(t)| = v(t)$ and

Distance traveled $= \int_0^{10} |v(t)|\, dt = \int_0^{10}\left(\frac{1}{2}t^2 + 4t + 5\right) dt = \left[\frac{1}{6}t^3 + 2t^2 + 5t\right]_0^{10} = \frac{500}{3} + 200 + 50 = 416\frac{2}{3}$ m.

**29.** $f_{\text{ave}} = \frac{1}{b-a}\int_a^b f(x)\, dx = \frac{1}{4-0}\int_0^4 (4x - x^2)\, dx = \frac{1}{4}\left[2x^2 - \frac{1}{3}x^3\right]_0^4 = \frac{1}{4}\left[\left(32 - \frac{64}{3}\right) - 0\right] = \frac{1}{4}\left(\frac{32}{3}\right) = \frac{8}{3}$

**31.** $g_{\text{ave}} = \frac{1}{b-a}\int_a^b g(x)\, dx = \frac{1}{8-1}\int_1^8 \sqrt[3]{x}\, dx = \frac{1}{7}\int_1^8 x^{1/3}\, dx = \frac{1}{7}\left[\frac{3}{4}x^{4/3}\right]_1^8 = \frac{1}{7}\cdot\frac{3}{4}[8^{4/3} - 1^{4/3}] = \frac{3}{28}(16 - 1) = \frac{45}{28}$

**33.** The period from 9 AM to 9 PM corresponds to $0 \leq t \leq 12$. Thus the average temperature is

$$T_{\text{ave}} = \frac{1}{b-a}\int_a^b T(t)\, dt = \frac{1}{12-0}\int_0^{12}(57 - 2.4t + 0.43t^2 - 0.014t^3)\, dt$$

$$= \frac{1}{12}\left[57t - 2.4\left(\frac{1}{2}t^2\right) + 0.43\left(\frac{1}{3}t^3\right) - 0.014\left(\frac{1}{4}t^4\right)\right]_0^{12}$$

$$= \frac{1}{12}\left[57(12) - 1.2(12)^2 + \frac{0.43}{3}(12)^3 - 0.0035(12)^4 - 0\right] \approx 57.2\,°\text{F}$$

**35.** $C_{\text{ave}} = \frac{1}{b-a}\int_a^b C(t)\, dt = \frac{1}{20-0}\int_0^{20} 8(e^{-0.05t} - e^{-0.4t})\, dt = \frac{1}{20}\left[8\left(\frac{1}{-0.05}e^{-0.05t} - \frac{1}{-0.4}e^{-0.4t}\right)\right]_0^{20}$

$= \frac{8}{20}\left[-20e^{-0.05t} + 2.5e^{-0.4t}\right]_0^{20} = 0.4[-20e^{-1} + 2.5e^{-8} + 20 - 2.5] \approx 4.06$ mg/L

**37.** By the Net Change Theorem, the change in the amount of water after four days is $\int_0^4 r(t)\, dt$. Using the Midpoint Rule with

$n = 4$ subintervals, we have $\Delta t = \frac{4-0}{4} = 1$ and

$$\int_0^4 r(t)\, dt \approx \Delta t\, [r(0.5) + r(1.5) + r(2.5) + r(3.5)] \approx (1)[1500 + 1770 + 740 + (-690)] = 3320 \text{ liters}$$

Thus the total amount of water four days later is $25{,}000 + 3320 = 28{,}320$ liters.

**39.** We use the Midpoint Rule with $n = 3$ subintervals to estimate the value of $\int_{20}^{50} f(x)\, dx$. Thus $\Delta x = \frac{50-20}{3} = 10$, and

$$\int_{20}^{50} f(x)\, dx \approx \Delta x[f(\overline{x}_1) + f(\overline{x}_2) + f(\overline{x}_3)] = 10\,[f(25) + f(35) + f(45)] = 10(38 + 29 + 48) = 1150$$

Then $f_{\text{ave}} = \frac{1}{50-20}\int_{20}^{50} f(x)\, dx \approx \frac{1}{30}(1150) \approx 38.3$.

**41.** The average value of the marginal cost function for $a \leq q \leq b$ is $\dfrac{1}{b-a}\displaystyle\int_a^b C'(q)\,dq = \dfrac{1}{b-a}[C(b) - C(a)]$ by the Net

Change Theorem. Because $C(q)$ is the total cost function, $\dfrac{1}{b-a}[C(b) - C(a)] = \dfrac{C(b) - C(a)}{b-a}$ is the average rate of change

of total cost for $a \leq q \leq b$.

## 5.4  The Substitution Rule

### Prepare Yourself

**1.** (a) $y = e^{x^3+1} \;\Rightarrow\; y' = e^{x^3+1} \cdot \dfrac{d}{dx}(x^3+1) = e^{x^3+1}(3x^2) = 3x^2 e^{x^3+1}$

(b) $Q(t) = \ln(3t + t^2) \;\Rightarrow\; Q'(t) = \dfrac{1}{3t + t^2} \cdot \dfrac{d}{dt}(3t + t^2) = \dfrac{3 + 2t}{3t + t^2}$

(c) $f(x) = (2x^2 + 3)^4 \;\Rightarrow\; f'(x) = 4(2x^2 + 3)^3 \cdot \dfrac{d}{dx}(2x^2 + 3) = 4(2x^2 + 3)^3(4x) = 16x(2x^2 + 3)^3$

(d) $g(z) = \sqrt{e^z + 5z} = (e^z + 5z)^{1/2} \;\Rightarrow\; g'(z) = \tfrac{1}{2}(e^z + 5z)^{-1/2} \cdot \dfrac{d}{dz}(e^z + 5z) = \dfrac{e^z + 5}{2\sqrt{e^z + 5z}}$

(e) $r = 3^{2t+2} \;\Rightarrow\; dr/dt = 3^{2t+2}(\ln 3) \cdot \dfrac{d}{dt}(2t + 2) = 3^{2t+2}(\ln 3) \cdot 2 = 3^{2t+2}(2\ln 3)$

**2.** (a) $\int (3x^5 + 4x - 1)\,dx = 3\left(\tfrac{1}{6}x^6\right) + 4\left(\tfrac{1}{2}x^2\right) - x + C = \tfrac{1}{2}x^6 + 2x^2 - x + C$

(b) $\int 8\sqrt{t}\,dt = \int 8t^{1/2}\,dt = 8 \cdot \dfrac{t^{3/2}}{3/2} + C = \tfrac{16}{3}t^{3/2} + C$

(c) $\displaystyle\int (5/v)\,dv = 5\int \dfrac{1}{v}\,dv = 5\ln|v| + C$

(d) $\int (5/v^2)\,dv = 5\int v^{-2}\,dv = 5(-v^{-1}) + C = -5/v + C$

(e) $\displaystyle\int 4^x\,dx = \dfrac{1}{\ln 4}4^x + C$

**3.** (a) One possibility is to let $g(x) = 3x^2 + 2$; then $f(x) = x^4$, so that $f(g(x)) = f(3x^2 + 2) = (3x^2 + 2)^4$.

(b) One possibility is $f(x) = \sqrt{x}$ and $g(x) = x^3 + 8$.

(c) One possibility is $f(x) = 1/x$ and $g(x) = x^3 - 2$.

(d) One possibility is $f(x) = e^x$ and $g(x) = x^2 + 1$.

### Exercises

**1.** Let $u = -x$. Then $du = -dx$, so $dx = -du$. Thus, $\int e^{-x}\,dx = \int e^u(-du) = -\int e^u\,du = -e^u + C = -e^{-x} + C$.

Don't forget that it is often very easy to check an indefinite integration by differentiating your answer. In this case,

$\dfrac{d}{dx}(-e^{-x} + C) = -[e^{-x}(-1)] = e^{-x}$, the desired result.

**3.** Let $u = x^3 + 1$. Then $du = 3x^2\,dx$ and $x^2\,dx = \frac{1}{3}\,du$, so

$$\int x^2 \sqrt{x^3 + 1}\,dx = \int \sqrt{u}\left(\tfrac{1}{3}\,du\right) = \frac{1}{3} \cdot \frac{u^{3/2}}{3/2} + C = \tfrac{1}{3} \cdot \tfrac{2}{3}u^{3/2} + C = \tfrac{2}{9}(x^3 + 1)^{3/2} + C.$$

**5.** Let $u = 1 + 4p^2$. Then $du = 8p\,dp$ and $p\,dp = \frac{1}{8}\,du$, so

$$\int \frac{p}{1 + 4p^2}\,dp = \int \frac{1}{u} \cdot \frac{1}{8}\,du = \frac{1}{8}\int \frac{1}{u}\,du = \tfrac{1}{8}\ln|u| + C = \tfrac{1}{8}\ln\left|1 + 4p^2\right| + C = \tfrac{1}{8}\ln(1 + 4p^2) + C \ \text{ (since } 1 + 4p^2 > 0\text{)}.$$

**7.** Let $u = 3 - t^2$. Then $du = -2t\,dt$ and $t\,dt = -\frac{1}{2}\,du$, so

$$\int t(3 - t^2)^4\,dt = \int (3 - t^2)^4 \cdot t\,dt = \int u^4\left(-\tfrac{1}{2}\,du\right) = -\tfrac{1}{2}\left(\tfrac{1}{5}u^5\right) + C = -\tfrac{1}{10}u^5 + C = -\tfrac{1}{10}(3 - t^2)^5 + C.$$

**9.** Let $u = 3x - 2$. Then $du = 3\,dx$ and $dx = \frac{1}{3}\,du$, so

$$\int (3x - 2)^{20}\,dx = \int u^{20}\left(\tfrac{1}{3}\,du\right) = \tfrac{1}{3}\int u^{20}\,du = \tfrac{1}{3} \cdot \tfrac{1}{21}u^{21} + C = \tfrac{1}{63}(3x - 2)^{21} + C.$$

**11.** Let $u = q^2 + 3.1$. Then $du = 2q\,dq$ and $q\,dq = \frac{1}{2}\,du$, so

$$\int q\sqrt{q^2 + 3.1}\,dq = \int \sqrt{q^2 + 3.1}\,(q\,dq) = \int u^{1/2}\left(\tfrac{1}{2}\,du\right) = \tfrac{1}{2}\int u^{1/2}\,du = \tfrac{1}{2}\left(\tfrac{2}{3}u^{3/2}\right) + C = \tfrac{1}{3}(q^2 + 3.1)^{3/2} + C.$$

**13.** Let $u = 0.4x^3 + 2.2$. Then $du = 1.2x^2\,dx = \frac{6}{5}x^2\,dx$ and $x^2\,dx = \frac{5}{6}\,du$, so

$$\int \frac{x^2}{\sqrt{0.4x^3 + 2.2}}\,dx = \int \frac{1}{\sqrt{0.4x^3 + 2.2}}\,(x^2\,dx) = \int \frac{1}{\sqrt{u}}\left(\frac{5}{6}\,du\right) = \frac{5}{6}\int u^{-1/2}\,du$$

$$= \frac{5}{6} \cdot \frac{u^{1/2}}{1/2} + C = \tfrac{5}{3}\sqrt{u} + C = \tfrac{5}{3}\sqrt{0.4x^3 + 2.2} + C$$

**15.** Let $u = z^3 + 2$. Then $du = 3z^2\,dz$ and $z^2\,dz = \frac{1}{3}\,du$, so

$$\int \frac{5z^2}{(z^3 + 2)^3}\,dz = 5\int \frac{1}{(z^3 + 2)^3}\,(z^2\,dz) = 5\int \frac{1}{u^3}\left(\frac{1}{3}\,du\right) = \frac{5}{3}\int u^{-3}\,du$$

$$= \tfrac{5}{3}\left(-\tfrac{1}{2}u^{-2}\right) + C = -\frac{5}{6u^2} + C = -\frac{5}{6(z^3 + 2)^2} + C$$

**17.** Let $u = 1 + e^x$. Then $du = e^x\,dx$, so $\int e^x \sqrt{1 + e^x}\,dx = \int \sqrt{u}\,du = \int u^{1/2}\,du = \tfrac{2}{3}u^{3/2} + C = \tfrac{2}{3}(1 + e^x)^{3/2} + C.$

**19.** Let $u = \ln x$. Then $du = \frac{1}{x}\,dx$, so $\int \frac{(\ln x)^2}{x}\,dx = \int u^2\,du = \tfrac{1}{3}u^3 + C = \tfrac{1}{3}(\ln x)^3 + C.$

**21.** Let $u = 2t^2$. Then $du = 4t\,dt$ and $t\,dt = \frac{1}{4}\,du$, so $\int te^{2t^2}\,dt = \int e^u\left(\tfrac{1}{4}\,du\right) = \tfrac{1}{4}\int e^u\,du = \tfrac{1}{4}e^u + C = \tfrac{1}{4}e^{2t^2} + C.$

**23.** Let $u = 5 - 3x$. Then $du = -3\,dx$ and $dx = -\frac{1}{3}\,du$, so

$$\int \frac{dx}{5 - 3x} = \int \frac{1}{u}\left(-\tfrac{1}{3}\,du\right) = -\frac{1}{3}\int \frac{1}{u}\,du = -\tfrac{1}{3}\ln|u| + C = -\tfrac{1}{3}\ln|5 - 3x| + C.$$

**25.** Let $u = \sqrt{t} + 1$. Then $du = \frac{1}{2}t^{-1/2}\,dt = \frac{1}{2\sqrt{t}}\,dt$ and $\frac{1}{\sqrt{t}}\,dt = 2\,du$, so

$$\int \frac{e^{\sqrt{t}+1}}{\sqrt{t}}\,dt = \int e^u(2\,du) = 2e^u + C = 2e^{\sqrt{t}+1} + C.$$

**27.** Let $u = 3 - 4t$. Then $du = -4\,dt$ and $dt = -\frac{1}{4}\,du$, so

$$\int 2^{3-4t}\,dt = \int 2^u\left(-\tfrac{1}{4}\,du\right) = -\frac{1}{4}\left(\frac{1}{\ln 2}\,2^u\right) + C = -\frac{1}{4\ln 2}\,2^{3-4t} + C.$$

**29.** Let $u = x^3 + 3x$. Then $du = (3x^2 + 3)\,dx = 3(x^2 + 1)\,dx$ and $\frac{1}{3}\,du = (x^2 + 1)\,dx$, so

$$\int (x^2 + 1)(x^3 + 3x)^4\,dx = \int u^4\left(\tfrac{1}{3}\,du\right) = \tfrac{1}{3}\cdot\tfrac{1}{5}u^5 + C = \tfrac{1}{15}(x^3 + 3x)^5 + C.$$

**31.** $f(x) = x(x^2 - 1)^3$. Let $u = x^2 - 1$; then $du = 2x\,dx$ and $\frac{1}{2}\,du = x\,dx$, so

$$\int x(x^2 - 1)^3\,dx = \int u^3\left(\tfrac{1}{2}\,du\right) = \tfrac{1}{2}\cdot\tfrac{1}{4}u^4 + C = \tfrac{1}{8}(x^2 - 1)^4 + C$$

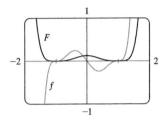

Where $f$ is positive, $F$ is increasing, and where $f$ is negative, $F$ is decreasing. Where $f$ changes from negative to positive, $F$ has a local minimum; $F$ has a local maximum where $f$ changes from positive to negative.

**33.** Let $u = 1 + 7x$. Then $du = 7\,dx$ and $\frac{1}{7}\,du = dx$. When $x = 0$, $u = 1$; when $x = 1$, $u = 8$. Thus

$$\int_0^1 \sqrt[3]{1 + 7x}\,dx = \int_1^8 \sqrt[3]{u}\left(\tfrac{1}{7}\,du\right) = \tfrac{1}{7}\int_1^8 u^{1/3}\,du = \tfrac{1}{7}\left[\tfrac{3}{4}u^{4/3}\right]_1^8 = \tfrac{3}{28}(8^{4/3} - 1^{4/3}) = \tfrac{3}{28}(16 - 1) = \tfrac{45}{28}$$

**35.** Let $u = 1 + 2x^3$, so $du = 6x^2\,dx \Rightarrow \frac{1}{6}\,du = x^2\,dx$. When $x = 0$, $u = 1$; when $x = 1$, $u = 3$. Thus

$$\int_0^1 x^2\left(1 + 2x^3\right)^5\,dx = \int_1^3 u^5\left(\tfrac{1}{6}\,du\right) = \tfrac{1}{6}\left[\tfrac{1}{6}u^6\right]_1^3 = \tfrac{1}{36}(3^6 - 1^6) = \tfrac{1}{36}(729 - 1) = \tfrac{728}{36} = \tfrac{182}{9}$$

**37.** Let $u = 3v + 1$, so $du = 3\,dv \Rightarrow \frac{1}{3}\,du = dv$. When $v = 0$, $u = 1$; when $v = 1$, $u = 4$. Thus

$$\int_0^1 \frac{1}{(3v + 1)^2}\,dv = \int_1^4 \frac{1}{u^2}\left(\tfrac{1}{3}\,du\right) = \tfrac{1}{3}\int_1^4 u^{-2}\,du = \left[\tfrac{1}{3}(-u^{-1})\right]_1^4 = \left[-\frac{1}{3u}\right]_1^4 = -\tfrac{1}{12} - \left(-\tfrac{1}{3}\right) = \tfrac{1}{4}$$

**39.** Let $u = z^2 - 1$, so $du = 2z\,dz \Rightarrow \frac{1}{2}\,du = z\,dz$. When $z = 1$, $u = 0$; when $z = 3$, $u = 8$. Thus

$$\int_1^3 4ze^{z^2-1}\,dz = 4\int_1^3 e^{z^2-1}(z\,dz) = 4\int_0^8 e^u\left(\tfrac{1}{2}\,du\right) = 2e^u\big]_0^8 = 2(e^8 - 1)$$

**41.** Let $u = \sqrt{x} = x^{1/2}$, so $du = \frac{1}{2}x^{-1/2}\,dx = \dfrac{1}{2\sqrt{x}}\,dx \Rightarrow 2\,du = \dfrac{1}{\sqrt{x}}\,dx$. When $x = 1$, $u = 1$; when $x = 4$, $u = 2$.

Thus, $\displaystyle\int_1^4 \frac{e^{\sqrt{x}}}{\sqrt{x}}\,dx = \int_1^2 e^u(2\,du) = 2\big[e^u\big]_1^2 = 2(e^2 - e).$

**43.** Let $u = \ln x$, so $du = \dfrac{1}{x}\,dx$. When $x = e$, $u = 1$; when $x = e^4$, $u = 4$. Thus

$$\int_e^{e^4} \frac{dx}{x\sqrt{\ln x}} = \int_1^4 \frac{1}{\sqrt{u}}\,du = \int_1^4 u^{-1/2}\,du = \left[\frac{u^{1/2}}{1/2}\right]_1^4 = 2\left[\sqrt{u}\right]_1^4 = 2\left(\sqrt{4} - \sqrt{1}\right) = 2(2 - 1) = 2.$$

**45.** If $f(x) = 3x^8 + x^4$, then $f(-x) = 3(-x)^8 + (-x)^4 = 3x^8 + x^4 = f(x)$, so $f(x)$ is even. Then, by Theorem 5,

$$\int_{-1}^1 f(x)\,dx = 2\int_0^1 f(x)\,dx = 2\int_0^1 (3x^8 + x^4)\,dx = 2\left[\tfrac{1}{3}x^9 + \tfrac{1}{5}x^5\right]_0^1 = 2\left[\left(\tfrac{1}{3} + \tfrac{1}{5}\right) - 0\right] = \tfrac{16}{15}.$$

**47.** If $f(t) = \dfrac{t^3}{t^6 + 1}$, then $f(-t) = \dfrac{(-t)^3}{(-t)^6 + 1} = \dfrac{-t^3}{t^6 + 1} = -f(t)$, so $f(t)$ is odd. By Theorem 5,

$$\int_{-2}^{2} f(t)\, dt = \int_{-2}^{2} \frac{t^3}{t^6 + 1}\, dt = 0.$$

**49.** The number of downloaded copies, in thousands, is $\displaystyle\int_0^6 50\left(\dfrac{t}{2t^2 + 5}\right) dt$. Let $u = 2t^2 + 5$, then $du = 4t\, dt \;\Rightarrow$

$\frac{1}{4}\, du = t\, dt$. When $t = 0$, $u = 5$; when $t = 6$, $u = 77$. Then

$$\int_0^6 50\left(\frac{t}{2t^2 + 5}\right) dt = 50\int_5^{77}\left(\frac{1}{u}\right)\left(\tfrac{1}{4}\, du\right) = \tfrac{25}{2}\Big[\ln|u|\Big]_5^{77} = \tfrac{25}{2}(\ln 77 - \ln 5) \approx 34.18$$

Thus about 34,200 copies were downloaded during the first six weeks.

**51.** The number of gallons added is $\int_0^{10} 8te^{-0.026t^2}\, dt$. Let $u = -0.026t^2$; then $du = -0.052t\, dt \;\Rightarrow\; -\dfrac{1}{0.052}\, du = t\, dt$.

When $t = 0$, $u = 0$; when $t = 10$, $u = -2.6$. Then

$$\int_0^{10} 8te^{-0.026t^2}\, dt = 8\int_0^{-2.6} e^u\left(-\frac{1}{0.052}\, du\right) = -\frac{8}{0.052}\Big[e^u\Big]_0^{-2.6} = -\frac{8}{0.052}(e^{-2.6} - 1) \approx 142.4$$

Thus about 142.4 gallons are added to the tank during the first 10 minutes.

**53.** Let $u = 2x + 5$. Then $x = \frac{1}{2}(u - 5)$ and $du = 2\, dx \;\Rightarrow\; \frac{1}{2}\, du = dx$.

$$\int x(2x + 5)^8\, dx = \int \tfrac{1}{2}(u - 5)u^8\left(\tfrac{1}{2}\, du\right) = \tfrac{1}{4}\int(u^9 - 5u^8)\, du$$

$$= \tfrac{1}{4}\left(\tfrac{1}{10}u^{10} - \tfrac{5}{9}u^9\right) + C = \tfrac{1}{40}(2x + 5)^{10} - \tfrac{5}{36}(2x + 5)^9 + C$$

**55.** Let $u = 2x$. Then $du = 2\, dx \;\Rightarrow\; \frac{1}{2}\, du = dx$. When $x = 0$, $u = 0$; when $x = 2$, $u = 4$. Thus

$\int_0^2 f(2x)\, dx = \int_0^4 f(u)\left(\frac{1}{2}\, du\right) = \frac{1}{2}\int_0^4 f(u)\, du = \frac{1}{2}(10) = 5$.

**57.** First write the integral as a sum of two integrals: $\int_{-2}^2 (x + 3)\sqrt{4 - x^2}\, dx = \int_{-2}^2 x\sqrt{4 - x^2}\, dx + \int_{-2}^2 3\sqrt{4 - x^2}\, dx$. The first

integral is 0 by Theorem 5(b), since $f(x) = x\sqrt{4 - x^2}$ is an odd function and we are integrating from $x = -2$ to $x = 2$. The

second integral can be interpreted as three times the area of a half-circle with radius 2, so its value is $3 \cdot \frac{1}{2}\left(\pi \cdot 2^2\right) = 6\pi$. Thus

$\int_{-2}^2 (x + 3)\sqrt{4 - x^2}\, dx = 0 + 6\pi = 6\pi$.

**59.** The area in the first figure is $\int_0^1 e^{\sqrt{x}}\, dx$. Let $u = \sqrt{x}$, so $x = u^2$ and $dx = 2u\, du$. When $x = 0$, $u = 0$; when $x = 1$, $u = 1$.

Thus, $\int_0^1 e^{\sqrt{x}}\, dx = \int_0^1 e^u(2u\, du) = \int_0^1 2ue^u\, du$. If we now change the variable $u$ to $x$, we get $\int_0^1 2xe^x\, dx$, the area

represented in the second figure.

**61.** Let $u = 1 - x$. Then $x = 1 - u$ and $dx = -du$. When $x = 0$, $u = 1$; when $x = 1$, $u = 0$. Thus

$$\int_0^1 x^a(1 - x)^b\, dx = \int_1^0 (1 - u)^a\, u^b(-du) = -\int_1^0 [-u^b(1 - u)^a]\, du = \int_0^1 u^b(1 - u)^a\, du$$

$$= \int_0^1 x^b(1 - x)^a\, dx \qquad \text{[changing the variable } u \text{ to } x]$$

## 5.5   Integration by Parts

**1.** Let $u = \ln x$, $dv = x\,dx$ $\Rightarrow$ $du = \dfrac{1}{x}\,dx$, $v = \frac{1}{2}x^2$. Then, by Equation 2,

$$\int x \ln x\,dx = (\ln x)\left(\tfrac{1}{2}x^2\right) - \int \left(\tfrac{1}{2}x^2\right)\left(\dfrac{1}{x}\right)dx = \tfrac{1}{2}x^2 \ln x - \tfrac{1}{2}\int x\,dx$$

$$= \tfrac{1}{2}x^2 \ln x - \tfrac{1}{2}\cdot\tfrac{1}{2}x^2 + C = \tfrac{1}{2}x^2 \ln x - \tfrac{1}{4}x^2 + C$$

**3.** Let $u = r$, $dv = e^{r/2}\,dr$ $\Rightarrow$ $du = dr$, $v = \dfrac{1}{1/2}\,e^{r/2} = 2e^{r/2}$. Then

$$\textstyle\int re^{r/2}\,dr = r\cdot 2e^{r/2} - \int 2e^{r/2}\,dr = 2re^{r/2} - 2\int e^{r/2}\,dr = 2re^{r/2} - 2\cdot 2e^{r/2} + C = 2(r-2)e^{r/2} + C.$$

**5.** Let $u = \ln 2x$, $dv = x^3\,dx$ $\Rightarrow$ $du = \dfrac{1}{2x}\cdot 2\,dx = \dfrac{1}{x}\,dx$, $v = \frac{1}{4}x^4$. Then

$$\int x^3 \ln 2x\,dx = (\ln 2x)\left(\tfrac{1}{4}x^4\right) - \int \left(\tfrac{1}{4}x^4\right)\left(\dfrac{1}{x}\right)dx = \tfrac{1}{4}x^4 \ln 2x - \tfrac{1}{4}\int x^3\,dx$$

$$= \tfrac{1}{4}x^4 \ln 2x - \tfrac{1}{4}\left(\tfrac{1}{4}x^4\right) + C = \tfrac{1}{4}x^4 \ln 2x - \tfrac{1}{16}x^4 + C$$

**7.** Let $u = 1 - 2z$, $dv = e^{-z}\,dz$ $\Rightarrow$ $du = -2\,dz$, $v = -e^{-z}$. Then

$$\textstyle\int (1 - 2z)\,e^{-z}\,dz = (1-2z)(-e^{-z}) - \int(-e^{-z})(-2)\,dz = -(1-2z)e^{-z} - 2\int e^{-z}\,dz$$

$$= (2z - 1)e^{-z} + 2e^{-z} + C = (2z - 1 + 2)e^{-z} + C = (2z + 1)e^{-z} + C$$

**9.** Let $u = \ln \sqrt[3]{x}$, $dv = dx$ $\Rightarrow$ $du = \dfrac{1}{\sqrt[3]{x}}\left(\tfrac{1}{3}x^{-2/3}\right)dx = \dfrac{1}{3x}\,dx$, $v = x$. Then

$$\int \ln \sqrt[3]{x}\,dx = x \ln \sqrt[3]{x} - \int x\cdot\dfrac{1}{3x}\,dx = x \ln \sqrt[3]{x} - \tfrac{1}{3}x + C.$$

*Another solution:* Rewrite $\int \ln \sqrt[3]{x}\,dx = \int \ln x^{1/3}\,dx = \int \tfrac{1}{3}\ln x\,dx = \tfrac{1}{3}\int \ln x\,dx$, and apply Example 3.

**11.** Let $u = r^2$, $dv = e^{-3r}\,dr$ $\Rightarrow$ $du = 2r\,dr$, $v = -\tfrac{1}{3}e^{-3r}$. Then

$$\textstyle\int r^2 e^{-3r}\,dr = (r^2)\left(-\tfrac{1}{3}e^{-3r}\right) - \int \left(-\tfrac{1}{3}e^{-3r}\right)(2r\,dr) = -\tfrac{1}{3}r^2 e^{-3r} + \tfrac{2}{3}\int re^{-3r}\,dr.$$ Using integration by parts again,

let $u = r$, $dv = e^{-3r}\,dr$ $\Rightarrow$ $du = dr$, $v = -\tfrac{1}{3}e^{-3r}$. Then

$$\textstyle\int re^{-3r}\,dr = r\left(-\tfrac{1}{3}e^{-3r}\right) - \int\left(-\tfrac{1}{3}e^{-3r}\right)dr = -\tfrac{1}{3}re^{-3r} + \tfrac{1}{3}\int e^{-3r}\,dr = -\tfrac{1}{3}re^{-3r} - \tfrac{1}{9}e^{-3r} + C.$$ Thus

$$\int r^2 e^{-3r}\,dr = -\tfrac{1}{3}r^2 e^{-3r} + \tfrac{2}{3}\int re^{-3r}\,dr = -\tfrac{1}{3}r^2 e^{-3r} + \tfrac{2}{3}\left[-\tfrac{1}{3}re^{-3r} - \tfrac{1}{9}e^{-3r} + C\right]$$

$$= -\tfrac{1}{3}r^2 e^{-3r} - \tfrac{2}{9}re^{-3r} - \tfrac{2}{27}e^{-3r} + \tfrac{2}{3}C = -\tfrac{1}{3}r^2 e^{-3r} - \tfrac{2}{9}re^{-3r} - \tfrac{2}{27}e^{-3r} + C_1 \quad \left[\text{where } C_1 = \tfrac{2}{3}C\right]$$

$$= -e^{-3r}\left(\tfrac{1}{3}r^2 + \tfrac{2}{9}r + \tfrac{2}{27}\right) + C \quad [\text{where we renamed } C_1 \text{ to } C]$$

**13.** Let $u = x$, $dv = e^{4x}\,dx$ $\Rightarrow$ $du = dx$, $v = \frac{1}{4}e^{4x}$. Then, by Formula 4,

$$\int_0^1 xe^{4x}\,dx = x\cdot\tfrac{1}{4}e^{4x}\Big]_0^1 - \int_0^1 \tfrac{1}{4}e^{4x}\,dx = \tfrac{1}{4}xe^{4x}\Big]_0^1 - \tfrac{1}{4}\int_0^1 e^{4x}\,dx = \tfrac{1}{4}e^4 - 0 - \tfrac{1}{4}\left[\tfrac{1}{4}e^{4x}\right]_0^1$$

$$= \tfrac{1}{4}e^4 - \tfrac{1}{16}(e^4 - e^0) = \tfrac{1}{4}e^4 - \tfrac{1}{16}e^4 + \tfrac{1}{16} = \tfrac{3}{16}e^4 + \tfrac{1}{16}$$

**15.** Let $u = y$, $dv = \dfrac{1}{e^{2y}}\,dy = e^{-2y}dy \;\;\Rightarrow\;\; du = dy$, $v = -\tfrac{1}{2}e^{-2y}$. Then

$$\int_0^1 \frac{y}{e^{2y}}\,dy = y\!\left(-\tfrac{1}{2}e^{-2y}\right)\Big]_0^1 - \int_0^1 \left(-\tfrac{1}{2}e^{-2y}\right)dy = -\tfrac{1}{2}ye^{-2y}\Big]_0^1 + \tfrac{1}{2}\int_0^1 e^{-2y}\,dy$$

$$= -\tfrac{1}{2}e^{-2} + 0 + \tfrac{1}{2}\int_0^1 e^{-2y}\,dy = -\tfrac{1}{2}e^{-2} + \tfrac{1}{2}\left[-\tfrac{1}{2}e^{-2y}\right]_0^1 = -\tfrac{1}{2}e^{-2} - \tfrac{1}{4}e^{-2} + \tfrac{1}{4} = \frac{1}{4} - \frac{3}{4e^2}$$

**17.** Let $u = \ln x$, $dv = x^{-2}\,dx \;\;\Rightarrow\;\; du = \dfrac{1}{x}\,dx$, $v = -x^{-1} = -\dfrac{1}{x}$. Then

$$\int_1^2 \frac{\ln x}{x^2}\,dx = \left[(\ln x)\!\left(-\frac{1}{x}\right)\right]_1^2 - \int_1^2 \left(-\frac{1}{x}\right)\cdot\frac{1}{x}\,dx = \left[-\frac{\ln x}{x}\right]_1^2 + \int_1^2 x^{-2}\,dx$$

$$= -\tfrac{1}{2}\ln 2 + \ln 1 + \left[-\frac{1}{x}\right]_1^2 = -\tfrac{1}{2}\ln 2 + 0 - \tfrac{1}{2} + 1 = \tfrac{1}{2} - \tfrac{1}{2}\ln 2$$

**19.** Let $u = (\ln x)^2$, $dv = dx \;\;\Rightarrow\;\; du = 2(\ln x)\cdot\dfrac{1}{x}\,dx = \dfrac{2}{x}\ln x\,dx$, $v = x$. Then $\int_1^2 (\ln x)^2\,dx = \left[x(\ln x)^2\right]_1^2 - 2\int_1^2 \ln x\,dx$.

From Example 3, an antiderivative of $\ln x$ is $x\ln x - x$, so

$$\int_1^2 (\ln x)^2\,dx = \left[x(\ln x)^2\right]_1^2 - 2\left[x\ln x - x\right]_1^2 = \left[x(\ln x)^2 - 2x\ln x + 2x\right]_1^2$$

$$= [2(\ln 2)^2 - 4\ln 2 + 4] - [0 - 0 + 2] = 2(\ln 2)^2 - 4\ln 2 + 2$$

**21.** Let $u = x$, $dv = e^{-2x}\,dx \;\;\Rightarrow\;\; du = dx$, $v = -\tfrac{1}{2}e^{-2x}$. Then

$\int xe^{-2x}\,dx = -\tfrac{1}{2}xe^{-2x} + \int \tfrac{1}{2}e^{-2x}\,dx = -\tfrac{1}{2}xe^{-2x} - \tfrac{1}{4}e^{-2x} + C$. We

see from the graph that this is reasonable, since $F$ has a minimum where $f$

changes from negative to positive. Also, $F$ increases where $f$ is positive and

$F$ decreases where $f$ is negative.

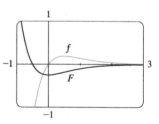

**23.** $\int_0^4 \sqrt{r^3 + r}\,dr \approx 14.133$

**25.** $\displaystyle\int_1^5 \frac{8}{1 + 2e^{0.8t}}\,dt \approx 1.936$

**27.** $2x^3 e^{-x} > 0$ for $1 \le x \le 5$, so the area under the curve $y = 2x^3 e^{-x}$ for $1 \le x \le 5$ is $\int_1^5 2x^3 e^{-x}\,dx \approx 8.592$.

**29.** (a) Since $v(t) > 0$ for all $t$, the distance traveled after $t$ seconds is $\int_0^t |v(w)|\,dw = \int_0^t w^2 e^{-w}\,dw$. (We changed the variable $t$

to $w$ in the function $v$ to avoid confusion with the limit of integration $t$.) Let $u = w^2$, $dv = e^{-w}\,dw \;\;\Rightarrow\;\; du = 2w\,dw$,

$v = -e^{-w}$. Then $\int_0^t w^2 e^{-w}\,dw = -w^2 e^{-w}\Big]_0^t - \int_0^t (-2we^{-w})\,dw = -t^2 e^{-t} + 0 + 2\int_0^t we^{-w}\,dw$. To evaluate

$\int_0^t we^{-w}\,dw$, let $u = w$, $dv = e^{-w}\,dw \;\;\Rightarrow\;\; du = dw$, $v = -e^{-w}$. Then

$$\int_0^t we^{-w}\,dw = -we^{-w}\Big]_0^t - \int_0^t (-e^{-w})\,dw = -te^{-t} + 0 + \left[-e^{-w}\right]_0^t$$

$$= -te^{-t} - e^{-t} + e^0 = -te^{-t} - e^{-t} + 1$$

Thus the distance traveled is

$$\int_0^t w^2 e^{-w}\, dw = -t^2 e^{-t} + 2(-te^{-t} - e^{-t} + 1) = -t^2 e^{-t} - 2te^{-t} - 2e^{-t} + 2$$

$$= 2 - e^{-t}(t^2 + 2t + 2) \text{ meters}$$

(b) The distanced traveled during the first ten seconds is

$$\int_0^{10} |v(w)|\, dw = 2 - e^{-10}(10^2 + 2\cdot 10 + 2) = 2 - 122e^{-10} \approx 1.99 \text{ meters}$$

**31.** Let $y = 1 + x \Rightarrow y - 1 = x$ and $dy = dx$. Then $\int x \ln(1+x)\, dx = \int (y-1)\ln y\, dy$. Now integrate by parts with

$u = \ln y$, $dv = (y - 1)\, dy$, $du = \dfrac{1}{y}\, dy$, $v = \frac{1}{2}y^2 - y$ to get

$$\int (y - 1)\ln y\, dy = \left(\tfrac{1}{2}y^2 - y\right)\ln y - \int \left(\tfrac{1}{2}y - 1\right) dy = \tfrac{1}{2}y(y - 2)\ln y - \tfrac{1}{4}y^2 + y + C$$

$$= \tfrac{1}{2}(1+x)(x-1)\ln(1+x) - \tfrac{1}{4}(1+x)^2 + 1 + x + C,$$

which can be written as $\frac{1}{2}(x^2 - 1)\ln(1+x) - \frac{1}{4}x^2 + \frac{1}{2}x + \frac{3}{4} + C$.

**33.** Let $u = (\ln x)^n$, $dv = dx \Rightarrow du = n(\ln x)^{n-1} \cdot \dfrac{1}{x}\, dx$, $v = x$. Then

$$\int (\ln x)^n\, dx = x(\ln x)^n - \int x \cdot n(\ln x)^{n-1} \cdot \frac{1}{x}\, dx = x(\ln x)^n - n\int (\ln x)^{n-1}\, dx$$

**35.** By repeated applications of the reduction formula found in Exercise 33,

$$\int (\ln x)^3\, dx = x(\ln x)^3 - 3\int (\ln x)^2\, dx = x(\ln x)^3 - 3\left[x(\ln x)^2 - 2\int (\ln x)^1\, dx\right]$$

$$= x(\ln x)^3 - 3x(\ln x)^2 + 6\left[x(\ln x)^1 - 1\int (\ln x)^0\, dx\right]$$

$$= x(\ln x)^3 - 3x(\ln x)^2 + 6x \ln x - 6\int 1\, dx = x(\ln x)^3 - 3x(\ln x)^2 + 6x \ln x - 6x + C$$

**37.** Let $u = x$, $dv = f''(x)\, dx \Rightarrow du = dx$, $v = f'(x)$. Then

$$\int_1^4 x f''(x)\, dx = \left[x f'(x)\right]_1^4 - \int_1^4 f'(x)\, dx = 4f'(4) - 1 \cdot f'(1) - \left[f(x)\right]_1^4$$

$$= 4f'(4) - f'(1) - [f(4) - f(1)] = 4 \cdot 3 - 5 - (7 - 2) = 12 - 5 - 5 = 2$$

(We used the fact that $f''$ is continuous to guarantee that the integral exists.)

# 5 Review

**1.** (a) We take batches of 10,000 units and use the initial marginal cost as the marginal cost for each unit in the batch. Then the estimated cost is

$$(\$17.48)(10{,}000) + (\$15.01)(10{,}000) + (\$13.84)(10{,}000) + (\$12.24)(10{,}000)$$

$$+ (\$11.56)(10{,}000) + (\$11.89)(10{,}000) = \$820{,}200$$

(b) Using batches of 5000 units and the ending marginal cost of each batch, the estimated cost is

$$(\$13.05)(5000) + (\$12.24)(5000) + (\$11.62)(5000) + (\$11.56)(5000) = \$242{,}350$$

**3.** (a) With 6 subintervals we have $\Delta x = 1$. Using left endpoints,

$$L_6 = f(x_0)\,\Delta x + f(x_1)\,\Delta x + f(x_2)\,\Delta x + f(x_3)\,\Delta x + f(x_4)\,\Delta x + f(x_5)\,\Delta x$$

$$= f(0)\cdot 1 + f(1)\cdot 1 + f(2)\cdot 1 + f(3)\cdot 1 + f(4)\cdot 1 + f(5)\cdot 1$$

$$\approx 2 + 3.5 + 4 + 2 + (-1) + (-2.5) = 8$$

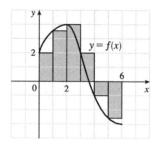

The Riemann sum represents the sum of the areas of the four rectangles above

the $x$-axis minus the sum of the areas of the two rectangles below the $x$-axis.

(b) $M_6 = f(\overline{x}_1)\,\Delta x + f(\overline{x}_2)\,\Delta x + f(\overline{x}_3)\,\Delta x + f(\overline{x}_4)\,\Delta x + f(\overline{x}_5)\,\Delta x + f(\overline{x}_6)\,\Delta x$

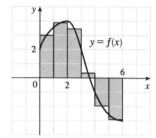

$$= f(0.5)\cdot 1 + f(1.5)\cdot 1 + f(2.5)\cdot 1 + f(3.5)\cdot 1 + f(4.5)\cdot 1 + f(5.5)\cdot 1$$

$$\approx 3 + 3.9 + 3.4 + 0.3 + (-2) + (-2.9) = 5.7$$

The Riemann sum represents the sum of the areas of the four rectangles above

the $x$-axis minus the sum of the areas of the two rectangles below the $x$-axis.

**5.** $\int_0^1 \left(x + \sqrt{1 - x^2}\right) dx = \int_0^1 x\,dx + \int_0^1 \sqrt{1 - x^2}\,dx$.

The first integral can be interpreted as the area of the triangle shown

in the figure and the second can be interpreted as the area of the

quarter-circle. Thus total area $= \frac{1}{2}(1)(1) + \frac{1}{4}(\pi)(1)^2 = \frac{1}{2} + \frac{\pi}{4}$.

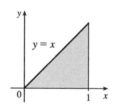

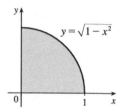

**7.** With $n = 6$ subintervals, we have $\Delta x = \frac{12 - 0}{6} = 2$ and the midpoints of the subintervals are $\overline{x}_1 = 1$, $\overline{x}_2 = 3$, $\overline{x}_3 = 5$,

$\overline{x}_4 = 7$, $\overline{x}_5 = 9$, and $\overline{x}_6 = 11$. Using the Midpoint Rule,

$$\int_0^{12} \ln(x^3 + 1)\,dx \approx M_6 = f(1)\,\Delta x + f(3)\,\Delta x + f(5)\,\Delta x + f(7)\,\Delta x + f(9)\,\Delta x + f(11)\,\Delta x$$

$$= \Delta x\,[\ln(1^3 + 1) + \ln(3^3 + 1) + \ln(5^3 + 1) + \ln(7^3 + 1) + \ln(9^3 + 1) + \ln(11^3 + 1)]$$

$$\approx 2(28.49) = 56.98$$

**9.** $f(x) = 2x^3 + 6x - 7 \;\Rightarrow\; F(x) = 2\left(\frac{1}{4}x^4\right) + 6\left(\frac{1}{2}x^2\right) - 7x + C = \frac{1}{2}x^4 + 3x^2 - 7x + C$

**11.** $p(r) = 4 + \dfrac{5}{r} = 4 + 5\cdot\dfrac{1}{r} \;\Rightarrow\; P(r) = 4r + 5\ln|r| + C$

**13.** $\int_1^2 (8x^3 + 3x^2)\,dx = \left[8\left(\frac{1}{4}x^4\right) + 3\left(\frac{1}{3}x^3\right)\right]_1^2 = \left[2x^4 + x^3\right]_1^2 = (2\cdot 2^4 + 2^3) - (2\cdot 1^4 + 1^3) = 40 - 3 = 37$

**15.** $\int_0^1 (1 - x^9)\,dx = \left[x - \frac{1}{10}x^{10}\right]_0^1 = \left(1 - \frac{1}{10}\right) - 0 = \frac{9}{10}$

**17.** $\int_0^2 5e^{2t}\,dt = \left[5\cdot\frac{1}{2}e^{2t}\right]_0^2 = \frac{5}{2}(e^4 - e^0) = \frac{5}{2}(e^4 - 1)$

**19.** $f(x) > 0$ for $1 \le x \le 8$, so

area $= \int_1^8 2\sqrt[3]{x}\,dx = 2\int_1^8 x^{1/3}\,dx = 2\left[\frac{3}{4}x^{4/3}\right]_1^8 = \frac{3}{2}(8^{4/3} - 1^{4/3}) = \frac{3}{2}(16 - 1) = \frac{45}{2}$.

**21.** $\int (7.2t^2 - 4.6t + 18.1)\,dt = 7.2\left(\frac{1}{3}t^3\right) - 4.6\left(\frac{1}{2}t^2\right) + 18.1t + C = 2.4t^3 - 2.3t^2 + 18.1t + C$

**23.** $\displaystyle\int\left(\frac{6}{x}+x\right)dx = 6\ln|x| + \tfrac{1}{2}x^2 + C$

**25.** Since $r(t)$ is the rate at which oil is consumed, by the Net Change Theorem, $\int_0^8 r(t)\,dt$ represents the net change in the amount of oil consumed (in barrels) from $t=0$ to $t=8$. In other words, it represents the number of barrels of oil consumed from January 1, 2000, through January 1, 2008.

**27.** By the Net Change Theorem, $\int_{500}^{1000} C'(q)\,dq = C(1000) - C(500)$ represents the change in cost when production is increased from 500 to 1000 laptops.

**29.** From the Net Change Theorem, the increase in revenue if production is raised from 1200 to 1800 pianos is

$$\int_{1200}^{1800}(4.3 - 0.002q)\,dq = \left[4.3q - 0.002\left(\tfrac{1}{2}q^2\right)\right]_{1200}^{1800} = \left[4.3q - 0.001q^2\right]_{1200}^{1800}$$

$$= [4.3(1800) - 0.001(1800)^2] - [4.3(1200) - 0.001(1200)^2]$$

$$= 4500 - 3720 = 780 \text{ thousand dollars or } \$780{,}000$$

**31.** $r(t)$ is the rate at which the gasoline leaks, so the amount of gasoline lost during the first 10 minutes is

$$\int_0^{10} r(t)\,dt = \int_0^{10} 2.1e^{-0.3t}\,dt = \left[2.1\left(\frac{1}{-0.3}e^{-0.3t}\right)\right]_0^{10} = -7(e^{-3} - e^0) = 7(1 - e^{-3}) \approx 6.65 \text{ gallons}$$

**33. (a)** Displacement $= \int_0^5 v(t)\,dt = \int_0^5 (t^2 - t)\,dt = \left[\tfrac{1}{3}t^3 - \tfrac{1}{2}t^2\right]_0^5 = \frac{125}{3} - \frac{25}{2} = \frac{175}{6} = 29.1\overline{6}$ meters

**(b)** The distance traveled is $\int_0^5 |v(t)|\,dt = \int_0^5 |t^2 - t|\,dt$. Now $t^2 - t = t(t-1) \geq 0$ for $t \geq 1$ and $t^2 - t < 0$ for $0 < t < 1$, so

$$\int_0^5 |v(t)|\,dt = \int_0^1 -(t^2 - t)\,dt + \int_1^5 (t^2 - t)\,dt = \int_0^1 (t - t^2)\,dt + \int_1^5 (t^2 - t)\,dt$$

$$= \left[\tfrac{1}{2}t^2 - \tfrac{1}{3}t^3\right]_0^1 + \left[\tfrac{1}{3}t^3 - \tfrac{1}{2}t^2\right]_1^5 = \tfrac{1}{2} - \tfrac{1}{3} - 0 + \left(\tfrac{125}{3} - \tfrac{25}{2}\right) - \left(\tfrac{1}{3} - \tfrac{1}{2}\right) = \frac{177}{6} = 29.5 \text{ meters}$$

**35.** $\displaystyle A_{\text{ave}} = \frac{1}{5-0}\int_0^5 A(t)\,dt = \frac{1}{5}\int_0^5 7.4e^{-0.12t}\,dt = \frac{1}{5}\left[7.4\left(\frac{1}{-0.12}e^{-0.12t}\right)\right]_0^5 = -\frac{1}{5}\cdot\frac{7.4}{0.12}\left(e^{-0.12(5)} - e^0\right)$

$\displaystyle = -\frac{37}{3}\left(e^{-0.6} - 1\right) \approx 5.56 \text{ oz.}$

**37.** Let $u = 1 + x^3$. Then $du = 3x^2\,dx$ and $x^2\,dx = \tfrac{1}{3}\,du$, so

$$\int x^2(1 + x^3)^6\,dx = \int u^6 \cdot \tfrac{1}{3}\,du = \tfrac{1}{3}\int u^6\,du = \tfrac{1}{3}\left(\tfrac{1}{7}u^7\right) + C = \tfrac{1}{21}(1 + x^3)^7 + C.$$

**39.** Let $u = x^2 + 1$. Then $du = 2x\,dx$ and $x\,dx = \tfrac{1}{2}\,du$. When $x = 0$, $u = 1$; when $x = 1$, $u = 2$. Thus

$$\int_0^1 \frac{x}{x^2 + 1}\,dx = \int_1^2 \frac{1}{u}\cdot\frac{1}{2}\,du = \frac{1}{2}\int_1^2 \frac{1}{u}\,du = \frac{1}{2}\Big[\ln|u|\Big]_1^2 = 12(\ln 2 - \ln 1) = \tfrac{1}{2}\ln 2.$$

**41.** Let $u = 4 - w^2$. Then $du = -2w\,dw$ and $w\,dw = -\tfrac{1}{2}\,du$. When $w = 0$, $u = 4$; when $w = 2$, $u = 0$. Thus

$$\int_0^2 we^{4-w^2}\,dw = \int_4^0 e^u\left(-\tfrac{1}{2}\right)du = -\tfrac{1}{2}\big[e^u\big]_4^0 = -\tfrac{1}{2}(e^0 - e^4) = \tfrac{1}{2}(e^4 - 1).$$

**43.** Let $u = e^x + 2$. Then $du = e^x\,dx$, so $\int e^x\sqrt{e^x + 2}\,dx = \int \sqrt{u}\,du = \int u^{1/2}\,du = \tfrac{2}{3}u^{3/2} + C = \tfrac{2}{3}(e^x + 2)^{3/2} + C.$

**45.** Let $u = x^2$. Then $du = 2x\,dx$ and $x\,dx = \frac{1}{2}\,du$, so

$$\int x5^{x^2}\,dx = \int 5^u \cdot \frac{1}{2}\,du = \frac{1}{2}\int 5^u\,du = \frac{1}{2}\left(\frac{1}{\ln 5}\,5^u\right) + C = \frac{1}{2\ln 5}\,5^{x^2} + C.$$

**47.** Let $u = \ln x$, $dv = x^{3/2}\,dx$ $\Rightarrow$ $du = \frac{1}{x}\,dx$, $v = \frac{2}{5}x^{5/2}$. Then

$$\int_1^4 x^{3/2}\ln x\,dx = (\ln x)\left(\tfrac{2}{5}x^{5/2}\right)\Big]_1^4 - \int_1^4 \left(\tfrac{2}{5}x^{5/2}\right)\left(\frac{1}{x}\right)dx = \left[\tfrac{2}{5}(4^{5/2}\ln 4) - 0\right] - \frac{2}{5}\int_1^4 x^{3/2}\,dx$$

$$= \tfrac{64}{5}\ln 4 - \tfrac{2}{5}\left[\tfrac{2}{5}x^{5/2}\right]_1^4 = \tfrac{64}{5}\ln 4 - \tfrac{4}{25}(4^{5/2} - 1^{5/2}) = \tfrac{64}{5}\ln 4 - \tfrac{4}{25}(32 - 1) = \tfrac{64}{5}\ln 4 - \tfrac{124}{25}$$

**49.** Let $u = t^2$, $dv = \dfrac{1}{e^{4t}}\,dt = e^{-4t}\,dt$ $\Rightarrow$ $du = 2t\,dt$, $v = -\frac{1}{4}e^{-4t}$. Then

$$\int \frac{t^2}{e^{4t}}\,dt = (t^2)\left(-\tfrac{1}{4}e^{-4t}\right) - \int \left(-\tfrac{1}{4}e^{-4t}\right)(2t\,dt) = -\tfrac{1}{4}t^2 e^{-4t} + \tfrac{1}{2}\int te^{-4t}\,dt$$

To evaluate $\displaystyle\int te^{-4t}\,dt$, we integrate by parts again: Let $U = t$, $dV = e^{-4t}\,dt$ $\Rightarrow$ $dU = dt$, $V = -\frac{1}{4}e^{-4t}$. Then

$$\int te^{-4t}\,dt = -\tfrac{1}{4}te^{-4t} - \int \left(-\tfrac{1}{4}e^{-4t}\right)dt = -\tfrac{1}{4}te^{-4t} + \tfrac{1}{4}\left(-\tfrac{1}{4}e^{-4t}\right) + C = -\tfrac{1}{4}te^{-4t} - \tfrac{1}{16}e^{-4t} + C.\text{ Thus}$$

$$\int \frac{t^2}{e^{4t}}\,dt = -\tfrac{1}{4}t^2 e^{-4t} + \tfrac{1}{2}\left[-\tfrac{1}{4}te^{-4t} - \tfrac{1}{16}e^{-4t} + C\right] = -\tfrac{1}{4}t^2 e^{-4t} - \tfrac{1}{8}te^{-4t} - \tfrac{1}{32}e^{-4t} + \tfrac{1}{2}C$$

$$= -e^{-4t}\left(\tfrac{1}{4}t^2 + \tfrac{1}{8}t + \tfrac{1}{32}\right) + C_1 \qquad \text{[where } C_1 = \tfrac{1}{2}C\text{]}$$

$$= -\frac{1}{e^{4t}}\left(\tfrac{1}{4}t^2 + \tfrac{1}{8}t + \tfrac{1}{32}\right) + C \qquad \text{[where we renamed } C_1 \text{ to } C\text{]}$$

**51.** $\displaystyle\int_0^3 3^t\,dt = \frac{3^t}{\ln 3}\Big]_0^3 = \frac{1}{\ln 3}(3^3 - 3^0) = \frac{1}{\ln 3}(27 - 1) = \frac{26}{\ln 3}$

**53.** Let $u = x^2 + 4x$. Then $du = (2x + 4)\,dx = 2(x + 2)\,dx$ and $x + 2 = \frac{1}{2}\,du$, so

$$\int \frac{x+2}{\sqrt{x^2+4x}}\,dx = \int \frac{1}{\sqrt{u}} \cdot \frac{1}{2}\,du = \frac{1}{2}\int u^{-1/2}\,du = \frac{1}{2}\cdot 2u^{1/2} + C = \sqrt{u} + C = \sqrt{x^2 + 4x} + C.$$

**55.** Integrate $f(x) = 8xe^{-x^2}$ by using the substitution $u = -x^2$, $du = -2x\,dx$. Then

$$F(x) = \int f(x)\,dx = \int 8xe^{-x^2}\,dx = 8\int e^u\left(-\tfrac{1}{2}\right)du = -4\int e^u\,du = -4e^u + C = -4e^{-x^2} + C.$$

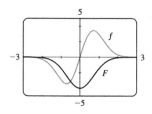

We see from the graph that this is reasonable, since $F$ has a minimum where $f$ changes from negative to positive. Also, $F$ increases where $f$ is positive and $F$ decreases where $f$ is negative.

**57.** $\displaystyle\int_0^{12} \frac{23}{1 + 3e^{-0.2t}}\,dt \approx 144.26$

**59.** $\int_0^6 f(x)\,dx = \int_0^4 f(x)\,dx + \int_4^6 f(x)\,dx$ $\Rightarrow$ $10 = 7 + \int_4^6 f(x)\,dx$ $\Rightarrow$ $\int_4^6 f(x)\,dx = 10 - 7 = 3$

# 6 APPLICATIONS OF INTEGRATION

## 6.1 Areas Between Curves

**Prepare Yourself**

1. (a) $\int_0^8 (0.4x^2 - 6x + 1.8)\,dx = \int_0^8 \left(\frac{2}{5}x^2 - 6x + \frac{9}{5}\right)dx = \left[\frac{2}{5}\cdot\frac{1}{3}x^3 - 6\cdot\frac{1}{2}x^2 + \frac{9}{5}x\right]_0^8 = \left[\frac{2}{15}x^3 - 3x^2 + \frac{9}{5}x\right]_0^8$

$$= \left[\frac{2}{15}(8)^3 - 3(8)^2 + \frac{9}{5}(8)\right] - 0 = \frac{1024}{15} - 192 + \frac{72}{5} = -\frac{328}{3}$$

(b) $\int_0^2 (e^t - t)\,dt = \left[e^t - \frac{1}{2}t^2\right]_0^2 = \left[e^2 - \frac{1}{2}(2)^2\right] - \left[e^0 - \frac{1}{2}(0)^2\right] = (e^2 - 2) - (1 - 0) = e^2 - 3$

(c) $\int_1^4 \left(1/x - 2\sqrt{x}\right)dx = \int_1^4 (1/x - 2x^{1/2})\,dx = \left[\ln|x| - 2\cdot\frac{2}{3}x^{3/2}\right]_1^4 = \left[\ln|x| - \frac{4}{3}x^{3/2}\right]_1^4$

$$= \left[\ln 4 - \frac{4}{3}(4)^{3/2}\right] - \left[\ln 1 - \frac{4}{3}(1)^{3/2}\right] = \left(\ln 4 - \frac{32}{3}\right) - \left(0 - \frac{4}{3}\right) = \ln 4 - \frac{28}{3}$$

(d) $\displaystyle\int_0^1 (2.5^x - 3x)\,dx = \left[\frac{1}{\ln 2.5}2.5^x - 3\cdot\frac{1}{2}x^2\right]_0^1 = \left[\frac{1}{\ln 2.5}2.5^x - \frac{3}{2}x^2\right]_0^1$

$$= \left[\frac{1}{\ln 2.5}(2.5)^1 - \frac{3}{2}(1)^2\right] - \left[\frac{1}{\ln 2.5}(2.5)^0 - \frac{3}{2}(0)^2\right] = \left(\frac{2.5}{\ln 2.5} - \frac{3}{2}\right) - \left(\frac{1}{\ln 2.5} - 0\right)$$

$$= \frac{1.5}{\ln 2.5} - \frac{3}{2}$$

(e) $\displaystyle\int_1^2 \left(\frac{1}{q} - \frac{4}{q^2}\right)dq = \int_1^2 \left(\frac{1}{q} - 4q^{-2}\right)dq = \left[\ln|q| - 4\cdot\frac{q^{-1}}{-1}\right]_1^2 = \left[\ln|q| + \frac{4}{q}\right]_1^2$

$$= \left(\ln 2 + \frac{4}{2}\right) - \left(\ln 1 + \frac{4}{1}\right) = (\ln 2 + 2) - (0 + 4) = \ln 2 - 2$$

2. With $n = 4$ subintervals, $\Delta x = (12 - 0)/4 = 3$. The midpoints of the subintervals are $\overline{x}_1 = 1.5$, $\overline{x}_2 = 4.5$, $\overline{x}_3 = 7.5$, and $\overline{x}_4 = 10.5$. Letting $f(x) = \ln(2x^2 + 5)$, we have

$$\int_0^{12} f(x)\,dx \approx f(\overline{x}_1)\,\Delta x + f(\overline{x}_2)\,\Delta x + f(\overline{x}_3)\,\Delta x + f(\overline{x}_4)\,\Delta x = \Delta x\,[f(1.5) + f(4.5) + f(7.5) + f(10.5)]$$
$$= 3[\ln(9.5) + \ln(45.5) + \ln(117.5) + \ln(225.5)] \approx 48.76$$

3. Find the point(s) of intersection by solving the pair of equations simultaneously. This gives $2x^2 - 8 = x^2 - x + 4 \;\Rightarrow$ $x^2 + x - 12 = 0 \;\Rightarrow\; (x+4)(x-3) = 0 \;\Rightarrow\; x + 4 = 0$ or $x - 3 = 0 \;\Rightarrow\; x = -4$ or $x = 3$. Evaluating either equation at $x = -4$ and $x = 3$ gives the $y$-coordinates of the points of intersection. Thus the points of intersection are $(-4, 24)$ and $(3, 10)$.

4. According to the Net Change Theorem, $\int_0^{1500} C'(q)\,dq = C(1500) - C(0)$. So, the integral represents the change in cost when production is raised from zero units to 1500 units.

5. Since $r(t)$ is the rate at which oil drains from the tank in qt/min, we can write $r(t) = V'(t)$, where $V(t)$ is volume of oil drained from the tank, in quarts, $t$ minutes after the draining began. Thus, by the Net Change Theorem, $\int_5^{20} r(t)\,dt = \int_5^{20} V'(t)\,dt = V(20) - V(5)$ is the volume of oil drained between the end of the 5th minute and the end of the 20th minute.

**Exercises**

**1.** The upper boundary curve is $y = 5x - x^2$ and the lower boundary curve is $y = x$. By Formula 1, the area is

$$A = \int_0^4 [(5x - x^2) - x]\,dx = \int_0^4 (4x - x^2)\,dx = \left[2x^2 - \tfrac{1}{3}x^3\right]_0^4 = \left[2(4)^2 - \tfrac{1}{3}(4)^3\right] - 0 = 32 - \tfrac{64}{3} = \tfrac{32}{3}$$

**3.** The top and bottom boundaries are $y_T = 2e^x$ and $y_B = 4x - 1$, so the area is

$$A = \int_0^2 (y_T - y_B)\,dx = \int_0^2 [(2e^x) - (4x - 1)]\,dx = \int_0^2 (2e^x - 4x + 1)\,dx$$
$$= \left[2e^x - 2x^2 + x\right]_0^2 = (2e^2 - 8 + 2) - (2e^0 - 0 + 0) = 2e^2 - 6 - 2 = 2e^2 - 8$$

**5.** The top curve is $y = 9 - x^2$ and the bottom curve is $y = x + 1$, so the area from $x = -1$ to $x = 2$ is

$$A = \int_{-1}^2 [(9 - x^2) - (x + 1)]\,dx = \int_{-1}^2 (8 - x - x^2)\,dx$$
$$= \left[8x - \frac{x^2}{2} - \frac{x^3}{3}\right]_{-1}^2$$
$$= \left(16 - 2 - \tfrac{8}{3}\right) - \left(-8 - \tfrac{1}{2} + \tfrac{1}{3}\right)$$
$$= 22 - 3 + \tfrac{1}{2} = \tfrac{39}{2} = 19.5$$

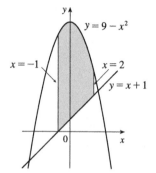

**7.** $A = \int_{-1}^1 [e^x - (x^2 - 1)]\,dx = \int_{-1}^1 (e^x - x^2 + 1)\,dx$
$$= \left[e^x - \tfrac{1}{3}x^3 + x\right]_{-1}^1$$
$$= \left(e - \tfrac{1}{3} + 1\right) - \left(e^{-1} + \tfrac{1}{3} - 1\right)$$
$$= e - \tfrac{1}{e} + \tfrac{4}{3}$$

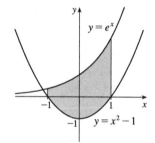

**9.** The curves intersect when $x = x^2 \Rightarrow x^2 - x = 0 \Rightarrow$
$x(x - 1) = 0 \Rightarrow x = 0$ or $1$.
$$A = \int_0^1 (x - x^2)\,dx$$
$$= \left[\tfrac{1}{2}x^2 - \tfrac{1}{3}x^3\right]_0^1$$
$$= \tfrac{1}{2} - \tfrac{1}{3} - 0 = \tfrac{1}{6}$$

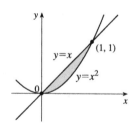

**11.** The curves intesect when $1/x = 1/x^2 \Rightarrow x^2 = x \Rightarrow$
$x(x - 1) = 0 \ (x \neq 0) \Rightarrow x = 1$.
$$A = \int_1^2 \left(\frac{1}{x} - \frac{1}{x^2}\right)\,dx = \left[\ln|x| + \frac{1}{x}\right]_1^2$$
$$= \left(\ln 2 + \tfrac{1}{2}\right) - (\ln 1 + 1)$$
$$= \ln 2 - \tfrac{1}{2} \approx 0.19$$

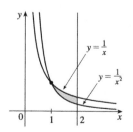

**13.** The curves intersect when $x^2 = \sqrt{x} \;\Rightarrow\; (x^2)^2 = \left(\sqrt{x}\right)^2 \;\Rightarrow$

$x^4 = x \;\Rightarrow\; x^4 - x = 0 \;\Rightarrow\; x^3(x-1) = 0 \;\Rightarrow\; x = 0 \text{ or } 1.$

$A = \int_0^1 \left(\sqrt{x} - x^2\right) dx = \int_0^1 (x^{1/2} - x^2)\, dx$

$\quad = \left[\frac{2}{3}x^{3/2} - \frac{1}{3}x^3\right]_0^1$

$\quad = \frac{2}{3} - \frac{1}{3} - 0 = \frac{1}{3}$

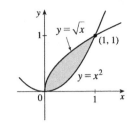

**15.** The curves intersect when $12 - x^2 = x^2 - 6 \;\Rightarrow\; 2x^2 = 18 \;\Rightarrow$

$x^2 = 9 \;\Rightarrow\; x = \pm 3.$

$A = \int_{-3}^{3}[(12 - x^2) - (x^2 - 6)]\, dx$

$\quad = \int_{-3}^{3}(18 - 2x^2)\, dx$

$\quad = \left[18x - \frac{2}{3}x^3\right]_{-3}^{3}$

$\quad = (54 - 18) - (-54 + 18)$

$\quad = 72$

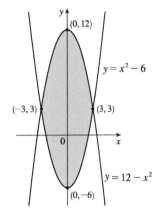

**17.** The curves intersect when $e^x = xe^x \;\Rightarrow\; e^x - xe^x = 0 \;\Rightarrow$

$e^x(1 - x) = 0 \;\Rightarrow\; x = 1.$

$A = \int_0^1 (e^x - xe^x)\, dx$

$\quad = \left[e^x - (xe^x - e^x)\right]_0^1 \qquad \text{[use parts with } u = x \text{ and } dv = e^x\, dx]$

$\quad = \left[2e^x - xe^x\right]_0^1 = (2e - e) - (2 - 0) = e - 2$

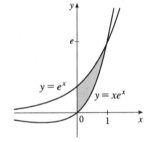

**19.** (a) Total area $= 12 + 27 = 39.$

(b) $f(x) \le g(x)$ for $0 \le x \le 2$ and $f(x) \ge g(x)$ for $2 \le x \le 5$, so

$$\int_0^5 [f(x) - g(x)]\, dx = \int_0^2 [f(x) - g(x)]\, dx + \int_2^5 [f(x) - g(x)]\, dx$$

$$= -\int_0^2 [g(x) - f(x)]\, dx + \int_2^5 [f(x) - g(x)]\, dx = -(12) + 27 = 15$$

**21.** The function $y = f(x) = \sqrt{x^2 - 1}$ is the upper curve and $y = g(x) = (\ln x)^2$ is the lower curve. Using $n = 4$ subintervals,

$\Delta x = (5 - 1)/4 = 1$. The midpoints of the subintervals are $\overline{x}_1 = 1.5$, $\overline{x}_2 = 2.5$, $\overline{x}_3 = 3.5$, and $\overline{x}_4 = 4.5$. The area of the

region is

$A = \int_1^5 [f(x) - g(x)]\, dx$

$\quad \approx [f(1.5) - g(1.5)]\,\Delta x + [f(2.5) - g(2.5)]\,\Delta x$

$\qquad + [f(3.5) - g(3.5)]\,\Delta x + [f(4.5) - g(4.5)]\,\Delta x$

$\quad = \left[\sqrt{(1.5)^2 - 1} - (\ln 1.5)^2\right] \cdot 1 + \left[\sqrt{(2.5)^2 - 1} - (\ln 2.5)^2\right] \cdot 1$

$\qquad + \left[\sqrt{(3.5)^2 - 1} - (\ln 3.5)^2\right] \cdot 1 + \left[\sqrt{(4.5)^2 - 1} - (\ln 4.5)^2\right] \cdot 1$

$\quad \approx 6.32$

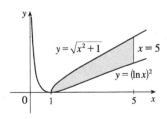

**23.**

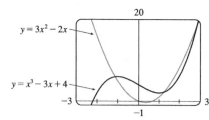

From the graph, we see that the curves intersect at $x \approx -1.11$,

$x \approx 1.25$, and $x \approx 2.86$. $y = x^3 - 3x + 4$ is the upper curve for

$-1.11 < x < 1.25$ and $y = 3x^2 - 2x$ is the upper curve for

$1.25 < x < 2.86$. So the area of the region bounded by the curves is

$$A \approx \int_{-1.11}^{1.25} \left[ (x^3 - 3x + 4) - (3x^2 - 2x) \right] dx + \int_{1.25}^{2.86} \left[ (3x^2 - 2x) - (x^3 - 3x + 4) \right] dx$$

$$= \int_{-1.11}^{1.25} (x^3 - 3x^2 - x + 4) \, dx + \int_{1.25}^{2.86} (-x^3 + 3x^2 + x - 4) \, dx$$

$$= \left[ \tfrac{1}{4}x^4 - x^3 - \tfrac{1}{2}x^2 + 4x \right]_{-1.11}^{1.25} + \left[ -\tfrac{1}{4}x^4 + x^3 + \tfrac{1}{2}x^2 - 4x \right]_{1.25}^{2.86} \approx 8.38$$

**25.**

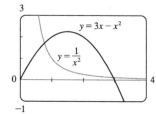

From the graph, we see that the curves intersect at $x \approx 0.76$ and $x \approx 2.96$.

$y = 3x - x^2$ is the upper curve, so the area of the region bounded by the curves is

$$A \approx \int_{0.76}^{2.96} \left[ (3x - x^2) - \frac{1}{x^2} \right] dx = \left[ \tfrac{3}{2}x^2 - \tfrac{1}{3}x^3 + \frac{1}{x} \right]_{0.76}^{2.96} \approx 2.80$$

**27.** $C'(q) > R'(q)$ for $0 \le q \le 200$, so the area between the graphs is

$$A = \int_0^{200} [C'(q) - R'(q)] \, dq = \int_0^{200} [(48 - 0.03q + 0.00002q^2) - (44 - 0.007q)] \, dq$$

$$= \int_0^{200} (4 - 0.023q + 0.00002q^2) \, dq = \left[ 4q - 0.0115q^2 + \frac{0.00002}{3}q^3 \right]_0^{200}$$

$$= 4(200) - 0.0115(200)^2 + \frac{0.00002}{3}(200)^3 - 0 \approx 393.33$$

The area between the graphs is the difference between total cost and revenue for $0 \le q \le 200$. Because cost is greater than revenue here, this means that the company's profit decreased by \$393.33 after producing and selling the first 200 heaters.

**29.** For $0 \le t \le 12$, $g(t) > f(t)$, so the area between the curves is given by

$$\int_0^{12} [g(t) - f(t)] \, dt = \int_0^{12} (5.3e^{0.0164t} - 4.3e^{0.0172t}) \, dt = \left[ \frac{5.3}{0.0164} e^{0.0164t} - \frac{4.3}{0.0172} e^{0.0172t} \right]_0^{12}$$

$$= \left( \frac{5.3}{0.0164} e^{0.1968} - \frac{4.3}{0.0172} e^{0.2064} \right) - \left( \frac{5.3}{0.0164} - \frac{4.3}{0.0172} \right) \approx 12.979$$

Thus, in 2011 the factory produced 12,979 more hard drives than in 2010.

**31.** For $0 \le t \le 10$, $r_1(t) > r_2(t)$, so the area between the curves is given by

$$\int_0^{10} [r_1(t) - r_2(t)] \, dt = \int_0^{10} [0.063(1.041^t) - 0.047(1.038^t)] \, dt = \left[ \frac{0.063}{\ln 1.041} 1.041^t - \frac{0.047}{\ln 1.038} 1.038^t \right]_0^{10}$$

$$= \left( \frac{0.063}{\ln 1.041} 1.041^{10} - \frac{0.047}{\ln 1.038} 1.038^{10} \right) - \left( \frac{0.063}{\ln 1.041} - \frac{0.047}{\ln 1.038} \right) \approx 0.206$$

Thus the value of one of the buildings increased about \$206,000 more than the other, 10 years after the purchase date.

**33.** Both velocity functions are positive so, as in Example 4, the area under a velocity curve is the distanced traveled.

(a) After one minute, the area under curve $A$ is greater than the area under curve $B$. So car A is ahead after one minute.

(b) The area of the shaded region is the difference between the areas under the velocity curves for $0 \le t \le 1$, which is the distance by which A is ahead of B after 1 minute.

(c) After two minutes, car B is traveling faster than car A and has gained some ground, but the area under curve A from $t = 0$ to $t = 2$ is still greater than the corresponding area for curve $B$, so car A is still ahead.

(d) From the graph, it appears that the area between curves $A$ and $B$ for $0 \le t \le 1$ (when car A is going faster), which corresponds to the distance by which car A is ahead, seems to be about 3 squares. Therefore the cars will be side by side at the time $x$ where the area between the curves for $1 \le t \le x$ (when car B is going faster) is the same as the area for $0 \le t \le 1$. From the graph, it appears that this time is $x \approx 2.2$. So the cars are side by side when $t \approx 2.2$ minutes.

**35.** 1 second $= \frac{1}{3600}$ hour, so 10 s $= \frac{1}{360}$ h. The difference in the distances traveled is the area between the graphs of $v_C$ and $v_K$, given by $\int_0^{1/360} (v_K - v_C)\, dt$. With the given data, we can take $n = 5$ to use the Midpoint Rule. $\Delta t = \frac{1/360 - 0}{5} = \frac{1}{1800}$, so

$$\text{distance}_{\text{Kelly}} - \text{distance}_{\text{Chris}} \approx M_5 = \frac{1}{1800} \left[ (v_K\,(1\text{ s}) - v_C\,(1\text{ s})) + (v_K\,(3\text{ s}) - v_C\,(3\text{ s})) + (v_K\,(5\text{ s}) - v_C\,(5\text{ s})) \right.$$
$$\left. + \; (v_K\,(7\text{ s}) - v_C\,(7\text{ s})) + (v_K\,(9\text{ s}) - v_C\,(9\text{ s})) \right]$$
$$= \frac{1}{1800} [(22 - 20) + (52 - 46) + (71 - 62) + (86 - 75) + (98 - 86)]$$
$$= \frac{1}{1800}(2 + 6 + 9 + 11 + 12) = \frac{1}{1800}(40) = \frac{1}{45} \text{ mile, or } 117\tfrac{1}{3} \text{ feet}$$

**37.** Let $h(x)$ denote the height of the wing at $x$ cm from the left end.

$$\text{Area} \approx M_5 = \frac{200 - 0}{5} [h(20) + h(60) + h(100) + h(140) + h(180)]$$
$$= 40(20.3 + 29.0 + 27.3 + 20.5 + 8.7) = 40(105.8) = 4232 \text{ cm}^2$$

**39.** The curves $y = 1/x$ and $y = x$ intersect when $1/x = x \;\Rightarrow\; 1 = x^2 \;\Rightarrow\; x = \pm 1$ and the curves $y = 1/x$ and $y = \frac{1}{4}x$ intersect when $1/x = \frac{1}{4}x \;\Rightarrow\; 4 = x^2 \;\Rightarrow\; x = \pm 2$. Thus for $x > 0$,

$$A = \int_0^1 \left( x - \frac{1}{4}x \right) dx + \int_1^2 \left( \frac{1}{x} - \frac{1}{4}x \right) dx$$
$$= \int_0^1 \left( \frac{3}{4}x \right) dx + \int_1^2 \left( \frac{1}{x} - \frac{1}{4}x \right) dx$$
$$= \left[ \frac{3}{8}x^2 \right]_0^1 + \left[ \ln |x| - \frac{1}{8}x^2 \right]_1^2 = \frac{3}{8} + \left( \ln 2 - \frac{1}{2} \right) - \left( 0 - \frac{1}{8} \right) = \ln 2$$

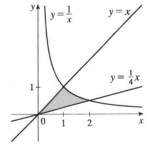

**41.**

By the symmetry of the problem, we consider only the first quadrant, where $y = x^2 \;\Rightarrow\; x = \sqrt{y}$. We integrate with respect to $y$, as in Exercise 73 in Section 5.2. We are looking for a number $b$ such that $\displaystyle\int_0^b \sqrt{y}\, dy = \int_b^4 \sqrt{y}\, dy \;\Rightarrow$

$$\frac{2}{3}\left[ y^{3/2} \right]_0^b = \frac{2}{3}\left[ y^{3/2} \right]_b^4 \;\Rightarrow\; b^{3/2} = 4^{3/2} - b^{3/2} \;\Rightarrow\; 2b^{3/2} = 8 \;\Rightarrow$$
$$b^{3/2} = 4 \;\Rightarrow\; b = 4^{2/3} \approx 2.52.$$

## 6.2 Applications to Economics

**1.** Since the number of products sold is $Q = 2000$, the corresponding price is $P = 1450 - 0.2(2000) = 1050$. By Definition 1, the consumer surplus is

$$\int_0^Q [D(q) - P]\, dq = \int_0^{2000}(1450 - 0.2q - 1050)\, dq = \int_0^{2000}(400 - 0.2q)\, dq$$

$$= 400q - 0.1q^2 \Big]_0^{2000} = 400(2000) - 0.1(2000)^2 - 0 = \$400{,}000$$

**3.** The number of products sold is $Q = 800$ and the corresponding price is $P = 760 - 0.1(800) - 0.0002(800)^2 = 552$. Therefore, the consumer surplus is

$$\int_0^Q [D(q) - P]\, dq = \int_0^{800}(760 - 0.1q - 0.0002q^2 - 552)\, dq = \int_0^{800}(208 - 0.1q - 0.0002q^2)\, dq$$

$$= 208q - 0.05q^2 - \frac{0.0002}{3}\, q^3 \Big]_0^{800} = 208(800) - 0.05(800)^2 - \frac{0.0002}{3}(800)^3$$

$$\approx \$100{,}266.67$$

**5.** The number of products sold is $Q = 45$ and the corresponding price is $P = 15{,}000e^{-0.03(45)} \approx 3888.60$. Therefore, the consumer surplus is

$$\int_0^Q [D(q) - P]\, dq \approx \int_0^{45}(15{,}000e^{-0.03q} - 3888.60)\, dq = -\frac{15{,}000}{0.03}\, e^{-0.03q} - 3888.60q \Big]_0^{45}$$

$$= \left[ -\frac{15{,}000}{0.03}\, e^{-0.03(45)} - 3888.60(45) \right] - \left[ -\frac{15{,}000}{0.03}\, e^0 - 0 \right] \approx \$195{,}392.87$$

**7.** The number of products sold is $Q = 100$ and the corresponding price is $P = 40 - 3.2\sqrt{100} = 8$. Therefore, the consumer surplus is

$$\int_0^Q [D(q) - P]\, dq = \int_0^{100}\left(40 - 3.2\sqrt{q} - 8\right) dq = \int_0^{100}(32 - 3.2q^{1/2})\, dq = 32q - 3.2\left(\tfrac{2}{3}\right)q^{3/2} \Big]_0^{100}$$

$$= 32(100) - 3.2\left(\tfrac{2}{3}\right)(100)^{3/2} \approx \$1066.67$$

**9.** $Q = 800$ and $P = D(800) = 2000 - 46\sqrt{800} = 2000 - 920\sqrt{2}$. The consumer surplus is

$$\int_0^Q [D(q) - P]\, dq = \int_0^{800}\left[2000 - 46\sqrt{q} - \left(2000 - 920\sqrt{2}\right)\right] dq$$

$$= \int_0^{800}\left[920\sqrt{2} - 46\sqrt{q}\right] dq = \left[920\sqrt{2}\, q - \tfrac{92}{3}q^{3/2}\right]_0^{800}$$

$$\approx \$1{,}040{,}861.18 - 693{,}907.45 = \$346{,}953.73$$

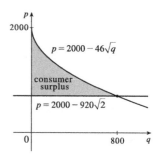

**11.** $p = 10 \quad\Rightarrow\quad \dfrac{450}{q + 8} = 10 \quad\Rightarrow\quad q + 8 = 45 \quad\Rightarrow\quad q = 37$.

$$\text{Consumer surplus} = \int_0^{37}\left(\frac{450}{q+8} - 10\right) dq = \Big[450\ln(q+8) - 10q\Big]_0^{37}$$

$$= (450\ln 45 - 370) - 450\ln 8 = 450\ln\left(\tfrac{45}{8}\right) - 370 \approx \$407.25$$

**13.** The demand function $p = D(q)$ is linear with slope $\dfrac{\Delta p}{\Delta q} = -\dfrac{1}{30}$ and $p = 18$ when $q = 210$, so an equation is

$p - 18 = -\frac{1}{30}(q - 210)$ or $p = 25 - \frac{1}{30}q$.

A selling price of $15 implies that $15 = 25 - \frac{1}{30}Q \Rightarrow -10 = -\frac{1}{30}Q \Rightarrow Q = 300$.

Consumer surplus $= \int_0^{300} \left(25 - \frac{1}{30}q - 15\right) dq = \int_0^{300} \left(10 - \frac{1}{30}q\right) dq = \left[10q - \frac{1}{60}q^2\right]_0^{300} = \$1500$.

**15.** Since the number of products sold is $Q = 500$, the corresponding price is $P = 16 + 0.03(500) = 31$. By Definition 2, the

producer surplus is

$$\int_0^Q [P - S(q)]\, dq = \int_0^{500} [31 - (16 + 0.03q)]\, dq = \int_0^{500} (15 - 0.03q)\, dq$$

$$= 15q - 0.015q^2 \Big]_0^{500} = 7500 - 3750 = \$3750$$

**17.** The number of products sold is $Q = 600$ and the corresponding price is $P = 22e^{0.002(600)} = 22e^{1.2}$. Therefore, the producer

surplus is

$$\int_0^Q [P - S(q)]\, dq = \int_0^{600} (22e^{1.2} - 22e^{0.002q})\, dq = 22e^{1.2}q - 11{,}000e^{0.002q}\Big]_0^{600}$$

$$= [22e^{1.2}(600) - 11{,}000e^{0.002(600)}] - [0 - 11{,}000e^0] \approx \$18{,}304.26$$

**19.** When $p = 400$, $400 = 200 + 0.2q^{3/2} \Rightarrow 200 = 0.2q^{3/2} \Rightarrow 1000 = q^{3/2} \Rightarrow q = 1000^{2/3} = 100$.

$$\text{Producer surplus} = \int_0^{100} [400 - (200 + 0.2q^{3/2})]\, dq = \int_0^{100} (200 - 0.2q^{3/2})\, dq$$

$$= \left[200q - 0.08q^{5/2}\right]_0^{100} = 20{,}000 - 8{,}000 = \$12{,}000$$

**21.** The graphs of $p = \dfrac{800{,}000e^{-q/5000}}{q + 20{,}000}$ and $p = 16$ intersect at $q \approx 3727$.

Using a graphing calculator (or computer software), the consumer surplus is

approximately $\displaystyle\int_0^{3727} \left[\dfrac{800{,}000e^{-q/5000}}{q + 20{,}000} - 16\right] dq \approx \$37{,}753$.

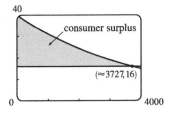

**23.** (a) At market equilibrium, $D(q) = S(q) \Rightarrow 228.4 - 18q = 27q + 57.4 \Rightarrow 171 = 45q \Rightarrow q = \frac{171}{45} = 3.8$. When

   $q = 3.8$, the corresponding price is $p = 228.4 - 18(3.8) = \$160$.

(b) From part (a), the market is in equilibrium for $q = 3.8$. Thus,

$$\text{Total surplus} = \int_0^{3.8} [D(q) - S(q)]\, dq = \int_0^{3.8} [(228.4 - 18q) - (27q + 57.4)]\, dq$$

$$= \int_0^{3.8} (171 - 45q)\, dq = \left[171q - 22.5q^2\right]_0^{3.8} = 324.9$$

   Because $q$ is measured in thousands, the total surplus is $324,900.

(c) The total surplus is maximized when the market is in equilibrium. So, from part (b), the maximum total surplus is

   $324,900.

**25.** The total surplus is maximized at market equilibrium. At equilibrium, $D(q) = S(q) \Rightarrow 312e^{-0.14q} = 26e^{0.2q} \Rightarrow$

$12 = e^{0.34q} \Rightarrow \ln 12 = 0.34q \Rightarrow q = \dfrac{1}{0.34}\ln 12 \approx 7.3085$. The corresponding price is

$p \approx 312e^{-0.14(7.3085)} \approx \$112.15$.

**27.** Assume that the income is continuous and is earned at the rate $1500/month $\times$ 12 months/year = $18,000 per year. The income earns 5% (compounded continuously), so $r = 0.05$, and the period is 10 years, so $T = 10$. By Equation 4, the future value is

$$\text{FV} = e^{rT} \int_0^T f(t)e^{-rt}\,dt = e^{0.05(10)} \int_0^{10} 18{,}000e^{-0.05t}\,dt = 18{,}000e^{0.5} \left(\frac{1}{-0.05}\right)e^{-0.05t}\Bigg]_0^{10}$$

$$= -360{,}000e^{0.5}(e^{-0.5} - e^0) \approx \$233{,}539.66$$

**29.** (a) $\text{FV} = e^{rT}\int_0^T f(t)e^{-rt}\,dt = e^{0.048(10)}\int_0^{10} 30{,}000e^{0.025t}e^{-0.048t}\,dt = 30{,}000e^{0.48}\int_0^{10}e^{-0.023t}\,dt$

$$= 30{,}000e^{0.48}\left(\frac{1}{-0.023}\right)e^{-0.023t}\Bigg]_0^{10} = -\frac{30{,}000}{0.023}e^{0.48}(e^{-0.23} - e^0) \approx \$433{,}107.37$$

(b) By Definition 5, the present value is

$$\text{PV} = \int_0^T f(t)e^{-rt}\,dt = \int_0^{10}30{,}000e^{0.025t}e^{-0.048t}\,dt = 30{,}000\int_0^{10}e^{-0.023t}\,dt$$

$$= 30{,}000\left(\frac{1}{-0.023}\right)e^{-0.023t}\Bigg]_0^{10} = -\frac{30{,}000}{0.023}(e^{-0.23} - e^0) \approx \$267{,}999.65$$

**31.** If we consider the income as a continuous stream, then the present value is

$$\text{PV} = \int_0^T f(t)e^{-rt}\,dt = \int_0^{20}50{,}000e^{-0.08t}\,dt = 50{,}000\left(\frac{1}{-0.08}\right)e^{-0.08t}\Bigg]_0^{20}$$

$$= -\frac{50{,}000}{0.08}(e^{-1.6} - e^0) \approx \$498{,}814.68$$

Since the present value of the installment plan exceeds the value of the one-time payment, the lottery winner should choose the installment plan rather than the one-time payment.

**33.** $\text{FV} = e^{rT}\int_0^T f(t)e^{-rt}\,dt = e^{0.052(15)}\int_0^{15}(12{,}000 + 500t)e^{-0.052t}\,dt = e^{0.78}\int_0^{15}(12{,}000 + 500t)e^{-0.052t}\,dt$. Using

integration by parts with $u = 12{,}000 + 500t \;\Rightarrow\; du = 500\,dt$ and $dv = e^{-0.052t}\,dt \;\Rightarrow\; v = -\dfrac{1}{0.052}e^{-0.052t}$, the

integral becomes

$$e^{0.78}\left[(12{,}000 + 500t)\left(-\frac{1}{0.052}\right)e^{-0.052t} - \int_0^{15}\left(-\frac{1}{0.052}\right)e^{-0.052t}(500)\,du\right]$$

$$= e^{0.78}\left[-\frac{1}{0.052}(12{,}000 + 500t)\,e^{-0.052t} + \frac{500}{0.052}\left(-\frac{1}{0.052}\right)e^{-0.052t}\right]_0^{15}$$

$$\approx \$346{,}884.18$$

Finally, note that $\text{FV} = e^{rT}\,\text{PV}$, so $\text{PV} = e^{-rT}\,\text{FV} = e^{-0.052(15)}(346{,}844.18) \approx \$159{,}013.79$.

**35.** The increase in capital is $f(8) - f(4) = \int_4^8 f'(t)\,dt = \int_4^8 \sqrt{t}\,dt = \left[\frac{2}{3}t^{3/2}\right]_4^8 = \frac{2}{3}(16\sqrt{2} - 8) \approx \$9.75$ million.

**37.** $N = \displaystyle\int_a^b Ax^{-k}\,dx = A\left[\frac{x^{-k+1}}{-k+1}\right]_a^b = \frac{A}{1-k}(b^{1-k} - a^{1-k})$.

Similarly, $\displaystyle\int_a^b Ax^{1-k}\,dx = A\left[\frac{x^{2-k}}{2-k}\right]_a^b = \frac{A}{2-k}(b^{2-k} - a^{2-k})$.

Thus, $\overline{x} = \dfrac{1}{N}\displaystyle\int_a^b Ax^{1-k}\,dx = \dfrac{[A/(2-k)](b^{2-k} - a^{2-k})}{[A/(1-k)](b^{1-k} - a^{1-k})} = \dfrac{(1-k)(b^{2-k} - a^{2-k})}{(2-k)(b^{1-k} - a^{1-k})}$.

## 6.3  Applications to Biology

**1.** (a) According to the survival function, the proportion of the population that survives is $S(4) = \frac{1}{5}$, so $\frac{1}{5}(7400) = 1480$ of the original members survive.

(b) $\int_0^4 R(t)\,dt = \int_0^4 (2240 + 60t)\,dt = \left[2240t + 30t^2\right]_0^4 = 8960 + 480 = 9440$ members are added.

(c) Not all of the 9440 new members survive.

**3.** According to Equation 1, the population $T = 12$ weeks from now is

$$P(12) = S(12) \cdot P_0 + \int_0^{12} S(12 - t)\,R(t)\,dt = e^{-0.2(12)}(22{,}500) + \int_0^{12} e^{-0.2(12-t)}(1225e^{0.14t})\,dt$$

$$= 22{,}500e^{-2.4} + 1225e^{-2.4}\int_0^{12} e^{0.34t}\,dt = 22{,}500e^{-2.4} + 1225e^{-2.4}\left[\frac{1}{0.34}e^{0.34t}\right]_0^{12}$$

$$= 22{,}500e^{-2.4} + \frac{1225}{0.34}e^{-2.4}(e^{4.08} - e^0) \approx 21{,}046 \text{ insects}$$

**5.** $t$ is measured in months, so the number of subscribers after 2 years is

$$P(24) = S(24) \cdot P_0 + \int_0^{24} S(24 - t)\,R(t)\,dt = e^{-0.06(24)}(8400) + \int_0^{24} e^{-0.06(24-t)}(180e^{0.04t})\,dt$$

$$= 8400e^{-1.44} + 180\int_0^{24} e^{-1.44+0.1t}\,dt = 8400e^{-1.44} + 180e^{-1.44}\int_0^{24} e^{0.1t}\,dt$$

$$= 8400e^{-1.44} + 180e^{-1.44}\left[\frac{1}{0.1}e^{0.1t}\right]_0^{24} = 8400e^{-1.44} + 1800e^{-1.44}(e^{2.4} - e^0) \approx 6265 \text{ subscribers}$$

**7.** $P(8) = S(8) \cdot P_0 + \int_0^8 S(8 - t)\,R(t)\,dt = e^{-0.25(8)}(50) + \int_0^8 e^{-0.25(8-t)}(12)\,dt$

$$= 50e^{-2} + 12e^{-2}\int_0^8 e^{0.25t}\,dt = 50e^{-2} + 12e^{-2}\left[\frac{1}{0.25}e^{0.25t}\right]_0^8$$

$$= 50e^{-2} + 48e^{-2}(e^2 - e^0) = 50e^{-2} + 48 - 48e^{-2} = 2e^{-2} + 48 \approx 48.3 \text{ mg}$$

**9.** $P(18) = S(18) \cdot P_0 + \int_0^{18} S(18 - t)\,R(t)\,dt = e^{-0.32(18)}(10{,}000) + \int_0^{18} e^{-0.32(18-t)}(1600e^{0.06t})\,dt$

$$= 10{,}000e^{-5.76} + 1600e^{-5.76}\int_0^{18} e^{0.38t}\,dt = 10{,}000e^{-5.76} + 1600e^{-5.76}\left[\frac{1}{0.38}e^{0.38t}\right]_0^{18}$$

$$= 10{,}000e^{-5.76} + \frac{1600}{0.38}e^{-5.76}(e^{6.84} - 1) \approx 12{,}417 \text{ gallons}$$

**11.** $F = \dfrac{\pi P R^4}{8\eta l} = \dfrac{\pi(4000)(0.008)^4}{8(0.027)(2)} \approx 0.000119 \text{ cm}^3/\text{s}$

**13.** From Equation 4, $F = \dfrac{A}{\int_0^T c(t)\,dt} = \dfrac{6}{20I}$, where

$$I = \int_0^{10} te^{-0.6t}\,dt = \left[\frac{1}{(-0.6)^2}(-0.6t - 1)e^{-0.6t}\right]_0^{10} \begin{bmatrix} \text{integrating} \\ \text{by parts} \end{bmatrix} = \frac{1}{0.36}(-7e^{-6} + 1)$$

Thus, $F = \dfrac{6(0.36)}{20(1 - 7e^{-6})} = \dfrac{0.108}{1 - 7e^{-6}} \approx 0.1099 \text{ L/s or } 6.594 \text{ L/min.}$

**15.** Divide the time interval into 8 subintervals of equal length $\Delta t = (16 - 0)/8 = 2$. Then the midpoints of the subintervals are

$\bar{t}_1 = 1, \bar{t}_2 = 3, \ldots, \bar{t}_8 = 15$. Then

$$\int_0^{16} c(t)\,dt \approx M_8 = c(1)\,\Delta t + c(3)\,\Delta t + c(5)\,\Delta t + c(7)\,\Delta t + c(9)\,\Delta t + c(11)\,\Delta t + c(13)\,\Delta t + c(15)\,\Delta t$$

$$\approx \Delta t(4.0 + 7.1 + 7.2 + 6.1 + 4.7 + 3.5 + 2.5 + 1.8) = (2)(36.9) = 73.8$$

So $F = \dfrac{A}{\int_0^{16} c(t)\,dt} \approx \dfrac{7}{73.8} \approx 0.0949 \text{ L/s} \approx 5.69 \text{ L/min}$.

**17.** Let $R(t) = R$ be the constant rate at which the drug is administered. The drug level $P(T)$ after $T$ hours is given by

Equation 1: $P(T) = S(T) \cdot P_0 + \int_0^T S(T - t)\,R(t)\,dt$, where $S(T) = e^{-0.4t}$ and $R(t) = R$. $P_0$ is the initial level, which

we want to remain constant, so $P(T) = P_0$ for any time $T$. Thus

$$P_0 = e^{-0.4T}P_0 + \int_0^T e^{-0.4(T-t)}R\,dt = P_0 e^{-0.4T} + Re^{-0.4T}\int_0^T e^{-0.4t}\,dt$$

$$= P_0 e^{-0.4T} + Re^{-0.4T}\left[\frac{1}{0.4}e^{0.4t}\right]_0^T = P_0 e^{-0.4T} + \frac{R}{0.4}e^{-0.4T}\left(e^{-0.4T} - 1\right)$$

Thus $P_0 - P_0 e^{-0.4T} = \dfrac{R}{0.4}(1 - e^{-0.4T})$ $\Rightarrow$ $P_0(1 - e^{-0.4T}) = \dfrac{R}{0.4}(1 - e^{-0.4T})$ $\Rightarrow$ $P_0 = \dfrac{R}{0.4}$ $\Rightarrow$ $R = 0.4P_0$.

Thus the drug should be administered at a constant rate of 0.4 times the desired level per hour.

## 6.4 Differential Equations

**Prepare Yourself**

**1.** $y = e^{-3x}$ $\Rightarrow$ $y' = -3e^{-3x}$ $\Rightarrow$ $y + y' = e^{-3x} - 3e^{-3x} = -2e^{-3x}$

**2.** $y = x^2 e^x$ $\Rightarrow$ $y' = x^2 \cdot e^x + e^x \cdot 2x = (x^2 + 2x)e^x$, so $xy' + y = x(x^2 + 2x)e^x + x^2 e^x = (x^3 + 3x^2)e^x$

**3.** (a) $\displaystyle\int \frac{1}{x}\,dx = \ln|x| + C$

(b) $\displaystyle\int \frac{1}{x^2}\,dx = \int x^{-2}\,dx = \frac{x^{-1}}{-1} + C = -\frac{1}{x} + C$

(c) Let $u = x + 4$ $\Rightarrow$ $du = dx$. Then $\displaystyle\int \frac{1}{x+4}\,dx = \int \frac{1}{u}\,du = \ln|u| + C = \ln|x+4| + C$.

(d) Let $u = x^2 + 4$ $\Rightarrow$ $du = 2x\,dx$ $\Rightarrow$ $x\,dx = \frac{1}{2}\,du$. Then

$$\int \frac{x}{x^2 + 4}\,dx = \int \frac{1}{u} \cdot \frac{1}{2}\,du = \tfrac{1}{2}\ln|u| + C = \tfrac{1}{2}\ln|x^2 + 4| + C = \tfrac{1}{2}\ln(x^2 + 4) + C \ \text{[since } x^2 + 4 > 0\text{]}.$$

(e) $\int e^{-2t}\,dt = -\frac{1}{2}e^{-2t} + C$

(f) $\int \sqrt{t}\,dt = \int t^{1/2}\,dt = \frac{2}{3}t^{3/2} + C$

**Exercises**

**1.** $y = \frac{2}{3}e^x + e^{-2x}$ $\Rightarrow$ $y' = \frac{2}{3}e^x - 2e^{-2x}$. Then

$y' + 2y = \frac{2}{3}e^x - 2e^{-2x} + 2\left(\frac{2}{3}e^x + e^{-2x}\right) = \frac{2}{3}e^x - 2e^{-2x} + \frac{4}{3}e^x + 2e^{-2x} = \frac{6}{3}e^x = 2e^x$ as desired.

**3.** $y = xe^x$ $\Rightarrow$ $y' = x \cdot e^x + e^x \cdot 1 = (x+1)e^x$ $\Rightarrow$ $y'/y = (x+1)e^x/(xe^x) = \dfrac{x+1}{x} = 1 + \dfrac{1}{x}$ as desired.

**5.** $\dfrac{dy}{dx} = xy^2$ $\Rightarrow$ $\dfrac{dy}{y^2} = x\,dx$ $[y \neq 0]$ $\Rightarrow$ $\int y^{-2}\,dy = \int x\,dx$ $\Rightarrow$ $-y^{-1} = \frac{1}{2}x^2 + C$ $\Rightarrow$

$\dfrac{1}{y} = -\frac{1}{2}x^2 - C$ $\Rightarrow$ $y = \dfrac{1}{-\frac{1}{2}x^2 - C} = \dfrac{2}{-x^2 - 2C} = \dfrac{2}{K - x^2}$, where $K = -2C$. $y = 0$ is also a solution.

**7.** $\dfrac{dy}{dx} = \dfrac{x+1}{y-1}$ $\Rightarrow$ $(y-1)\,dy = (x+1)\,dx$ $\Rightarrow$ $\int (y-1)\,dy = \int (x+1)\,dx$ $\Rightarrow$ $\frac{1}{2}y^2 - y = \frac{1}{2}x^2 + x + C$ $\Rightarrow$

$2\left(\frac{1}{2}y^2 - y\right) = 2\left(\frac{1}{2}x^2 + x + C\right)$ $\Rightarrow$ $y^2 - 2y = x^2 + 2x + 2C$ $\Rightarrow$ $y^2 - 2y = x^2 + 2x + K$, where $K = 2C$.

**9.** $\dfrac{dy}{dx} = \dfrac{y}{x}$ $\Rightarrow$ $\dfrac{1}{y}\,dy = \dfrac{1}{x}\,dx$ $[y \neq 0]$ $\Rightarrow$ $\displaystyle\int \frac{1}{y}\,dy = \int \frac{1}{x}\,dx$ $\Rightarrow$ $\ln|y| = \ln|x| + C$ $\Rightarrow$ $e^{\ln|y|} = e^{\ln|x|+C}$ $\Rightarrow$

$|y| = e^{\ln|x|}e^C = e^C|x|$ $\Rightarrow$ $y = \pm Kx$, where $K = e^C$. Letting $K$ be an arbitrary constant (either positive or negative), the solution can be written as $y = Kx$.

**11.** $(x^2 + 1)y' = xy$ $\Rightarrow$ $\dfrac{dy}{dx} = \dfrac{xy}{x^2 + 1}$ $\Rightarrow$ $\dfrac{1}{y}\,dy = \dfrac{x}{x^2+1}\,dx$ $[y \neq 0]$ $\Rightarrow$ $\displaystyle\int \frac{1}{y}\,dy = \int \frac{x}{x^2+1}\,dx$ $\Rightarrow$

$\ln|y| = \frac{1}{2}\ln(x^2 + 1) + C$ $[u = x^2 + 1, du = 2x\,dx]$ $= \ln(x^2+1)^{1/2} + C$ $\Rightarrow$

$e^{\ln|y|} = e^{\ln(x^2+1)^{1/2}+C} = e^{\ln(x^2+1)^{1/2}}e^C$ $\Rightarrow$ $|y| = e^C(x^2+1)^{1/2}$ $\Rightarrow$ $y = \pm e^C\sqrt{x^2+1}$. Since $y = 0$ is also a

solution, we can write $y = K\sqrt{x^2+1}$, where $K$ is any constant.

**13.** $\dfrac{du}{dt} = (1+u)(2+t)$ $\Rightarrow$ $\displaystyle\int \frac{du}{1+u} = \int (2+t)\,dt$ $[u \neq -1]$ $\Rightarrow$ $\ln|1+u| = \frac{1}{2}t^2 + 2t + C$ $\Rightarrow$

$|1+u| = e^{t^2/2 + 2t + C} = Ke^{t^2/2 + 2t}$ where $K = e^C$ $\Rightarrow$ $1 + u = \pm Ke^{t^2/2 + 2t}$ $\Rightarrow$ $u = -1 \pm Ke^{t^2/2 + 2t}$

where $K > 0$. $u = -1$ is also a solution, so $u = Ae^{t^2/2 + 2t} - 1$ where $A$ is an arbitrary constant.

**15.** $\dfrac{dy}{dx} = \dfrac{x}{y}$ $\Rightarrow$ $y\,dy = x\,dx$ $\Rightarrow$ $\int y\,dy = \int x\,dx$ $\Rightarrow$ $\frac{1}{2}y^2 = \frac{1}{2}x^2 + C$. $y(0) = -3$ $\Rightarrow$

$\frac{1}{2}(-3)^2 = \frac{1}{2}(0)^2 + C$ $\Rightarrow$ $C = \frac{9}{2}$, so $\frac{1}{2}y^2 = \frac{1}{2}x^2 + \frac{9}{2}$ $\Rightarrow$ $y^2 = x^2 + 9$ $\Rightarrow$ $y = -\sqrt{x^2+9}$ since $y(0) = -3 < 0$.

**17.** $y' = \dfrac{dy}{dx} = \dfrac{e^{2x}}{y^2}$ $\Rightarrow$ $y^2\,dy = e^{2x}\,dx$ $\Rightarrow$ $\int y^2\,dy = \int e^{2x}\,dx$ $\Rightarrow$ $\frac{1}{3}y^3 = \frac{1}{2}e^{2x} + C$ $\Rightarrow$ $y^3 = \frac{3}{2}e^{2x} + 3C$ $\Rightarrow$

$y = \sqrt[3]{\frac{3}{2}e^{2x} + K}$, where $K = 3C$. Then $y(0) = 3$ $\Rightarrow$ $3 = \sqrt[3]{\frac{3}{2}e^0 + K}$ $\Rightarrow$ $3 = \sqrt[3]{\frac{3}{2} + K}$ $\Rightarrow$ $3^3 = \frac{3}{2} + K$ $\Rightarrow$

$K = 27 - \frac{3}{2} = \frac{51}{2}$ $\Rightarrow$ $y = \sqrt[3]{\frac{3}{2}e^{2x} + \frac{51}{2}}$.

**19.** $\dfrac{dA}{dt} = \dfrac{1}{At}$ $\Rightarrow$ $A\,dA = \dfrac{1}{t}\,dt$ $\Rightarrow$ $\displaystyle\int A\,dA = \int \dfrac{1}{t}\,dt$ $\Rightarrow$ $\tfrac{1}{2}A^2 = \ln|t| + C$ $\Rightarrow$

$A = \pm\sqrt{2\ln|t| + 2C} = \pm\sqrt{2\ln|t| + K}$ where $K = 2C$. Since $A > 0$ and $t > 0$, we have $A = \sqrt{2\ln t + K}$.

Then $A(1) = 4$ $\Rightarrow$ $4 = \sqrt{2\ln 1 + K}$ $\Rightarrow$ $4 = \sqrt{K}$ $\Rightarrow$ $K = 16$ $\Rightarrow$ $A = \sqrt{2\ln t + 16}$.

**21.** (a) Since the derivative $y' = -y^2$ is always negative (or 0 if $y = 0$), the function $y$ must be decreasing (or equal to 0) on any

interval on which it is defined.

(b) $y = \dfrac{1}{x + C} = (x + C)^{-1}$ $\Rightarrow$ $y' = -(x + C)^{-2} = -\dfrac{1}{(x + C)^2}$. Then $-y^2 = -\left(\dfrac{1}{x + C}\right)^2 = -\dfrac{1}{(x + C)^2} = y'$ as

desired.

(c) $y = 0$ is a solution of $y' = -y^2$ that is not a member of the family in part (b).

(d) If $y = \dfrac{1}{x + C}$, then $y(0) = 0.5$ $\Rightarrow$ $\dfrac{1}{0 + C} = 0.5$ $\Rightarrow$ $\dfrac{1}{C} = \dfrac{1}{2}$ $\Rightarrow$ $C = 2$, so $y = \dfrac{1}{x + 2}$.

**23.** If the slope at the point $(x, y)$ is $xy$, then we have $\dfrac{dy}{dx} = xy$ $\Rightarrow$ $\dfrac{dy}{y} = x\,dx$ $[y \neq 0]$ $\Rightarrow$ $\displaystyle\int \dfrac{dy}{y} = \int x\,dx$ $\Rightarrow$

$\ln|y| = \tfrac{1}{2}x^2 + C$. $y(0) = 1$ $\Rightarrow$ $\ln 1 = 0 + C$ $\Rightarrow$ $C = 0$. Thus, $|y| = e^{x^2/2}$ $\Rightarrow$ $y = \pm e^{x^2/2}$, so $y = e^{x^2/2}$

since $y(0) = 1 > 0$. Note that $y = 0$ is not a solution because it doesn't satisfy the initial condition $y(0) = 1$.

**25.** Since $B(t)$ always increases at a rate inversely proportional to the value of $B(t)$, we write $B' = \dfrac{dB}{dt} = \dfrac{k}{B}$, where $k$ is the

constant of proportionality.

**27.** Let $P(t)$ represent the population $t$ years from now. Then $\dfrac{dP}{dt} = \dfrac{1}{100}P$ $\Rightarrow$ $\dfrac{1}{P}\,dP = \dfrac{1}{100}\,dt$ $\Rightarrow$

$\displaystyle\int \dfrac{1}{P}\,dP = \int \dfrac{1}{100}\,dt$ $\Rightarrow$ $\ln|P| = \dfrac{1}{100}t + C$ $\Rightarrow$ $|P| = e^C e^{t/100}$ $\Rightarrow$ $P = Ke^{t/100}$ where $K$ is a positive constant.

$P(0) = 20{,}000$ $\Rightarrow$ $20{,}000 = Ke^{0/100} = K$ $\Rightarrow$ $P(t) = 20{,}000e^{t/100}$.

**29.** Let $P(t)$ represent the ant population $t$ months from now. Then $\dfrac{dP}{dt} = \dfrac{1}{50}P + 500$ $\Rightarrow$ $\dfrac{1}{\frac{1}{50}P + 500}\,dP = dt$ $\Rightarrow$

$\dfrac{50}{P + 25{,}000}\,dP = dt$ $\Rightarrow$ $50\displaystyle\int \dfrac{1}{P + 25{,}000}\,dP = \int dt$ $\Rightarrow$ $50\ln(P + 25{,}000) = t + C$. $[P > 0$, so no absolute

value is needed.] Then $\ln(P + 25{,}000) = \tfrac{1}{50}t + \tfrac{1}{50}C$ $\Rightarrow$ $P + 25{,}000 = e^{C/50}e^{t/50}$ $\Rightarrow$ $P = Ke^{t/50} - 25{,}000$, where

$K$ is a positive constant.

$P(0) = 300{,}000$ $\Rightarrow$ $300{,}000 = K - 25{,}000$ $\Rightarrow$ $K = 325{,}000$ and $P(t) = 325{,}000e^{t/50} - 25{,}000$. In one year there

will be $P(12) = 325{,}000e^{12/50} - 25{,}000 \approx 388{,}156$ ants.

**31.** (a) $P$ increases most rapidly at the beginning, since there are usually many simple, easily-learned sub-skills associated with

learning a skill. As $t$ increases, we would expect $dP/dt$ to remain positive, but decrease. This is because as time

progresses, the only points left to learn are the more difficult ones.

(b) $\dfrac{dP}{dt} = k(M - P)$ is always positive, so the level of performance $P$

is increasing. As $P$ gets close to $M$, $dP/dt$ gets close to 0; that is,

the performance levels off, as explained in part (a).

(c)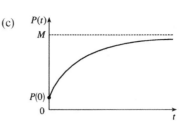

(d) $\dfrac{dP}{dt} = k(M - P)$ $\Rightarrow$ $\dfrac{1}{M - P}\,dP = k\,dt$ $\Rightarrow$ $\displaystyle\int \dfrac{1}{M - P}\,dP = \int k\,dt = -\ln|M - P| = kt + C$ $\Rightarrow$

$\ln|M - P| = -kt - C$ $\Rightarrow$ $|M - P| = e^{-kt-C}$ $\Rightarrow$ $M - P = e^{-C}e^{-kt}$ [since $M - P > 0$] $\Rightarrow$

$P = M - e^{-C}e^{-kt} = M + Ae^{-kt}$ where $A = -e^{-C}$ is a negative constant. If we assume that performance is at level 0

when $t = 0$, then $P(0) = 0$ $\Rightarrow$ $0 = M + A$ $\Rightarrow$ $A = -M$ $\Rightarrow$ $P(t) = M - Me^{-kt}$.

$\displaystyle\lim_{t\to\infty} P(t) = M - M\cdot 0 = M$.

**33.** $\dfrac{dy}{dt} = \dfrac{te^t}{y\sqrt{1 + y^2}}$ $\Rightarrow$ $y\sqrt{1 + y^2}\,dy = te^t\,dt$ $\Rightarrow$ $\displaystyle\int y\sqrt{1 + y^2}\,dy = \int te^t\,dt$ $\Rightarrow$ $\tfrac{1}{3}\left(1 + y^2\right)^{3/2} = te^t - e^t + C$

[where the first integral is evaluated by substitution and the second by parts] $\Rightarrow$ $1 + y^2 = [3(te^t - e^t + C)]^{2/3}$ $\Rightarrow$

$y = \pm\sqrt{[3(te^t - e^t + C)]^{2/3} - 1}$.

**35.** $u = x + y$ $\Rightarrow$ $\dfrac{d}{dx}(u) = \dfrac{d}{dx}(x + y)$ $\Rightarrow$ $\dfrac{du}{dx} = 1 + \dfrac{dy}{dx}$, but $\dfrac{dy}{dx} = x + y = u$, so $\dfrac{du}{dx} = 1 + u$ $\Rightarrow$

$\dfrac{du}{1 + u} = dx$ $[u \neq -1]$ $\Rightarrow$ $\displaystyle\int \dfrac{du}{1 + u} = \int dx$ $\Rightarrow$ $\ln|1 + u| = x + C$ $\Rightarrow$ $|1 + u| = e^{x+C}$ $\Rightarrow$

$1 + u = \pm e^C e^x$ $\Rightarrow$ $u = \pm e^C e^x - 1$ $\Rightarrow$ $x + y = \pm e^C e^x - 1$ $\Rightarrow$ $y = Ke^x - x - 1$, where $K = \pm e^C \neq 0$.

If $u = -1$, then $-1 = x + y$ $\Rightarrow$ $y = -x - 1$, which is just $y = Ke^x - x - 1$ with $K = 0$. Thus, the general solution

is $y = Ke^x - x - 1$, where $K$ is any constant.

**37.** (a) $\dfrac{dC}{dt} = r - kC$ $\Rightarrow$ $\dfrac{dC}{dt} = -(kC - r)$ $\Rightarrow$ $\displaystyle\int \dfrac{dC}{kC - r} = \int - dt$ $\Rightarrow$ $(1/k)\ln|kC - r| = -t + M_1$

[substitute $u = kC - r$ $\Rightarrow$ $du = k\,dC$] $\Rightarrow$ $\ln|kC - r| = -kt + M_2$ $\Rightarrow$ $|kC - r| = e^{-kt+M_2}$ $\Rightarrow$

$kC - r = M_3 e^{-kt}$ $\Rightarrow$ $kC = M_3 e^{-kt} + r$ $\Rightarrow$ $C(t) = M_4 e^{-kt} + r/k$. $C(0) = C_0$ $\Rightarrow$ $C_0 = M_4 + r/k$ $\Rightarrow$

$M_4 = C_0 - r/k$ $\Rightarrow$ $C(t) = (C_0 - r/k)e^{-kt} + r/k$.

(b) If $C_0 < r/k$, then $C_0 - r/k < 0$ and the formula for $C(t)$ shows that $C(t)$ increases and $\displaystyle\lim_{t\to\infty} C(t) = r/k$.

As $t$ increases, the formula for $C(t)$ shows how the role of $C_0$ steadily diminishes as that of $r/k$ increases.

**39.** The rate of growth of the area is jointly proportional to $\sqrt{A(t)}$ and $M - A(t)$; that is, the rate is proportional to the product of

those two quantities. So for some constant $k$, $dA/dt = k\sqrt{A}\,(M - A)$. We are interested in the maximum of the function

$dA/dt$ (when the tissue grows the fastest), so we differentiate, using the Chain Rule and then substituting for $dA/dt$ from the differential equation:

$$\frac{d}{dt}\left(\frac{dA}{dt}\right) = k\left[\sqrt{A}(-1)\frac{dA}{dt} + (M-A)\cdot\frac{1}{2}A^{-1/2}\frac{dA}{dt}\right] = \frac{1}{2}kA^{-1/2}\frac{dA}{dt}\left[-2A + (M-A)\right]$$

$$= \frac{1}{2}kA^{-1/2}\left[k\sqrt{A}(M-A)\right][M-3A] = \frac{1}{2}k^2(M-A)(M-3A)$$

This is 0 when $M - A = 0$ [this situation never actually occurs, since the graph of $A(t)$ approaches but never reaches the line $y = M$] and when $M - 3A = 0 \iff A(t) = M/3$. This represents a maximum by the First Derivative Test, since $\frac{d}{dt}\left(\frac{dA}{dt}\right)$ goes from positive to negative when $A(t) = M/3$.

## 6.5  Improper Integrals

---

**Prepare Yourself**

**1.** (a) $e^{-t} = \dfrac{1}{e^t}$ and $e^t \to \infty$ as $t \to \infty$, so its reciprocal $1/e^t \to 0$. Thus $\lim\limits_{t\to\infty} e^{-t} = 0$.

(b) $\lim\limits_{x\to\infty}(x + e^{-x}) = \lim\limits_{x\to\infty} x + \lim\limits_{x\to\infty} e^{-x}$ and $\lim\limits_{x\to\infty} e^{-x} = 0$ from part (a). But $\lim\limits_{x\to\infty} x = \infty$, so $\lim\limits_{x\to\infty}(x + e^{-x}) = \infty$.

(c) $\lim\limits_{x\to-\infty}(4 + e^{-x}) = \lim\limits_{x\to-\infty} 4 + \lim\limits_{x\to-\infty} e^{-x} = 4 + \lim\limits_{x\to-\infty}\dfrac{1}{e^x}$. As $x \to -\infty$, $e^x \to 0^+$ so $\dfrac{1}{e^x} \to \infty$. Thus $\lim\limits_{x\to-\infty}(4 + e^{-x}) = \infty$.

(d) $\lim\limits_{t\to\infty}\left(1 + 5\sqrt{t}\right) = \lim\limits_{t\to\infty} 1 + 5\left(\lim\limits_{t\to\infty}\sqrt{t}\right) = 1 + 5\lim\limits_{t\to\infty}\sqrt{t}$. But $\sqrt{t} \to \infty$ as $t \to \infty$, so $\lim\limits_{t\to\infty}\left(1 + 5\sqrt{t}\right) = \infty$.

(e) $\lim\limits_{t\to\infty}\left(1 + 5/\sqrt{t}\right) = \lim\limits_{t\to\infty} 1 + 5\left(\lim\limits_{t\to\infty}\dfrac{1}{\sqrt{t}}\right) = 1 + 5\lim\limits_{t\to\infty}\dfrac{1}{\sqrt{t}}$. As $t \to \infty$, $\sqrt{t} \to \infty$ so its reciprocal $1/\sqrt{t} \to 0$. Thus $\lim\limits_{t\to\infty}\left(1 + 5/\sqrt{t}\right) = 1 + 5\cdot 0 = 1$.

(f) As $x \to \infty$, $\ln x \to \infty \implies 1/\ln x \to 0$. So $\lim\limits_{x\to\infty}(1/\ln x) = 0$.

(g) As $x \to -\infty$, $1 + x^2 \to \infty \implies \ln(1 + x^2) \to \infty$. So $\lim\limits_{x\to-\infty}\ln(1 + x^2) = \infty$.

**2.** (a) $\displaystyle\int_2^5 \frac{1}{(x+1)^{5/2}}\,dx$. Let $u = x + 1 \implies du = dx$. Also, when $x = 2$, $u = 3$; when $x = 5$, $u = 6$. Then the integral becomes $\displaystyle\int_3^6 \frac{1}{u^{5/2}}\,du = \int_3^6 u^{-5/2}\,du = -\frac{2}{3}u^{-3/2}\Big]_3^6 = -\frac{2}{3}(6^{-3/2} - 3^{-3/2}) = -\frac{2}{3}\left(\frac{1}{6^{3/2}} - \frac{1}{3^{3/2}}\right)$.

(b) $\int_1^w e^{-x/3}\,dx = -3e^{-x/3}\Big]_1^w = -3(e^{-w/3} - e^{-1/3}) = 3e^{-1/3} - 3e^{-w/3}$

(c) Let $u = t^2 \implies du = 2t\,dt \implies t\,dt = \frac{1}{2}\,du$. Then $\int te^{t^2}\,dt = \int e^{t^2}(t\,dt) = \int e^u\left(\frac{1}{2}\,du\right) = \frac{1}{2}e^u + C = \frac{1}{2}e^{t^2} + C$.

(d) Let $u = \ln x \implies du = \dfrac{1}{x}\,dx$. Then $\displaystyle\int\frac{1}{x\ln x}\,dx = \int\frac{1}{\ln x}\left(\frac{1}{x}\,dx\right) = \int\frac{1}{u}\,du = \ln|u| + C = \ln|\ln x| + C$.

(e) $\displaystyle\int\frac{1}{t^{1.4}}\,dt = \int t^{-1.4}\,dt = \frac{1}{-0.4}t^{-0.4} + C = -2.5/t^{0.4} + C$

## Exercises

**1.** The area under the graph of $y = 1/x^3 = x^{-3}$ between $x = 1$ and $x = t$ is

$A(t) = \int_1^t x^{-3}\, dx = \left[-\frac{1}{2}x^{-2}\right]_1^t = -\frac{1}{2}t^{-2} - \left(-\frac{1}{2}\right) = \frac{1}{2} - 1/(2t^2)$. So the area for $1 \le x \le 10$ is

$A(10) = 0.5 - 0.005 = 0.495$, the area for $1 \le x \le 100$ is $A(100) = 0.5 - 0.00005 = 0.49995$, and the area for

$1 \le x \le 1000$ is $A(1000) = 0.5 - 0.0000005 = 0.4999995$. The total area under the curve for $x \ge 1$ is

$\displaystyle\lim_{t\to\infty} A(t) = \lim_{t\to\infty}\left[\frac{1}{2} - 1/(2t^2)\right] = \frac{1}{2} - 0 = \frac{1}{2}$.

**3.** $\displaystyle\int_1^\infty \frac{2}{x^3}\, dx = \lim_{t\to\infty}\int_1^t \frac{2}{x^3}\, dx = \lim_{t\to\infty} 2\int_1^t x^{-3}\, dx = \lim_{t\to\infty} 2\left[\frac{x^{-2}}{-2}\right]_1^t = \lim_{t\to\infty}\left[-\frac{1}{x^2}\right]_1^t$

$\displaystyle = \lim_{t\to\infty}\left(-\frac{1}{t^2} + \frac{1}{1}\right) = 0 + 1 = 1.$   Convergent

**5.** $\displaystyle\int_3^\infty \frac{1}{(x-2)^{3/2}}\, dx = \lim_{t\to\infty}\int_3^t (x-2)^{-3/2}\, dx = \lim_{t\to\infty}\left[-2(x-2)^{-1/2}\right]_3^t$   [substitute $u = x - 2,\, du = dx$]

$\displaystyle = \lim_{t\to\infty}\left[-2(t-2)^{-1/2} + 2(1)^{-1/2}\right] = \lim_{t\to\infty}\left(\frac{-2}{\sqrt{t-2}} + 2\right) = 0 + 2 = 2.$   Convergent

**7.** $\displaystyle\int_{-\infty}^{-1} \frac{1}{\sqrt{2-w}}\, dw = \lim_{t\to-\infty}\int_t^{-1} (2-w)^{-1/2}\, dw = \lim_{t\to-\infty}\left[-2(2-w)^{1/2}\right]_t^{-1}$   [$u = 2 - w,\, du = -dw$]

$\displaystyle = \lim_{t\to-\infty}\left[-2\sqrt{2-w}\right]_t^{-1} = \lim_{t\to-\infty}\left[-2\sqrt{3} + 2\sqrt{2-t}\right] = \infty$

since $2 - t \to \infty \;\Rightarrow\; \sqrt{2-t} \to \infty$ as $t \to -\infty$.   Divergent

**9.** $\displaystyle\int_4^\infty e^{-y/2}\, dy = \lim_{t\to\infty}\int_4^t e^{-y/2}\, dy = \lim_{t\to\infty}\left[-2e^{-y/2}\right]_4^t = \lim_{t\to\infty}\left(-2e^{-t/2} + 2e^{-2}\right) = 0 + 2e^{-2} = 2e^{-2}$ since $-t/2 \to -\infty$

as $t \to \infty$ and then $e^{-t/2} \to 0$.   Convergent

**11.** First we split the integral, following Definition 3: $\int_{-\infty}^\infty xe^{-x^2}\, dx = \int_{-\infty}^0 xe^{-x^2}\, dx + \int_0^\infty xe^{-x^2}\, dx$.

$\int_{-\infty}^0 xe^{-x^2}\, dx = \lim_{t\to-\infty}\left(-\frac{1}{2}\right)\left[e^{-x^2}\right]_t^0$   [substitute $u = -x^2$] $= \lim_{t\to-\infty}\left(-\frac{1}{2}\right)\left(1 - e^{-t^2}\right) = -\frac{1}{2}(1 - 0) = -\frac{1}{2}$, and

$\int_0^\infty xe^{-x^2}\, dx = \lim_{t\to\infty}\left(-\frac{1}{2}\right)\left[e^{-x^2}\right]_0^t = \lim_{t\to\infty}\left(-\frac{1}{2}\right)\left(e^{-t^2} - 1\right) = -\frac{1}{2}(0 - 1) = \frac{1}{2}$.

Therefore $\int_{-\infty}^\infty xe^{-x^2}\, dx = -\frac{1}{2} + \frac{1}{2} = 0.$   Convergent

**13.** $\displaystyle\int_1^\infty \frac{\ln x}{x}\, dx = \lim_{t\to\infty}\left[\frac{(\ln x)^2}{2}\right]_1^t \quad \begin{bmatrix} \text{by substitution with} \\ u = \ln x,\, du = dx/x \end{bmatrix} = \lim_{t\to\infty}\left[\frac{1}{2}(\ln t)^2 - 0\right] = \infty.$   Divergent

**15.** $\displaystyle\int_e^\infty \frac{1}{x(\ln x)^3}\, dx = \lim_{t\to\infty}\int_e^t (\ln x)^{-3}\cdot\frac{1}{x}\, dx = \lim_{t\to\infty}\left[-\frac{1}{2}(\ln x)^{-2}\right]_e^t \quad \begin{bmatrix} u = \ln x, \\ du = dx/x \end{bmatrix}$

$\displaystyle = \lim_{t\to\infty}\left[-\frac{1}{2}(\ln t)^{-2} + \frac{1}{2}(\ln e)^{-2}\right] = -\frac{1}{2}(0) + \frac{1}{2}(1) = \frac{1}{2}$

because $\ln t \to \infty$ as $t \to \infty$ and then $(\ln t)^{-2} = \dfrac{1}{(\ln t)^2} \to 0.$   Convergent

**17.**

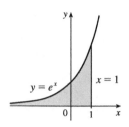

$$\text{Area} = \int_{-\infty}^{1} e^x \, dx = \lim_{t \to -\infty} \int_{t}^{1} e^x \, dx = \lim_{t \to -\infty} \left[ e^x \right]_{t}^{1}$$

$$= \lim_{t \to -\infty} (e^1 - e^t) = e - \lim_{t \to -\infty} e^t = e - 0 = e$$

**19.**

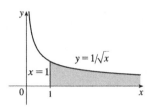

$$\text{Area} = \int_{1}^{\infty} \frac{1}{\sqrt{x}} \, dx = \lim_{t \to \infty} \int_{1}^{\infty} x^{-1/2} \, dx = \lim_{t \to \infty} \left[ 2x^{1/2} \right]_{1}^{t}$$

$$= \lim_{t \to \infty} (2\sqrt{t} - 2) = \infty$$

The area is infinite.

**21.** The yearly rate of the income is \$400/month $\times$ 12 months/year = \$4800/year. By Equation 4, the present value is

$$\text{PV} = \int_{0}^{\infty} f(t)e^{-rt} \, dt = \lim_{T \to \infty} \int_{0}^{T} 4800e^{-0.05t} \, dt = \lim_{T \to \infty} \left[ \frac{4800}{-0.05} e^{-0.05t} \right]_{0}^{T} = \lim_{T \to \infty} [-96{,}000(e^{-0.05T} - 1)]$$

$$= -96{,}000(0 - 1) = \$96{,}000$$

**23.** The annual revenue is 3% of \$10,000, or \$300. Thus

$$\text{PV} = \int_{0}^{\infty} f(t)e^{-rt} \, dt = \lim_{T \to \infty} \int_{0}^{T} 300e^{-0.042t} \, dt = \lim_{T \to \infty} \left[ \frac{300}{-0.042} e^{-0.042t} \right]_{0}^{T} = \lim_{T \to \infty} \left[ -\frac{300}{0.042} (e^{-0.042T} - 1) \right]$$

$$= -\frac{300}{0.042} (0 - 1) \approx \$7142.86$$

**25.** $\text{PV} = \int_{0}^{\infty} f(t)e^{-rt} \, dt = \lim_{T \to \infty} \int_{0}^{T} (8000e^{-0.01t})e^{-0.055t} \, dt = \lim_{T \to \infty} \int_{0}^{T} 8000e^{-0.045t} \, dt = \lim_{T \to \infty} \left[ \frac{8000}{-0.045} e^{-0.045t} \right]_{0}^{T}$

$$= \lim_{T \to \infty} \left[ -\frac{8000}{0.045} (e^{-0.045T} - 1) \right] = -\frac{8000}{0.045} (0 - 1) \approx \$177{,}777.78$$

**27.** We would expect a small percentage of bulbs to burn out in the first few hundred hours, most of the bulbs to burn out after close to 700 hours, and a few overachievers to burn on and on.

(a)

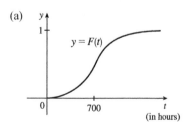

(b) $r(t) = F'(t)$ is the rate at which the fraction $F(t)$ of burnt-out bulbs increases as $t$ increases. This could be interpreted as a fractional burnout rate.

(c) $\int_{0}^{\infty} r(t) \, dt = \lim_{x \to \infty} F(x) = 1$, since all of the bulbs will eventually burn out.

## 6.6  Probability

**1.** (a) $\int_{30,000}^{40,000} f(x)\,dx$ is the probability that a randomly chosen tire will have a lifetime between 30,000 and 40,000 miles.

(b) $\int_{25,000}^{\infty} f(x)\,dx$ is the probability that a randomly chosen tire will have a lifetime of at least 25,000 miles.

**3.** (a) In general, we must satisfy the two conditions that are mentioned before Example 1—namely, (1) $f(x) \geq 0$ for all $x$,

and (2) $\int_{-\infty}^{\infty} f(x)\,dx = 1$. For $0 \leq x \leq 4$, we have $f(x) = \frac{3}{64}x\sqrt{16 - x^2} \geq 0$, so $f(x) \geq 0$ for all $x$. Also,

$$\int_{-\infty}^{\infty} f(x)\,dx = \int_0^4 \frac{3}{64}x\sqrt{16 - x^2}\,dx = \int_{16}^0 \frac{3}{64}\sqrt{u}\left(-\frac{1}{2}\,du\right) \qquad [\text{Let } u = 16 - x^2, du = -2x\,dx]$$

$$= -\frac{3}{128} \cdot \frac{2}{3}u^{3/2}\Big]_{16}^0 = -\frac{1}{64}(0 - 16^{3/2}) = -\frac{1}{64}(-64) = 1.$$

Therefore $f$ is a probability density function.

(b) $P(X < 2) = \int_{-\infty}^2 f(x)\,dx = \int_0^2 \frac{3}{64}x\sqrt{16 - x^2}\,dx = \int_{16}^{12} \frac{3}{64}\sqrt{u}\left(-\frac{1}{2}\,du\right) = -\frac{1}{64}u^{3/2}\Big]_{16}^{12}$

$= -\frac{1}{64}(12^{3/2} - 16^{3/2}) = -\frac{1}{64}(24\sqrt{3} - 64) = 1 - \frac{3}{8}\sqrt{3} \approx 0.350$

**5.** (a) The constant $c$ should be chosen so that $\int_{-\infty}^{\infty} f(x)\,dx = 1$ and so that $f(x) \geq 0$ for all $x$.

$$\int_{-\infty}^{\infty} f(x)\,dx = \int_{-\infty}^0 0\,dx + \int_0^{\infty} cxe^{-x^2}\,dx = 0 + \lim_{t \to \infty} \int_0^t cxe^{-x^2}\,dx$$

$$= \lim_{t \to \infty}\left[-\frac{c}{2}e^{-x^2}\right]_0^t \quad [\text{substitute } u = -x^2] \quad = -\frac{c}{2}\lim_{t \to \infty}\left(e^{-t^2} - 1\right) = -\frac{c}{2}(0 - 1) = \frac{c}{2}$$

Then $\int_{-\infty}^{\infty} f(x)\,dx = 1 \quad \Rightarrow \quad c = 2.$

(b) For $c = 2$, $P(1 < X < 4) = \int_1^4 f(x)\,dx = \int_1^4 2xe^{-x^2}\,dx = \left[-e^{-x^2}\right]_1^4 = -e^{-16} - (-e^{-1}) = \dfrac{1}{e} - \dfrac{1}{e^{16}} \approx 0.368.$

**7.** (a) In general, we must satisfy the two conditions that are mentioned before Example 1—namely, (1) $f(x) \geq 0$ for all $x$,

and (2) $\int_{-\infty}^{\infty} f(x)\,dx = 1$. Since $f(x) = 0$ or $f(x) = 0.1$, condition (1) is satisfied. For condition (2), we see that

$\int_{-\infty}^{\infty} f(x)\,dx = \int_0^{10} 0.1\,dx = \left[\frac{1}{10}x\right]_0^{10} = 1$. Thus, $f(x)$ is a probability density function for the spinner's values.

(b) Since all the numbers between 0 and 10 are equally likely to be selected, we expect the mean to be halfway between the endpoints of the interval; that is, $x = 5$.

$$\mu = \int_{-\infty}^{\infty} xf(x)\,dx = \int_0^{10} x(0.1)\,dx = \left[\frac{1}{20}x^2\right]_0^{10} = \frac{100}{20} = 5, \quad \text{as expected.}$$

**9.** We need to find $m$ so that $\int_m^{\infty} f(t)\,dt = \frac{1}{2} \quad \Rightarrow \quad \lim_{x \to \infty} \int_m^x \frac{1}{5}e^{-t/5}\,dt = \frac{1}{2} \quad \Rightarrow \quad \lim_{x \to \infty}\left[\frac{1}{5}(-5)e^{-t/5}\right]_m^x = \frac{1}{2} \quad \Rightarrow$

$\lim_{x \to \infty}(-1)(e^{-x/5} - e^{-m/5}) = \frac{1}{2} \quad \Rightarrow \quad (-1)(0 - e^{-m/5}) = \frac{1}{2} \quad \Rightarrow \quad e^{-m/5} = \frac{1}{2} \quad \Rightarrow \quad -m/5 = \ln\frac{1}{2} \quad \Rightarrow$

$m = -5\ln\frac{1}{2} \approx 3.47$ min.

**11.** We use an exponential density function with $\mu = 2.5$ min.

(a) $P(X > 4) = \int_4^{\infty} f(t)\,dt = \lim_{x \to \infty} \int_4^x \frac{1}{2.5}e^{-t/2.5}\,dt = \lim_{x \to \infty}\left[-e^{-t/2.5}\right]_4^x = \lim_{x \to \infty}\left[-e^{-x/2.5} + e^{-4/2.5}\right]$

$= 0 + e^{-4/2.5} \approx 0.202$

(b) $P(0 \le X \le 2) = \int_0^2 f(t)\,dt = \left[-e^{-t/2.5}\right]_0^2 = -e^{-2/2.5} + 1 \approx 0.551$

(c) We need to find a value $a$ so that $P(X \ge a) = 0.02$, or, equivalently, $P(0 \le X \le a) = 0.98 \Rightarrow$

$\int_0^a f(t)\,dt = 0.98 \Rightarrow \left[-e^{-t/2.5}\right]_0^a = 0.98 \Rightarrow -e^{-a/2.5} + 1 = 0.98 \Rightarrow e^{-a/2.5} = 0.02 \Rightarrow$

$-a/2.5 = \ln 0.02 \Rightarrow a = -2.5 \ln \frac{1}{50} = 2.5 \ln 50 \approx 9.78 \text{ min} \approx 10 \text{ min}$. The ad should say that if you aren't served

within 10 minutes, you get a free hamburger.

**13.** $P(X \ge 10) = \int_{10}^{\infty} \dfrac{1}{4.2\sqrt{2\pi}} e^{-(x-9.4)^2/[2(4.2)^2]}\,dx$. To avoid the improper integral we approximate it

by the integral from 10 to 100. (The area to the right of $x = 100$ is not significant.) Thus

$P(X \ge 10) \approx \int_{10}^{100} \dfrac{1}{4.2\sqrt{2\pi}} e^{-(x-9.4)^2/[2(4.2)^2]}\,dx \approx 0.443$ (using a calculator or computer to estimate the integral), so

about 44 percent of the households throw out at least 10 lb of paper a week.

*Note:* We can't evaluate $1 - P(0 \le X \le 10)$ for this problem since a significant amount of area lies to the left of $X = 0$.

**15.** (a) $P(0 \le X \le 100) = \int_0^{100} \dfrac{1}{8\sqrt{2\pi}} e^{-(x-112)^2/(2\cdot 8^2)}\,dx \approx 0.0668$ (using a calculator or computer to estimate the

integral), so there is about a 6.68% chance that a randomly chosen vehicle is traveling at a legal speed.

(b) $P(X \ge 125) = \int_{125}^{\infty} \dfrac{1}{8\sqrt{2\pi}} e^{-(x-112)^2/(2\cdot 8^2)}\,dx = \int_{125}^{\infty} f(x)\,dx$. In this case, we could use a calculator or computer to

estimate either $\int_{125}^{300} f(x)\,dx$ or $1 - \int_0^{125} f(x)\,dx$. Both are approximately 0.0521, so about 5.21% of the motorists are

targeted.

**17.** $P(\mu - 2\sigma \le X \le \mu + 2\sigma) = \int_{\mu-2\sigma}^{\mu+2\sigma} \dfrac{1}{\sigma\sqrt{2\pi}} e^{-(x-\mu)^2/(2\sigma^2)}\,dx$. Substituting $t = \dfrac{x-\mu}{\sigma}$ and $dt = \dfrac{1}{\sigma}\,dx$ gives us

$\int_{-2}^{2} \dfrac{1}{\sigma\sqrt{2\pi}} e^{-t^2/2}(\sigma\,dt) = \dfrac{1}{\sqrt{2\pi}} \int_{-2}^{2} e^{-t^2/2}\,dt \approx 0.9545.$

## 6  Review

**1.**

$y = 5 + 2x$

$x = 2$

$x = 0$

$y = x^2 + 1$

Area $= \int_0^2 [(5 + 2x) - (x^2 + 1)]\,dx = \int_0^2 (4 + 2x - x^2)\,dx$

$\quad = \left[4x + x^2 - \frac{1}{3}x^3\right]_0^2 = 8 + 4 - \frac{8}{3} - 0 = \frac{28}{3}$

**3.**

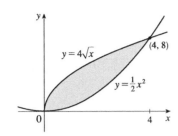

The curves intersect when $\frac{1}{2}x^2 = 4\sqrt{x} \Rightarrow x^2 = 8\sqrt{x} \Rightarrow$
$x^4 = 64x \Rightarrow x(x^3 - 64) = 0 \Rightarrow x = 0, 4$.

$\text{Area} = \int_0^4 \left(4\sqrt{x} - \frac{1}{2}x^2\right) dx = \left[4 \cdot \frac{2}{3}x^{3/2} - \frac{1}{2} \cdot \frac{1}{3}x^3\right]_0^4$

$= \left[\frac{8}{3}x^{3/2} - \frac{1}{6}x^3\right]_0^4 = \frac{64}{3} - \frac{64}{6} - 0 = \frac{32}{3}$

**5.** The curves intersect when $x^2 = 4x - x^2 \Rightarrow 2x^2 - 4x = 0 \Rightarrow$

$2x(x - 2) = 0 \Rightarrow x = 0$ or $2$. The upper boundary curve is $y = 4x - x^2$

and the lower boundary curve is $y = x^2$. Thus the area of the region is

$A = \int_0^2 \left[(4x - x^2) - x^2\right] dx = \int_0^2 (4x - 2x^2)\, dx$

$= \left[2x^2 - \frac{2}{3}x^3\right]_0^2 = \left(8 - \frac{16}{3}\right) - 0 = \frac{8}{3}$

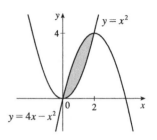

**7.** If $x \geq 0$, then $|x| = x$, and the graphs intersect when $x = 1 - 2x^2 \Rightarrow 2x^2 + x - 1 = 0 \Rightarrow (2x - 1)(x + 1) = 0 \Rightarrow$

$x = \frac{1}{2}$ or $-1$, but $-1 < 0$. By symmetry, we can double the area from $x = 0$ to $x = \frac{1}{2}$.

$A = 2\int_0^{1/2} \left[(1 - 2x^2) - x\right] dx = 2\int_0^{1/2} (-2x^2 - x + 1)\, dx$

$= 2\left[-\frac{2}{3}x^3 - \frac{1}{2}x^2 + x\right]_0^{1/2} = 2\left[\left(-\frac{1}{12} - \frac{1}{8} + \frac{1}{2}\right) - 0\right]$

$= 2\left(\frac{7}{24}\right) = \frac{7}{12}$

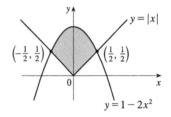

**9.** Note that $R'(q) > C'(q)$ for $0 \leq q \leq 500$, so the area between the graphs is

$\int_0^{500} [R'(q) - C'(q)]\, dq = \int_0^{500} [(22 - 0.005q) - (16 - 0.02q + 0.00004q^2)]\, dq$

$= \int_0^{500} (6 + 0.015q - 0.00004q^2)\, dq = \left[6q + 0.0075q^2 - \frac{0.00004}{3}q^3\right]_0^{500}$

$= 3000 + 1875 - \frac{0.00004}{3}(500)^3 \approx 3208.33$

The area represents the increase in profit ($3208.33) for producing and selling the first 500 backpacks.

**11.** The number of products sold is $Q = 100$ and the corresponding price is $P = 2000 - 0.1(100) - 0.01(100)^2 = 1890$

$\text{Consumer surplus} = \int_0^{100} [p(q) - P]\, dq = \int_0^{100} (2000 - 0.1q - 0.01q^2 - 1890)\, dq$

$= \left[110q - 0.05q^2 - \frac{0.01}{3}q^3\right]_0^{100} = 11{,}000 - 500 - \frac{10{,}000}{3} \approx \$7166.67$

**13.** Since the number of products sold is $Q = 500$, the corresponding price is $P = 18e^{0.003(500)} = 18e^{1.5}$. Therefore the producer surplus is

$\int_0^{500} (18e^{1.5} - 18e^{0.003q})\, dq = \left[18e^{1.5}q - \frac{18}{0.003}e^{0.003q}\right]_0^{500}$

$= (9000e^{1.5} - 6000e^{1.5}) - (0 - 6000) = 6000 + 3000e^{1.5} \approx \$19{,}455.07$

**15.** Assume that the income is continuous and is earned at the rate $2000/month × 12 months/year = $24,000 per year. Then

$$\text{FV} = e^{rT} \int_0^T f(t)e^{-rt}\, dt = e^{0.062(6)} \int_0^6 24{,}000 e^{-0.062t}\, dt = 24{,}000 e^{0.372} \left( \frac{1}{-0.062} \right) e^{-0.062t} \Big]_0^6$$

$$= -\frac{24{,}000}{0.062}\, e^{0.372}(e^{-0.372} - 1) \approx \$174{,}438.57$$

**17.** The population 10 years from now is

$$P(10) = S(10) \cdot P_0 + \int_0^{10} S(10 - t)\, R(t)\, dt = e^{-0.1(10)}(75{,}000) + \int_0^{10} e^{-0.1(10-t)}(3200 e^{0.05t})\, dt$$

$$= 75{,}000 e^{-1} + 3200 e^{-1} \int_0^{10} e^{0.15t}\, dt = 75{,}000 e^{-1} + 3200 e^{-1} \left[ \frac{1}{0.15}\, e^{0.15t} \right]_0^{10}$$

$$= 75{,}000 e^{-1} + \frac{3200}{0.15}\, e^{-1}(e^{1.5} - 1) \approx 54{,}916$$

**19.** $P(8) = S(8) \cdot P_0 + \displaystyle\int_0^8 S(8 - t)\, R(t)\, dt = \frac{2}{8+2}(26{,}500) + \int_0^8 \left( \frac{2}{(8-t)+2} \right)(1720)\, dt$

$$= 5300 + 1720 \int_0^8 \frac{2}{10 - t}\, dt = 5300 + 3440 \left[ -\ln|10 - t| \right]_0^8 = 5300 - 3440(\ln 2 - \ln 10)$$

$$\approx 10{,}836$$

**21.** $A' = \dfrac{dA}{dt} = \dfrac{k}{A^2}$, where $k$ is the constant of proportionality.

**23.** $\dfrac{dB}{dt} = \dfrac{e^{2B}}{\sqrt{t}} \;\Rightarrow\; e^{-2B}\, dB = t^{-1/2}\, dt \;\Rightarrow\; \int e^{-2B}\, dB = \int t^{-1/2}\, dt \;\Rightarrow\; -\tfrac{1}{2} e^{-2B} = 2t^{1/2} + C \;\Rightarrow$

$e^{-2B} = -4t^{1/2} - 2C \;\Rightarrow\; -2B = \ln\left(-4\sqrt{t} - 2C\right) \;\Rightarrow\; B = -\tfrac{1}{2}\ln\left(K - 4\sqrt{t}\right)$, where $K = -2C$.

**25.** $y' = xy^3 \;\Rightarrow\; \dfrac{dy}{dx} = xy^3 \;\Rightarrow\; y^{-3}\, dy = x\, dx \;(y \neq 0) \;\Rightarrow\; \int y^{-3}\, dy = \int x\, dx \;\Rightarrow\; -\tfrac{1}{2} y^{-2} = \tfrac{1}{2} x^2 + C \;\Rightarrow$

$y^{-2} = -x^2 - 2C \;\Rightarrow\; y^2 = \dfrac{1}{K - x^2}$, where $K = -2C$.

$y\left(\tfrac{1}{4}\right) = 4 \;\Rightarrow\; 4^2 = \dfrac{1}{K - \left(\tfrac{1}{4}\right)^2} \;\Rightarrow\; 16 = \dfrac{1}{K - \frac{1}{16}} \;\Rightarrow\; K - \tfrac{1}{16} = \tfrac{1}{16} \;\Rightarrow\; K = \tfrac{1}{8}$, so $y^2 = \dfrac{1}{\frac{1}{8} - x^2} \;\Rightarrow$

$y = \pm\sqrt{1/\left(\tfrac{1}{8} - x^2\right)}$. Only the positive solution satisfies the initial condition $y\left(\tfrac{1}{4}\right) = 4$, so the solution is

$y = \sqrt{1/\left(\tfrac{1}{8} - x^2\right)}$.

**27.** $\displaystyle\int_1^\infty \frac{1}{(2x + 1)^3}\, dx = \lim_{t \to \infty} \int_1^t (2x + 1)^{-3}\, dx = \lim_{t \to \infty} \left[ -\tfrac{1}{4}(2x + 1)^{-2} \right]_1^t$ \quad [substitute $u = 2x + 1,\, du = 2\, dx$]

$$= \lim_{t \to \infty} \left[ -\frac{1}{4(2x + 1)^2} \right]_1^t = -\frac{1}{4} \lim_{t \to \infty} \left[ \frac{1}{(2t + 1)^2} - \frac{1}{9} \right] = -\frac{1}{4}\left( 0 - \frac{1}{9} \right) = \frac{1}{36}.$$

Converges

**29.** $\int_{-\infty}^{0} e^{-x/2}\,dx = \lim\limits_{t\to-\infty}\int_{t}^{0} e^{-x/2}\,dx = \lim\limits_{t\to-\infty}\left[-2e^{-x/2}\right]_{t}^{0} = -2\lim\limits_{t\to-\infty}\left(1 - e^{-t/2}\right)$. $-t/2 \to \infty$ as $t \to -\infty$, so

$e^{-t/2} \to \infty$ and the integral diverges.

**31.** (a) For $f(x)$ to qualify as a probability density function, two conditions must be satisfied: (1) $f(x) \geq 0$ and

(2) $\int_{-\infty}^{\infty} f(x)\,dx = 1$. Since $f(x) = \frac{1}{288}(12x - x^2) \geq 0$ for $0 \leq x \leq 12$ and $f(x) = 0$ otherwise, condition (1) is

satisfied. Also,

$$\int_{-\infty}^{\infty} f(x)\,dx = \int_{0}^{12} \tfrac{1}{288}(12x - x^2)\,dx = \tfrac{1}{288}\left[6x^2 - \tfrac{1}{3}x^3\right]_{0}^{12} = \tfrac{1}{288}\left[6(12)^2 - \tfrac{1}{3}(12)^3 - 0\right] = \tfrac{1}{288}(288) = 1.$$

Therefore $f$ is a probability density function.

(b) $P(x < 4) = \int_{-\infty}^{4} f(x)\,dx = \int_{0}^{4} \tfrac{1}{288}(12x - x^2)\,dx = \tfrac{1}{288}\left[6x^2 - \tfrac{1}{3}x^3\right]_{0}^{4} = \tfrac{1}{288}\left[6(4)^2 - \tfrac{1}{3}(4)^3 - 0\right] = \tfrac{7}{27} \approx 0.259$

(c) $\mu = \int_{-\infty}^{\infty} x f(x)\,dx = \int_{0}^{12} (x)\left(\tfrac{1}{288}\right)(12x - x^2)\,dx = \tfrac{1}{288}\int_{0}^{12}(12x^2 - x^3)\,dx$

$= \tfrac{1}{288}\left[4x^3 - \tfrac{1}{4}x^4\right]_{0}^{12} = \tfrac{1}{288}\left[4(12)^3 - \tfrac{1}{4}(12)^4 - 0\right] = 6$

Since $f(x)$ is symmetric about $x = 6$, we would expect the mean to be 6.

**33.** (a) The probability density function is $f(t) = \begin{cases} 0 & \text{if } t < 0 \\ \tfrac{1}{8}e^{-t/8} & \text{if } t \geq 0 \end{cases}$

$$P(0 \leq X \leq 3) = \int_{0}^{3} \tfrac{1}{8}e^{-t/8}\,dt = \left[-e^{-t/8}\right]_{0}^{3} = -e^{-3/8} + 1 \approx 0.3127$$

(b) $P(X > 10) = \int_{10}^{\infty} \tfrac{1}{8}e^{-t/8}\,dt = \lim\limits_{x\to\infty}\left[-e^{-t/8}\right]_{10}^{x} = \lim\limits_{x\to\infty}\left(-e^{-x/8} + e^{-10/8}\right) = 0 + e^{-5/4} \approx 0.2865$

(c) We need to find $m$ such that $P(X \geq m) = \tfrac{1}{2}$ $\Rightarrow$ $\int_{m}^{\infty} \tfrac{1}{8}e^{-t/8}\,dt = \tfrac{1}{2}$ $\Rightarrow$ $\lim\limits_{x\to\infty}\left[-e^{-t/8}\right]_{m}^{x} = \tfrac{1}{2}$ $\Rightarrow$

$\lim\limits_{x\to\infty}\left(-e^{-x/8} + e^{-m/8}\right) = \tfrac{1}{2}$ $\Rightarrow$ $e^{-m/8} = \tfrac{1}{2}$ $\Rightarrow$ $-m/8 = \ln\tfrac{1}{2}$ $\Rightarrow$ $m = -8\ln\tfrac{1}{2} = 8\ln 2 \approx 5.55$ minutes.

# 7 FUNCTIONS OF SEVERAL VARIABLES

## 7.1 Functions of Several Variables

---

**Prepare Yourself**

**1.** (a) $f(x) = \sqrt{x+5}$ is defined only when $x + 5 \geq 0 \;\Rightarrow\; x \geq -5$, so the domain of $f$ is $\{x \mid x \geq -5\} = [-5, \infty)$.

(b) $g(x) = \sqrt{4 - x^2}$ is defined only when $4 - x^2 \geq 0 \;\Rightarrow\; x^2 \leq 4 \;\Rightarrow\; -2 \leq x \leq 2$, so the domain of $g$ is $\{x \mid -2 \leq x \leq 2\} = [-2, 2]$.

(c) $F(a) = \dfrac{a^2}{a-2}$ is defined only when $a - 2 \neq 0 \;\Rightarrow\; a \neq 2$, so the domain of $F$ is $\{a \mid a \neq 2\}$.

(d) $L(t) = \ln(1 - 2t)$ is defined only when $1 - 2t > 0 \;\Rightarrow\; 2t < 1 \;\Rightarrow\; t < \frac{1}{2}$, so the domain of $L$ is $\{t \mid t < \frac{1}{2}\} = \left(-\infty, \frac{1}{2}\right)$.

**2.** (a) $3x + y = 6 \;\Rightarrow\; y = -3x + 6$, a line with slope $-3$ and $y$-intercept 6.

(b) $x^2 + y = -1 \;\Rightarrow\; y = -x^2 - 1$, a parabola centered on the $y$-axis opening downward with vertex $(0, -1)$.

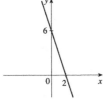

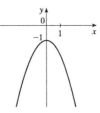

(c) $xy = 4 \;\Rightarrow\; y = 4/x$, a multiple of the reciprocal function.

(d) $\sqrt{8 - x^2 - y^2} = 2 \;\Rightarrow\; 8 - x^2 - y^2 = 4 \;\Rightarrow\; x^2 + y^2 = 4$, a circle with center the origin and radius 2.

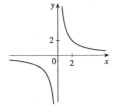

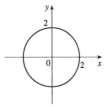

---

**Exercises**

**1.** $f(x, y) = 1 + 4xy - 3y^2$

(a) $f(6, 2) = 1 + 4(6)(2) - 3(2)^2 = 1 + 48 - 12 = 37$

(b) $f(-1, 4) = 1 + 4(-1)(4) - 3(4)^2 = 1 - 16 - 48 = -63$

(c) $f(0, -3) = 1 + 4(0)(-3) - 3(-3)^2 = 1 + 0 - 27 = -26$

(d) $f(x, 2) = 1 + 4(x)(2) - 3(2)^2 = 1 + 8x - 12 = 8x - 11$

**3.** $g(x, y) = x^2 e^{3y}$

(a) $g(-3, 0) = (-3)^2 e^{3(0)} = 9 \cdot 1 = 9$

(b) $g\left(3, \frac{1}{3}\right) = (3)^2 e^{3(1/3)} = 9 \cdot e^1 = 9e$

(c) $g(1, -1) = (1)^2 e^{3(-1)} = 1 \cdot e^{-3} = 1/e^3$

(d) $g(-2, y) = (-2)^2 e^{3y} = 4e^{3y}$

**5.** $f(x, y, z) = \dfrac{x}{y - z}$

(a) $f(12, 2, -2) = \dfrac{12}{2 - (-2)} = \dfrac{12}{4} = 3$

(b) $f(6, 5, 1) = \dfrac{6}{5 - 1} = \dfrac{6}{4} = \dfrac{3}{2}$

(c) $f\left(\frac{1}{6}, \frac{1}{2}, \frac{1}{3}\right) = \dfrac{\frac{1}{6}}{\frac{1}{2} - \frac{1}{3}} = \dfrac{\frac{1}{6}}{\frac{1}{6}} = 1$

(d) $f(x, 2, 3) = \dfrac{x}{2 - 3} = \dfrac{x}{-1} = -x$

**7.** $f(x, y) = e^{xy}$

(a) $f(x + h, y) - f(x, y) = e^{(x+h)y} - e^{xy}$

(b) $f(x, y + h) - f(x, y) = e^{x(y+h)} - e^{xy}$

**9.** $f(x, y) = x^2 e^{3xy}$

(a) $f(2, 0) = (2)^2 e^{3(2)(0)} = 4e^0 = 4(1) = 4$

(b) Since both $x^2$ and the exponential function are defined everywhere, $x^2 e^{3xy}$ is defined for all choices of values for $x$ and $y$. Thus the domain of $f$ is the entire $xy$-plane.

**11.** $f(x, y) = \sqrt{x + y}$ is defined whenever $x + y \geq 0 \Rightarrow$ $y \geq -x$, so the domain is $\{(x, y) \mid y \geq -x\}$ which is the set of points that lie on or above the line $y = -x$.

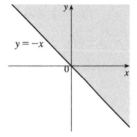

**13.** $f(x, y) = \dfrac{xy}{x - y}$ is defined whenever $x - y \neq 0 \Rightarrow$ $y \neq x$, so the domain is $\{(x, y) \mid y \neq x\}$ which is the set of all points not on the line $y = x$.

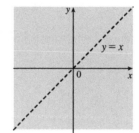

**15.** $g(x, y) = \ln(4 - x^2 - y^2)$ is defined whenever $4 - x^2 - y^2 > 0 \Rightarrow$ $x^2 + y^2 < 4$, so the domain is $\{(x, y) \mid x^2 + y^2 < 4\}$, the interior of a circle of radius 2 with center the origin.

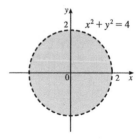

**17.** We start at the origin, which has coordinates $(0, 0, 0)$. First we move 4 units along the positive $x$-axis, affecting only the $x$-coordinate, bringing us to the point $(4, 0, 0)$. We then move 3 units straight downward, in the negative $z$-direction. Thus only the $z$-coordinate is affected, and we arrive at $(4, 0, -3)$.

**19.**

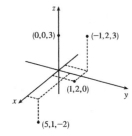

**21.** The graph of $f(x, y) = 1 - x - y$ has equation

$z = 1 - x - y$ $\Rightarrow$ $x + y + z = 1$, a plane with

$x$-intercept 1 (put $y = z = 0$), $y$-intercept 1, and

$z$-intercept 1.

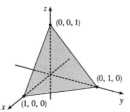

**23.** The level curves of $f(x, y) = 2x - y$ are $2x - y = k$

or $y = 2x - k$, a family of lines with slope 2 and

$y$-intercept $-k$.

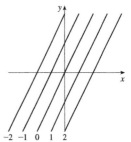

**25.** The level curves are $xy = k$ or $y = k/x$. For $k = 0$ the curves are the

coordinate axes; if $k > 0$, they are hyperbolas in the first and third quadrants

(multiples of the reciprocal function); if $k < 0$, they are hyperbolas in the

second and fourth quadrants.

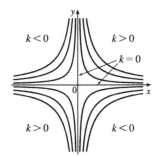

**27.** The level curves are $ye^x = k$ or $y = ke^{-x}$, a family of exponential curves.

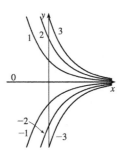

**29.** $z = (1 - x^2)(1 - y^2)$     (a) B     (b) III

Reasons: $z = 0$ when $1 - x^2 = 0$ $\Rightarrow$ $x = \pm 1$ or $1 - y^2 = 0$ $\Rightarrow$ $y = \pm 1$, thus the function is 0 along the lines $x = \pm 1$

and $y = \pm 1$. The only contour map in which this could occur is III. Also when $x = 0$ and $y = 0$ we have $z = 1$; the only

graph where this is possible is B.

**31.** $z = \dfrac{1}{x^2 + 4y^2}$     (a) A     (b) IV

Reasons: $z$ is undefined at the origin, and gets larger and larger as we use points closer to the origin. This behavior is satisfied

only by graph A which matches contour map IV.

**33.** $P(L, K) = 1.47L^{0.65}K^{0.35}$ $\Rightarrow$ $P(120, 20) = 1.47(120)^{0.65}(20)^{0.35} \approx 94.2$. This means that when the manufacturer invests \$20 million and 120,000 labor hours are spent, its yearly production is about \$94.2 million.

**35.** (a) $C = f(x, y, z) = 8000 + 2.5x + 4y + 4.5z$, measured in dollars.

(b) $f(3000, 5000, 4000) = 8000 + 2.5(3000) + 4(5000) + 4.5(4000) = 8000 + 7500 + 20,000 + 18,000 = 53,500$.
This means that it costs the company \$53,500 to make 3000 small boxes, 5000 medium boxes, and 4000 large boxes.

(c) $x$, $y$, or $z$ cannot be negative so the domain is $\{(x, y, z) \mid x \geq 0, y \geq 0, z \geq 0\}$. (In practice, only integer values would be used.)

**37.** (a) $f(160, 70) = 0.1091(160)^{0.425}(70)^{0.725} \approx 20.52$. This means that a person who weighs 160 pounds and is 70 inches tall has surface area about 20.52 square feet.

(b) Clearly answers will vary!

**39.** $f(x, y) = e^{-x^2} + e^{-2y^2}$

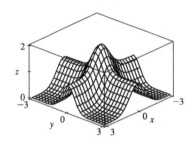
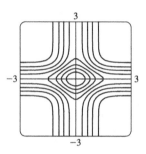

**41.** $f(x, y) = xy^2 - x^3$

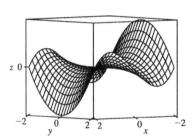
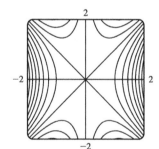

The surface is called a monkey saddle because a monkey sitting on the surface near the origin has places for both legs and tail to rest.

## 7.2 Partial Derivatives

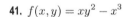

**Prepare Yourself**

**1.** $f'(1.5) = 0.6$ means that rain is falling at a rate of 0.6 inches per hour 1.5 hours after the start of the storm.

**2.** (a) $g(x) = 5x^3 - 8x^2 + 13x - 4$ $\Rightarrow$ $g'(x) = 5 \cdot 3x^2 - 8 \cdot 2x + 13 - 0 = 15x^2 - 16x + 13$

(b) $f(x) = (x + 2)^8$ $\Rightarrow$ $f'(x) = 8(x + 2)^7 \cdot 1 = 8(x + 2)^7$

(c) $K(v) = 3^v$ $\Rightarrow$ $K'(v) = 3^v \ln 3$

(d) $B(u) = u^3 e^u$ $\Rightarrow$ $B'(u) = u^3 \cdot e^u + e^u \cdot 3u^2 = (u^3 + 3u^2)e^u$

(e) $H(t) = \dfrac{e^t}{t^3} \;\Rightarrow\; H'(t) = \dfrac{t^3 \cdot e^t - e^t \cdot 3t^2}{(t^3)^2} = \dfrac{t^2 e^t (t-3)}{t^6} = \dfrac{(t-3)e^t}{t^4}$

(f) $f(x) = \dfrac{7x}{x^2+1} \;\Rightarrow\; f'(x) = \dfrac{(x^2+1)\cdot 7 - 7x \cdot 2x}{(x^2+1)^2} = \dfrac{7 - 7x^2}{(x^2+1)^2}$

(g) $g(y) = \sqrt{y} + \ln y + 1 \;\Rightarrow\; g'(y) = \tfrac{1}{2} y^{-1/2} + \dfrac{1}{y} + 0 = \dfrac{1}{2\sqrt{y}} + \dfrac{1}{y}$

(h) $y = e^{x^2+2} \;\Rightarrow\; dy/dx = e^{x^2+2}(2x+0) = 2x e^{x^2+2}$

(i) $y = \ln(t^2 - 5t) \;\Rightarrow\; \dfrac{dy}{dt} = \dfrac{1}{t^2 - 5t}(2t - 5) = \dfrac{2t - 5}{t^2 - 5t}$

(j) $A(t) = t\sqrt{t^3 - 1} = t(t^3 - 1)^{1/2} \;\Rightarrow\;$

$\quad A'(t) = t \cdot \tfrac{1}{2}(t^3 - 1)^{-1/2}(3t^2) + (t^3 - 1)^{1/2} \cdot 1 = \dfrac{3t^3}{2\sqrt{t^3 - 1}} + \sqrt{t^3 - 1}$

**3.** (a) $z = a^2 + x^3$ and $a$ is a constant, so $dz/dx = 0 + 3x^2 = 3x^2$.

(b) $z = ae^x$ and $a$ is a constant, so $dz/dx = ae^x$.

(c) $z = \ln(x + ax^2 + b)$ and $a$ and $b$ are constants, so $\dfrac{dz}{dx} = \dfrac{1}{x + ax^2 + b}(1 + a \cdot 2x + 0) = \dfrac{1 + 2ax}{x + ax^2 + b}$.

(d) $z = \dfrac{x}{2x - c}$ and $c$ is a constant, so $\dfrac{dz}{dx} = \dfrac{(2x - c)\cdot 1 - x \cdot (2 - 0)}{(2x - c)^2} = -\dfrac{c}{(2x - c)^2}$.

**4.** (a) $y = \dfrac{x}{x - 1} \;\Rightarrow\; \dfrac{dy}{dx} = \dfrac{(x - 1)\cdot 1 - x \cdot 1}{(x - 1)^2} = -\dfrac{1}{(x - 1)^2} = -(x - 1)^{-2} \;\Rightarrow\;$

$\dfrac{d^2 y}{dx^2} = -(-2)(x - 1)^{-3}(1) = \dfrac{2}{(x - 1)^3}$.

(b) $y = \sqrt{x^2 + 1} = (x^2 + 1)^{1/2} \;\Rightarrow\;$

$\dfrac{dy}{dx} = \tfrac{1}{2}(x^2 + 1)^{-1/2} \cdot 2x = \dfrac{x}{(x^2 + 1)^{1/2}} \;\Rightarrow\;$

$\dfrac{d^2 y}{dx^2} = \dfrac{(x^2 + 1)^{1/2} \cdot 1 - x \cdot \tfrac{1}{2}(x^2 + 1)^{-1/2}(2x)}{[(x^2 + 1)^{1/2}]^2} = \dfrac{(x^2 + 1)^{-1/2}\left[(x^2 + 1) - x^2\right]}{(x^2 + 1)} = \dfrac{1}{(x^2 + 1)^{3/2}}$

## Exercises

**1.** $C_w(150, 80)$ represents the rate at which the cost changes, in dollars per pound, to deliver a package 150 miles if the weight changes from 80 pounds (but the distance does not change).

**3.** (a) $\partial T/\partial x$ represents the rate of change of $T$ when we hold $y$ and $t$ constant and consider $T$ as a function of the single variable $x$, which describes how quickly the temperature changes when longitude changes but latitude and time are constant. $\partial T/\partial y$ represents the rate of change of $T$ when we hold $x$ and $t$ constant and consider $T$ as a function of $y$, which describes how quickly the temperature changes when latitude changes but longitude and time are constant. $\partial T/\partial t$ represents the rate of change of $T$ when we hold $x$ and $y$ constant and consider $T$ as a function of $t$, which describes how quickly the temperature changes over time for a constant longitude and latitude.

(b) $f_x(158, 21, 9)$ represents the rate of change of temperature at longitude $158°$W, latitude $21°$N at 9:00 AM when only longitude varies. Since the air is warmer to the west than to the east, increasing longitude results in an increased air temperature, so we would expect $f_x(158, 21, 9)$ to be positive. $f_y(158, 21, 9)$ represents the rate of change of temperature at the same time and location when only latitude varies. Since the air is warmer to the south and cooler to the north, increasing latitude results in a decreased air temperature, so we would expect $f_y(158, 21, 9)$ to be negative. $f_t(158, 21, 9)$ represents the rate of change of temperature at the same time and location when only time varies. Since typically air temperature increases from the morning to the afternoon as the sun warms it, we would expect $f_t(158, 21, 9)$ to be positive.

**5.** (a) By Definition 2, $f_T(-15, 30) = \lim\limits_{h \to 0} \dfrac{f(-15 + h, 30) - f(-15, 30)}{h}$, which we can approximate by considering $h = 5$

and $h = -5$ and using the values given in the table:

$$f_T(-15, 30) \approx \frac{f(-10, 30) - f(-15, 30)}{5} = \frac{-20 - (-26)}{5} = \frac{6}{5} = 1.2,$$

$$f_T(-15, 30) \approx \frac{f(-20, 30) - f(-15, 30)}{-5} = \frac{-33 - (-26)}{-5} = \frac{-7}{-5} = 1.4. \text{ Averaging these values, we estimate}$$

$f_T(-15, 30)$ to be approximately 1.3. Thus, when the actual temperature is $-15°$C and the wind speed is 30 km/h, the apparent temperature rises by about $1.3°$C for every degree that the actual temperature rises.

Similarly, $f_v(-15, 30) = \lim\limits_{h \to 0} \dfrac{f(-15, 30 + h) - f(-15, 30)}{h}$ which we can approximate by considering $h = 10$

and $h = -10$: $f_v(-15, 30) \approx \dfrac{f(-15, 40) - f(-15, 30)}{10} = \dfrac{-27 - (-26)}{10} = \dfrac{-1}{10} = -0.1,$

$$f_v(-15, 30) \approx \frac{f(-15, 20) - f(-15, 30)}{-10} = \frac{-24 - (-26)}{-10} = \frac{2}{-10} = -0.2. \text{ Averaging these values, we estimate}$$

$f_v(-15, 30)$ to be approximately $-0.15$. Thus, when the actual temperature is $-15°$C and the wind speed is 30 km/h, the apparent temperature decreases by about $0.15°$C for every km/h that the wind speed increases.

(b) For a fixed wind speed $v$, the values of the wind-chill index $W$ increase as temperature $T$ increases (look at a column of

the table), so $\dfrac{\partial W}{\partial T}$ is positive. For a fixed temperature $T$, the values of $W$ decrease (or remain constant) as $v$ increases

(look at a row of the table), so $\dfrac{\partial W}{\partial v}$ is negative (or perhaps 0).

(c) For fixed values of $T$, the function values $f(T, v)$ appear to become constant (or nearly constant) as $v$ increases, so the

corresponding rate of change is 0 or near 0 as $v$ increases. This suggests that $\lim\limits_{v \to \infty} (\partial W/\partial v) = 0$.

**7.** (a) If we start at $(1, 2)$ and move in the positive $x$-direction, the graph of $f$ increases. Thus $f_x(1, 2)$ is positive.

(b) If we start at $(1, 2)$ and move in the positive $y$-direction, the graph of $f$ decreases. Thus $f_y(1, 2)$ is negative.

(c) The graph of $f$ decreases if we start at $(-1, 2)$ and move in the positive $x$-direction, so $f_x(-1, 2)$ is negative.

(d) The graph of $f$ decreases if we start at $(-1, 2)$ and move in the positive $y$-direction, so $f_y(-1, 2)$ is negative.

**9.** $f(x, y) = 16 - 4x^2 - y^2$. According to Definition 1, we find $f_x(x, 2)$ by considering the function of one variable that we get

by letting $y = 2$ in the expression for $f(x, y)$: $f(x, 2) = 16 - 4x^2 - 2^2 = 12 - 4x^2$. Then we differentiate with respect

to $x$: $f_x(x, 2) = -8x$. Then with $x = 1$, $f_x(1, 2) = -8(1) = -8$.

Similarly, $f(1, y) = 16 - 4(1)^2 - y^2 = 12 - y^2$, so $f_y(1, y) = -2y$. Putting $y = 2$ in this equation gives

$f_y(1, 2) = -2(2) = -4$.

The graph of $f$ is the paraboloid $z = 16 - 4x^2 - y^2$ and the vertical plane $y = 2$ intersects it in the parabola $z = 12 - 4x^2$,

$y = 2$ (the curve $C_1$ in the first figure). The slope of the tangent line to this parabola at $(1, 2, 8)$ is $f_x(1, 2) = -8$. Similarly

the plane $x = 1$ intersects the paraboloid in the parabola $z = 12 - y^2$, $x = 1$ (the curve $C_2$ in the second figure) and the slope

of the tangent line at $(1, 2, 8)$ is $f_y(1, 2) = -4$.

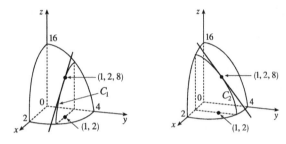

**11.** (a) $f(x, y) = x^2 y - 5xy^2 \ \Rightarrow \ f(x, -2) = x^2(-2) - 5x(-2)^2 = -2x^2 - 20x$, a function of the single variable $x$. Then

we differentiate with respect to $x$ to get $f_x(x, -2) = -4x - 20$ (see Definition 1).

(b) $f(3, y) = (3)^2 y - 5(3)y^2 = 9y - 15y^2$. Differentiating with respect to $y$, we have $f_y(3, y) = 9 - 30y$.

**13.** $f(x, y) = y^5 - 3xy$. To find $f_x$, we regard $y$ as a constant and differentiate $f$ with respect to $x$:

$f_x(x, y) = 0 - 3y = -3y$. To find $f_y$, we regard $x$ as a constant and differentiate $f$ with respect to $y$:

$f_y(x, y) = 5y^4 - 3x$.

**15.** $f(x, y) = x^4 + x^2 y^2 + y^4 \ \Rightarrow \ f_x(x, y) = 4x^3 + 2xy^2 + 0 = 4x^3 + 2xy^2, \quad f_y(x, y) = 0 + x^2 \cdot 2y + 4y^3 = 2x^2 y + 4y^3$

**17.** $z = (2x + 3y)^{10} \ \Rightarrow \ \dfrac{\partial z}{\partial x} = 10(2x + 3y)^9 \cdot 2 = 20(2x + 3y)^9, \dfrac{\partial z}{\partial y} = 10(2x + 3y)^9 \cdot 3 = 30(2x + 3y)^9$

**19.** $f(x, y) = \dfrac{x - y}{x + y} \ \Rightarrow \ f_x(x, y) = \dfrac{(x + y) \cdot 1 - (x - y) \cdot 1}{(x + y)^2} = \dfrac{2y}{(x + y)^2}$,

$f_y(x, y) = \dfrac{(x + y)(-1) - (x - y)(1)}{(x + y)^2} = -\dfrac{2x}{(x + y)^2}$

**21.** $f(r, s) = r \ln(r^2 + s^2) \ \Rightarrow \ f_r(r, s) = r \cdot \dfrac{1}{r^2 + s^2}(2r) + \ln(r^2 + s^2) \cdot 1 = \dfrac{2r^2}{r^2 + s^2} + \ln(r^2 + s^2)$,

$f_s(r, s) = r \cdot \dfrac{1}{r^2 + s^2}(2s) = \dfrac{2rs}{r^2 + s^2}$

**23.** $u = te^{w/t} \ \Rightarrow \ \dfrac{\partial u}{\partial t} = t \cdot e^{w/t}(-wt^{-2}) + e^{w/t} \cdot 1 = e^{w/t} - \dfrac{w}{t}e^{w/t} = e^{w/t}\left(1 - \dfrac{w}{t}\right), \dfrac{\partial u}{\partial w} = te^{w/t} \cdot \dfrac{1}{t} = e^{w/t}$

**25.** $f(s,t) = \sqrt{2 - 3s^2 - 5t^2} = (2 - 3s^2 - 5t^2)^{1/2}$ $\Rightarrow$

$f_s(s,t) = \frac{1}{2}(2 - 3s^2 - 5t^2)^{-1/2}(-6s) = -\dfrac{3s}{\sqrt{2 - 3s^2 - 5t^2}}$,

$f_t(s,t) = \frac{1}{2}(2 - 3s^2 - 5t^2)^{-1/2}(-10t) = -\dfrac{5t}{\sqrt{2 - 3s^2 - 5t^2}}$

**27.** $f(x,y,z) = xz - 5x^2y^3z^4$ $\Rightarrow$ $f_x(x,y,z) = z - 10xy^3z^4$, $f_y(x,y,z) = 0 - 5x^2 \cdot 3y^2 \cdot z^4 = -15x^2y^2z^4$,

$f_z(x,y,z) = x - 5x^2y^3 \cdot 4z^3 = x - 20x^2y^3z^3$

**29.** $w = \ln(x + 2y + 3z)$ $\Rightarrow$ $\dfrac{\partial w}{\partial x} = \dfrac{1}{x + 2y + 3z}$, $\dfrac{\partial w}{\partial y} = \dfrac{2}{x + 2y + 3z}$, $\dfrac{\partial w}{\partial z} = \dfrac{3}{x + 2y + 3z}$

**31.** $f(x,y,z) = \dfrac{x}{y + z} = x(y + z)^{-1}$ $\Rightarrow$ $f_x(x,y,z) = \dfrac{1}{y + z}$, $f_y(x,y,z) = -x(y + z)^{-2} \cdot 1 = -\dfrac{x}{(y + z)^2}$,

$f_z(x,y,z) = -x(y + z)^{-2} \cdot 1 = -\dfrac{x}{(y + z)^2}$

**33.** $f(x,y,z) = x^{yz}$ $\Rightarrow$ $f_x(x,y,z) = yzx^{yz-1}$, $f_y(x,y,z) = x^{yz}\ln x \cdot \dfrac{\partial}{\partial y}(yz) = x^{yz}\ln x \cdot z = zx^{yz}\ln x$,

$f_z(x,y,z) = x^{yz}\ln x \cdot \dfrac{\partial}{\partial z}(yz) = x^{yz}\ln x \cdot y = yx^{yz}\ln x$

**35.** $f(x,y,z,t) = xy^2z^3t^4$ $\Rightarrow$ $f_x(x,y,z,t) = y^2z^3t^4$, $f_y(x,y,z,t) = 2xyz^3t^4$,

$f_z(x,y,z,t) = 3xy^2z^2t^4$, $f_t(x,y,z,t) = 4xy^2z^3t^3$

**37.** $f(x,y) = x^3y^5$ $\Rightarrow$ $f_x(x,y) = 3x^2y^5$, so $f_x(3,-1) = 3(3)^2(-1)^5 = -27$

**39.** $f(x,y) = \ln\left(x + \sqrt{x^2 + y^2}\right)$ $\Rightarrow$

$f_x(x,y) = \dfrac{1}{x + \sqrt{x^2 + y^2}}\left[1 + \frac{1}{2}(x^2 + y^2)^{-1/2}(2x)\right] = \dfrac{1}{x + \sqrt{x^2 + y^2}}\left(1 + \dfrac{x}{\sqrt{x^2 + y^2}}\right)$,

so $f_x(3,4) = \dfrac{1}{3 + \sqrt{3^2 + 4^2}}\left(1 + \dfrac{3}{\sqrt{3^2 + 4^2}}\right) = \frac{1}{8}\left(1 + \frac{3}{5}\right) = \frac{1}{5}$.

**41.** $f(x,y,z) = \dfrac{y}{x + y + z}$ $\Rightarrow$ $f_y(x,y,z) = \dfrac{(x + y + z) \cdot 1 - y \cdot 1}{(x + y + z)^2} = \dfrac{x + z}{(x + y + z)^2}$,

so $f_y(2,1,-1) = \dfrac{2 + (-1)}{(2 + 1 + (-1))^2} = \dfrac{1}{4}$.

**43.** $f(x,y) = x^3y^5 + 2x^4y$ $\Rightarrow$ $f_x(x,y) = 3x^2y^5 + 8x^3y$, $f_y(x,y) = 5x^3y^4 + 2x^4$. Then $f_{xx}(x,y) = 6xy^5 + 24x^2y$,

$f_{xy}(x,y) = 15x^2y^4 + 8x^3$, $f_{yx}(x,y) = 15x^2y^4 + 8x^3$, and $f_{yy}(x,y) = 20x^3y^3$.

**45.** $w = \sqrt{u^2 + v^2}$ $\Rightarrow$ $w_u = \frac{1}{2}(u^2 + v^2)^{-1/2} \cdot 2u = \dfrac{u}{\sqrt{u^2 + v^2}}$, $w_v = \frac{1}{2}(u^2 + v^2)^{-1/2} \cdot 2v = \dfrac{v}{\sqrt{u^2 + v^2}}$. Then

$w_{uu} = \dfrac{\sqrt{u^2 + v^2} \cdot 1 - u \cdot \frac{1}{2}(u^2 + v^2)^{-1/2}(2u)}{\left(\sqrt{u^2 + v^2}\right)^2} = \dfrac{\sqrt{u^2 + v^2} - u^2/\sqrt{u^2 + v^2}}{u^2 + v^2} = \dfrac{u^2 + v^2 - u^2}{(u^2 + v^2)^{3/2}} = \dfrac{v^2}{(u^2 + v^2)^{3/2}}$,

$$w_{uv} = u\left(-\tfrac{1}{2}\right)\left(u^2 + v^2\right)^{-3/2}(2v) = -\frac{uv}{(u^2+v^2)^{3/2}}, \; w_{vu} = v\left(-\tfrac{1}{2}\right)\left(u^2 + v^2\right)^{-3/2}(2u) = -\frac{uv}{(u^2+v^2)^{3/2}},$$

$$w_{vv} = \frac{\sqrt{u^2+v^2}\cdot 1 - v\cdot\tfrac{1}{2}(u^2+v^2)^{-1/2}(2v)}{\left(\sqrt{u^2+v^2}\right)^2} = \frac{\sqrt{u^2+v^2} - v^2/\sqrt{u^2+v^2}}{u^2+v^2} = \frac{u^2+v^2-v^2}{(u^2+v^2)^{3/2}} = \frac{u^2}{(u^2+v^2)^{3/2}}.$$

**47.** $u = 3x^2y - 8x^3 + 2y^2 \;\Rightarrow\; u_x = 6xy - 24x^2, \; u_{xy} = 6x$ and $u_y = 3x^2 + 4y \;\Rightarrow\; u_{yx} = 6x$. Thus $u_{xy} = u_{yx}$.

**49.** $u = \ln\sqrt{x^2+y^2} = \ln(x^2+y^2)^{1/2} = \tfrac{1}{2}\ln(x^2+y^2) \;\Rightarrow\; u_x = \frac{1}{2}\frac{1}{x^2+y^2}\cdot 2x = \frac{x}{x^2+y^2} = x(x^2+y^2)^{-1}$,

$$u_{xy} = x(-1)(x^2+y^2)^{-2}(2y) = -\frac{2xy}{(x^2+y^2)^2} \quad\text{and}\quad u_y = \frac{1}{2}\frac{1}{x^2+y^2}\cdot 2y = \frac{y}{x^2+y^2} = y(x^2+y^2)^{-1},$$

$$u_{yx} = y(-1)(x^2+y^2)^{-2}(2x) = -\frac{2xy}{(x^2+y^2)^2}. \text{ Thus } u_{xy} = u_{yx}.$$

**51.** (a) $P(L, K) = 1.47L^{0.65}K^{0.35} \;\Rightarrow\; P_L(L, K) = 1.47(0.65L^{0.65-1})K^{0.35} = 0.9555L^{-0.35}K^{0.35}$,

   $P_K(L, K) = 1.47L^{0.65}(0.35K^{0.35-1}) = 0.5145L^{0.65}K^{-0.65}$

(b) The marginal productivity of labor is $P_L(120, 20) = 0.9555(120)^{-0.35}(20)^{0.35} \approx 0.51$ and the marginal productivity of

   capital is $P_K(120, 20) = 0.5145(120)^{0.65}(20)^{-0.65} \approx 1.65$.

(c) If capital investment is held constant at $20 million and labor is increased from 120,000 labor hours, then production

   would increase at a rate of about 0.51 units per thousand labor hours. If labor is held constant at 120,000 labor hours and

   capital investment is increased from $20 million, then production would increase at a rate of about 1.65 units per million

   dollars.

**53.** $q_1 = 8000 - 25p_1 - 10p_2 \;\Rightarrow\; \partial q_1/\partial p_2 = -10$ and $q_2 = 15{,}000 - 120p_1 - 50p_2 \;\Rightarrow\; \partial q_2/\partial p_1 = -120$.

Because both mixed partials are negative, the products are complementary. (See the paragraph following Example 10.)

**55.** $\dfrac{1}{R} = \dfrac{1}{R_1} + \dfrac{1}{R_2} + \dfrac{1}{R_3} \;\Rightarrow\; R = \dfrac{1}{\dfrac{1}{R_1} + \dfrac{1}{R_2} + \dfrac{1}{R_3}} = \left(\dfrac{1}{R_1} + \dfrac{1}{R_2} + \dfrac{1}{R_3}\right)^{-1}$, so

$$\frac{\partial R}{\partial R_1} = (-1)\left(\frac{1}{R_1} + \frac{1}{R_2} + \frac{1}{R_3}\right)^{-2}\left(-\frac{1}{R_1^2} + 0 + 0\right) = \frac{1}{R_1^2}\left(\frac{1}{R_1} + \frac{1}{R_2} + \frac{1}{R_3}\right)^{-2}. \text{ But } \frac{1}{R_1} + \frac{1}{R_2} + \frac{1}{R_3} = \frac{1}{R}, \text{ so we}$$

have $\dfrac{\partial R}{\partial R_1} = \dfrac{1}{R_1^2}\left(\dfrac{1}{R}\right)^{-2} = \dfrac{R^2}{R_1^2}$.

**57.** $\dfrac{\partial K}{\partial m} = \tfrac{1}{2}v^2, \; \dfrac{\partial K}{\partial v} = mv, \; \dfrac{\partial^2 K}{\partial v^2} = m$. Thus $\dfrac{\partial K}{\partial m}\cdot\dfrac{\partial^2 K}{\partial v^2} = \tfrac{1}{2}v^2 m = K$.

## 7.3   Maximum and Minimum Values

**1.** (a) First we compute $D(1, 1) = f_{xx}(1, 1)f_{yy}(1, 1) - [f_{xy}(1, 1)]^2 = (4)(2) - (1)^2 = 7$. Since $D(1, 1) > 0$ and

   $f_{xx}(1, 1) > 0$, $f$ has a local minimum at $(1, 1)$ by the Second Derivatives Test.

(b) $D(1,1) = f_{xx}(1,1)\, f_{yy}(1,1) - [f_{xy}(1,1)]^2 = (4)(2) - (3)^2 = -1$. Since $D(1,1) < 0$, $f$ has a saddle point at $(1,1)$ by the Second Derivatives Test.

**3.** $f(x,y) = 9 - 2x + 4y - x^2 - 4y^2 \;\Rightarrow\; f_x = -2 - 2x,\ f_y = 4 - 8y,\ f_{xx} = -2,\ f_{xy} = 0,\ f_{yy} = -8.$ $\;\; f_x = 0$ only when $x = -1$ and $f_y = 0$ only when $y = \frac{1}{2}$, so the only critical point is $\left(-1, \frac{1}{2}\right)$.

$D(x,y) = f_{xx} f_{yy} - (f_{xy})^2 = (-2)(-8) - 0^2 = 16$, and since $D\left(-1, \frac{1}{2}\right) = 16 > 0$ and $f_{xx}\left(-1, \frac{1}{2}\right) = -2 < 0$,

$f\left(-1, \frac{1}{2}\right) = 11$ is a local maximum by the Second Derivatives Test.

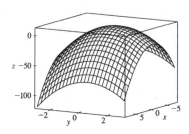

**5.** $f(x,y) = (x+1)^2 + (y-2)^2 - 4 \;\Rightarrow\; f_x = 2(x+1),\ f_y = 2(y-2),\ f_{xx} = 2,\ f_{xy} = 0,\ f_{yy} = 2.$ $\;\; f_x = 0$ only when $x = -1$ and $f_y = 0$ only when $y = 2$, so the only critical point is $(-1, 2)$.

$D(x,y) = f_{xx} f_{yy} - (f_{xy})^2 = (2)(2) - 0^2 = 4$, and since $D(-1,2) = 4 > 0$ and $f_{xx}(-1,2) = 2 > 0$, $f(-1,2) = -4$ is a local minimum by the Second Derivatives Test.

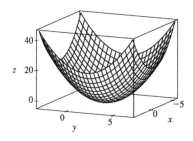

**7.** $f(x,y) = x^2 + xy + y^2 + y \;\Rightarrow\; f_x = 2x + y,\ f_y = x + 2y + 1,\ f_{xx} = 2,\ f_{xy} = 1,\ f_{yy} = 2.$ Then $f_x = 0$ implies

$y = -2x$, and substitution into $f_y = x + 2y + 1 = 0$ gives $x + 2(-2x) + 1 = 0 \;\Rightarrow\; -3x = -1 \;\Rightarrow\; x = \frac{1}{3}$.

Then $y = -2\left(\frac{1}{3}\right) = -\frac{2}{3}$ and the only critical point is $\left(\frac{1}{3}, -\frac{2}{3}\right)$. $\;\; D(x,y) = f_{xx} f_{yy} - (f_{xy})^2 = (2)(2) - (1)^2 = 3$, and

since $D\left(\frac{1}{3}, -\frac{2}{3}\right) = 3 > 0$ and $f_{xx}\left(\frac{1}{3}, -\frac{2}{3}\right) = 2 > 0$, $f\left(\frac{1}{3}, -\frac{2}{3}\right) = -\frac{1}{3}$ is a local minimum by the Second Derivatives Test.

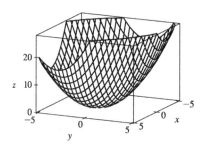

**9.** $f(x, y) = (x - y)(1 - xy) = x - y - x^2 y + xy^2$ $\Rightarrow$ $f_x = 1 - 2xy + y^2$, $f_y = -1 - x^2 + 2xy$, $f_{xx} = -2y$,

$f_{xy} = -2x + 2y$, $f_{yy} = 2x$. Then $f_x = 0$ implies $1 - 2xy + y^2 = 0$ and $f_y = 0$ implies $-1 - x^2 + 2xy = 0$. Adding the

two equations gives $1 + y^2 - 1 - x^2 = 0$ $\Rightarrow$ $y^2 = x^2$ $\Rightarrow$ $y = \pm x$, but if $y = -x$ then $f_x = 0$ implies

$1 + 2x^2 + x^2 = 0$ $\Rightarrow$ $3x^2 = -1$ which has no real solution. If $y = x$ then substitution into $f_x = 0$ gives

$1 - 2x^2 + x^2 = 0$ $\Rightarrow$ $x^2 = 1$ $\Rightarrow$ $x = \pm 1$, so the critical points are $(1, 1)$ and $(-1, -1)$. Now

$D(1, 1) = (-2)(2) - 0^2 = -4 < 0$ and $D(-1, -1) = (2)(-2) - 0^2 = -4 < 0$, so $(1, 1)$ and $(-1, -1)$ are saddle points.

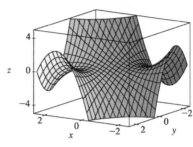

**11.** $f(x, y) = xy(1 - x - y) = xy - x^2 y - xy^2$ $\Rightarrow$ $f_x = y - 2xy - y^2$, $f_y = x - x^2 - 2xy$, $f_{xx} = -2y$,

$f_{xy} = 1 - 2x - 2y$, $f_{yy} = -2x$. Then $f_x = 0$ implies $y(1 - 2x - y) = 0$, so $y = 0$ or $1 - 2x - y = 0$ $\Rightarrow$ $y = 1 - 2x$.

If $y = 0$ then substitution into $f_y = 0$ gives $x - x^2 = 0$ $\Rightarrow$ $x(1 - x) = 0$ $\Rightarrow$ $x = 0$ or $x = 1$, so we have critical

points $(0, 0)$ and $(1, 0)$. If $y = 1 - 2x$, substitution into $f_y = 0$ gives $x - x^2 - 2x(1 - 2x) = 0$ $\Rightarrow$ $3x^2 - x = 0$ $\Rightarrow$

$x(3x - 1) = 0$ $\Rightarrow$ $x = 0$ or $x = \frac{1}{3}$. If $x = 0$ then $y = 1$, and if $x = \frac{1}{3}$ then $y = \frac{1}{3}$, so $(0, 1)$ and $\left(\frac{1}{3}, \frac{1}{3}\right)$ are critical points.

$D(0, 0) = (0)(0) - 1^2 = -1 < 0$, $D(1, 0) = (0)(-2) - (-1)^2 = -1 < 0$, and $D(0, 1) = (-2)(0) - (-1)^2 = -1 < 0$, so

$(0, 0)$, $(1, 0)$, and $(0, 1)$ are saddle points. $D\left(\frac{1}{3}, \frac{1}{3}\right) = \left(-\frac{2}{3}\right)\left(-\frac{2}{3}\right) - \left(-\frac{1}{3}\right)^2 = \frac{1}{3} > 0$ and $f_{xx}\left(\frac{1}{3}, \frac{1}{3}\right) = -\frac{2}{3} < 0$, so

$f\left(\frac{1}{3}, \frac{1}{3}\right) = \frac{1}{27}$ is a local maximum.

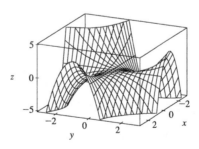

**13.** $f(x, y) = y^3 + 3x^2 y - 6x^2 - 6y^2 + 2$ $\Rightarrow$ $f_x = 6xy - 12x$, $f_y = 3y^2 + 3x^2 - 12y$, $f_{xx} = 6y - 12$, $f_{xy} = 6x$,

$f_{yy} = 6y - 12$. Then $f_x = 0$ implies $6x(y - 2) = 0$, so $x = 0$ or $y = 2$. If $x = 0$ then substitution into $f_y = 0$ gives

$3y^2 - 12y = 0$ $\Rightarrow$ $3y(y - 4) = 0$ $\Rightarrow$ $y = 0$ or $y = 4$, so we have critical points $(0, 0)$ and $(0, 4)$. If $y = 2$, substitution

into $f_y = 0$ gives $12 + 3x^2 - 24 = 0$ $\Rightarrow$ $x^2 = 4$ $\Rightarrow$ $x = \pm 2$, so we have critical points $(\pm 2, 2)$.

$D(0, 0) = (-12)(-12) - 0^2 = 144 > 0$ and $f_{xx}(0, 0) = -12 < 0$, so $f(0, 0) = 2$ is a local maximum.

[continued]

$D(0,4) = (12)(12) - 0^2 = 144 > 0$ and $f_{xx}(0,4) = 12 > 0$, so $f(0,4) = -30$ is a local minimum.

$D(\pm 2, 2) = (0)(0) - (\pm 12)^2 = -144 < 0$, so $(-2,2)$ and $(2,2)$ are saddle points.

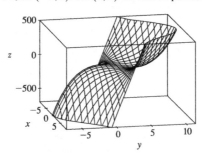

**15.** $f(x,y) = x^3 - 12xy + 8y^3 \quad \Rightarrow \quad f_x = 3x^2 - 12y, f_y = -12x + 24y^2, f_{xx} = 6x, f_{xy} = -12, f_{yy} = 48y$. Then $f_x = 0$

implies $x^2 = 4y$ and $f_y = 0$ implies $x = 2y^2$. Substituting the second equation into the first gives $(2y^2)^2 = 4y \quad \Rightarrow$

$4y^4 = 4y \quad \Rightarrow \quad 4y(y^3 - 1) = 0 \quad \Rightarrow \quad y = 0$ or $y = 1$. If $y = 0$ then $x = 0$ and if $y = 1$ then $x = 2$, so the critical points

are $(0,0)$ and $(2,1)$.   $D(0,0) = (0)(0) - (-12)^2 = -144 < 0$, so $(0,0)$ is a saddle point.

$D(2,1) = (12)(48) - (-12)^2 = 432 > 0$ and $f_{xx}(2,1) = 12 > 0$ so $f(2,1) = -8$ is a local minimum.

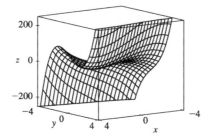

**17.** $f(x,y) = xe^{-2x^2 - 2y^2} \quad \Rightarrow \quad f_x = x \cdot e^{-2x^2 - 2y^2}(-4x) + e^{-2x^2 - 2y^2} \cdot 1 = (1 - 4x^2)e^{-2x^2 - 2y^2}, \quad f_y = -4xye^{-2x^2 - 2y^2},$

$f_{xx} = (1 - 4x^2) \cdot e^{-2x^2 - 2y^2}(-4x) + e^{-2x^2 - 2y^2} \cdot (-8x) = (16x^2 - 12)xe^{-2x^2 - 2y^2},$

$f_{xy} = (1 - 4x^2)e^{-2x^2 - 2y^2}(-4y) = (16x^2 - 4)ye^{-2x^2 - 2y^2},$

$f_{yy} = -4\left[xy \cdot e^{-2x^2 - 2y^2}(-4y) + e^{-2x^2 - 2y^2} \cdot x\right] = (16y^2 - 4)xe^{-2x^2 - 2y^2}$. Then $f_x = 0$ implies $1 - 4x^2 = 0 \quad \Rightarrow$

$x = \pm\frac{1}{2}$, and substitution into $f_y = 0$ implies $-4xy = -4\left(\pm\frac{1}{2}\right)y = 0 \quad \Rightarrow \quad y = 0$, so the critical points are $\left(\pm\frac{1}{2}, 0\right)$. Now

$D\left(\frac{1}{2}, 0\right) = (-4e^{-1/2})(-2e^{-1/2}) - 0^2 = 8e^{-1} > 0$ and $f_{xx}\left(\frac{1}{2}, 0\right) = -4e^{-1/2} < 0$, so $f\left(\frac{1}{2}, 0\right) = \frac{1}{2}e^{-1/2}$ is a local

maximum.   $D\left(-\frac{1}{2}, 0\right) = (4e^{-1/2})(2e^{-1/2}) - 0^2 = 8e^{-1} > 0$ and $f_{xx}\left(-\frac{1}{2}, 0\right) = 4e^{-1/2} > 0$, so $f\left(-\frac{1}{2}, 0\right) = -\frac{1}{2}e^{-1/2}$

is a local minimum.

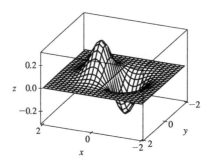

**19.** $f(x,y) = (x^2 + y^2)e^{y^2 - x^2}$   $\Rightarrow$

$f_x = (x^2 + y^2) \cdot e^{y^2 - x^2}(-2x) + e^{y^2 - x^2} \cdot 2x = 2xe^{y^2 - x^2}(1 - x^2 - y^2),$

$f_y = (x^2 + y^2) \cdot e^{y^2 - x^2}(2y) + e^{y^2 - x^2} \cdot 2y = 2ye^{y^2 - x^2}(1 + x^2 + y^2),$

$f_{xx} = 2xe^{y^2 - x^2}(-2x) + (1 - x^2 - y^2)\left(2x\left(-2xe^{y^2 - x^2}\right) + 2e^{y^2 - x^2}\right) = 2e^{y^2 - x^2}((1 - x^2 - y^2)(1 - 2x^2) - 2x^2),$

$f_{xy} = 2xe^{y^2 - x^2}(-2y) + 2x(2y)e^{y^2 - x^2}(1 - x^2 - y^2) = -4xye^{y^2 - x^2}(x^2 + y^2),$

$f_{yy} = 2ye^{y^2 - x^2}(2y) + (1 + x^2 + y^2)\left(2y\left(2ye^{y^2 - x^2}\right) + 2e^{y^2 - x^2}\right) = 2e^{y^2 - x^2}((1 + x^2 + y^2)(1 + 2y^2) + 2y^2).$

$f_y = 0$ implies $y = 0$, and substituting into $f_x = 0$ gives $2xe^{-x^2}(1 - x^2) = 0$   $\Rightarrow$   $x = 0$ or $x = \pm 1$. Thus the critical

points are $(0,0)$ and $(\pm 1, 0)$. Now $D(0,0) = (2)(2) - 0 > 0$ and $f_{xx}(0,0) = 2 > 0$, so $f(0,0) = 0$ is a local minimum.

$D(\pm 1, 0) = (-4e^{-1})(4e^{-1}) - 0 < 0$ so $(1,0)$ and $(-1,0)$ are saddle points.

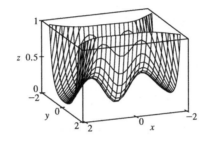

**21.** Let $x$, $y$, and $z$ be the three positive numbers. Then $x + y + z = 100$   $\Rightarrow$   $z = 100 - x - y$, and we wish to maximize the

product $xyz = xy(100 - x - y) = 100xy - x^2y - xy^2 = f(x,y)$ for $0 < x, y < 100$.   $f_x = 100y - 2xy - y^2,$

$f_y = 100x - x^2 - 2xy$, $f_{xx} = -2y$, $f_{yy} = -2x$, $f_{xy} = 100 - 2x - 2y$. Then $f_x = y(100 - 2x - y) = 0$ implies $y = 0$

or $y = 100 - 2x$. Substituting $y = 0$ into $f_y = x(100 - x - 2y) = 0$ gives $x = 0$ or $x = 100$ and substituting $y = 100 - 2x$

into $f_y = 0$ gives $100x - x^2 - 2x(100 - 2x) = 3x^2 - 100x = x(3x - 100) = 0$ so $x = 0$ or $\frac{100}{3}$. If $x = 0$ then

$y = 100 - 2(0) = 100$, and if $x = \frac{100}{3}$, then $y = 100 - 2 \cdot \frac{100}{3} = \frac{100}{3}$. Thus the critical points are $(0,0)$, $(100,0)$, $(0,100)$,

and $\left(\frac{100}{3}, \frac{100}{3}\right)$.   $D(0,0) = D(100,0) = D(0,100) = -10{,}000$ while $D\left(\frac{100}{3}, \frac{100}{3}\right) = \frac{10{,}000}{3}$ and

$f_{xx}\left(\frac{100}{3}, \frac{100}{3}\right) = -\frac{200}{3} < 0$. Thus $(0,0)$, $(100,0)$, and $(0,100)$ are saddle points and $f\left(\frac{100}{3}, \frac{100}{3}\right)$ is a local maximum. Here

the local maximum is also the absolute maximum [compare to the values of $f$ as $x, y \to 0$ or $100$], so the numbers are

$x = y = z = \frac{100}{3}$.

**23.** $R = 40q_A + 60q_B - 6q_A^2 - 9q_B^2 - 4q_Aq_B$   $\Rightarrow$   $R_{q_A} = 40 - 12q_A - 4q_B$,   $R_{q_B} = 60 - 18q_B - 4q_A$,   $R_{q_Aq_A} = -12$,

$R_{q_Aq_B} = -4$,   $R_{q_Bq_B} = -18$.   $R_{q_A} = 0$ implies $3q_A + q_B = 10$   $\Rightarrow$   $q_B = 10 - 3q_A$ and $R_{q_B} = 0$ implies

$2q_A + 9q_B = 30$. Substituting the first equation into the second gives $2q_A + 9(10 - 3q_A) = 30$   $\Rightarrow$

$90 - 25q_A = 30$   $\Rightarrow$   $q_A = 2.4$, and then $q_B = 10 - 3(2.4) = 2.8$, so the only critical point is $(2.4, 2.8)$.

$D(2.4, 2.8) = (-12)(-18) - (-4)^2 = 200 > 0$ and $R_{q_Aq_A}(2.4, 2.8) = -12 < 0$, so $R(2.4, 2.8)$ is a (local) maximum.

Thus revenue is maximized for $q_A = 2.4$ and $q_B = 2.8$.

**25.** The joint cost function is

$$C = 12q_1 + 10q_2 + q_1q_2 = 12(60 - 2p_1) + 10(40 - p_2) + (60 - 2p_1)(40 - p_2)$$

$$= 3520 - 104p_1 - 70p_2 + 2p_1p_2$$

Revenue is

$$R = p_1q_1 + p_2q_2 = p_1(60 - 2p_1) + p_2(40 - p_2)$$

$$= 60p_1 - 2p_1^2 + 40p_2 - p_2^2$$

and the profit function is

$$P = R - C$$

$$= 60p_1 - 2p_1^2 + 40p_2 - p_2^2 - (3520 - 104p_1 - 70p_2 + 2p_1p_2)$$

$$= 164p_1 - 2p_1^2 + 110p_2 - p_2^2 - 2p_1p_2 - 3520$$

The critical numbers of the profit function occur when its partial derivatives are 0:

$$\frac{\partial P}{\partial p_1} = 164 - 4p_1 - 2p_2 = 2(82 - 2p_1 - p_2) = 0$$

$$\frac{\partial P}{\partial p_2} = 110 - 2p_2 - 2p_1 = 2(55 - p_1 - p_2) = 0$$

Then $\dfrac{\partial P}{\partial p_1} = 0$ implies $p_2 = 82 - 2p_1$ and substituting into $\dfrac{\partial P}{\partial p_2} = 0$ implies $55 - p_1 - (82 - 2p_1) = 0 \Rightarrow$

$p_1 - 27 = 0 \Rightarrow p_1 = 27$. Then $p_2 = 82 - 2(27) = 28$, so the only critical point is $(27, 28)$. The second partial derivatives

are $\dfrac{\partial^2 P}{\partial p_1^2} = -4$, $\dfrac{\partial^2 P}{\partial p_2^2} = -2$, $\dfrac{\partial^2 P}{\partial p_2 \partial p_1} = -2$, so $D(27, 28) = (-4)(-2) - (-2)^2 = 4 > 0$. Since $\partial^2 P/\partial p_1^2$ is

negative, the Second Derivatives Test says that $p_1 = 27$ and $p_2 = 28$ give a maximum. The corresponding quantities are

$q_1 = 60 - 2p_1 = 60 - 2(27) = 6$ and $q_2 = 40 - p_2 = 40 - 28 = 12$.

**27.** Let the dimensions of the box be $x$ (width), $y$ (length), and $z$ (height). We are given that

volume $= xyz = 32{,}000$ cm$^3 \Rightarrow z = \dfrac{32{,}000}{xy}$, and we wish to minimize the surface area

$xy$ (base) $+2xz$ (2 sides) $+2yz$ (the other 2 sides) $= xy + 2x \cdot \dfrac{32{,}000}{xy} + 2y \cdot \dfrac{32{,}000}{xy} = xy + \dfrac{64{,}000}{y} + \dfrac{64{,}000}{x}$. Thus we

minimize $f(x, y) = xy + 64{,}000(x^{-1} + y^{-1})$. $f_x = y - 64{,}000x^{-2}$, $f_y = x - 64{,}000y^{-2}$, $f_{xx} = 128{,}000x^{-3}$,

$f_{xy} = 1$, $f_{yy} = 128{,}000y^{-3}$. $f_x = 0$ implies $y = 64{,}000/x^2$; substituting into $f_y = 0$ implies

$x - \dfrac{64{,}000}{(64{,}000/x^2)^2} = 0 \Rightarrow x = \dfrac{x^4}{64{,}000} \Rightarrow x^3 = 64{,}000$ or $x = 40$ and then $y = 40$. Now

$D(40, 40) = (2)(2) - 1^2 > 0$ and $f_{xx}(40, 40) > 0$ so this is indeed a minimum. Thus the dimensions of the box are

$x = y = 40$ cm, $z = 20$ cm.

**29.** Let the dimensions of the box be $x$ (width), $y$ (length), and $z$ (height). The sum of the 12 edges is 30 ft, so

$4x + 4y + 4z = 30 \Rightarrow 4z = 30 - 4x - 4y \Rightarrow z = 7.5 - x - y$. The volume is

$xyz = xy(7.5 - x - y) = 7.5xy - x^2y - xy^2$, so we maximize $f(x, y) = 7.5xy - x^2y - xy^2$ for $0 < x, y, z < 7.5$. (If any

edge measures 7.5 or more, there won't be 12 edges.) Then $f_x = 7.5y - 2xy - y^2$, $f_y = 7.5x - x^2 - 2xy$, $f_{xx} = -2y$,

$f_{xy} = 7.5 - 2x - 2y$, $f_{yy} = -2x$. Setting $f_x = 0$ implies $y(7.5 - 2x - y) = 0$ $\Rightarrow$ $y = 7.5 - 2x$ [note $y \neq 0$], and

substituting into $f_y = 0$ gives $7.5x - x^2 - 2x(7.5 - 2x) = 0$ $\Rightarrow$ $3x^2 - 7.5x = x(3x - 7.5) = 0$ $\Rightarrow$ $x = 2.5$, and

then $y = 7.5 - 2(2.5) = 2.5$ and $z = 7.5 - 2.5 - 2.5 = 2.5$. $D(2.5, 2.5) = (-5)(-5) - (-2.5)^2 > 0$ and

$f_{xx}(2.5, 2.5) < 0$, so this is a maximum. (We can also argue that because we know there is a maximum volume, it must occur

at the one critical point.) Thus the box is a cube with edge length 2.5 ft.

## 7.4 Lagrange Multipliers

**1.** We wish to find the maximum and minimum values of $f(x, y) = xy$ subject to the constraint $x - 2y = 1$, so

$g(x, y) = x - 2y$. Following the method outlined in the section, we solve the system of equations $f_x = \lambda g_x$ $\Rightarrow$

$y = \lambda \cdot 1$ $\Rightarrow$ $y = \lambda$ **(1)**, $f_y = \lambda g_y$ $\Rightarrow$ $x = \lambda(-2)$ $\Rightarrow$ $x = -2\lambda$ **(2)**, and $g(x, y) = 1$ $\Rightarrow$ $x - 2y = 1$ **(3)**.

Substituting (1) into (2) gives $x = -2\lambda = -2y$. Substituting this into (3), we have $-2y - 2y = 1$ $\Rightarrow$ $y = -\frac{1}{4}$, and then

$x = -2y = -2\left(-\frac{1}{4}\right) = \frac{1}{2}$. Thus $f$ has a possible extreme value at only one point, $\left(\frac{1}{2}, -\frac{1}{4}\right)$. $f\left(\frac{1}{2}, -\frac{1}{4}\right) = -\frac{1}{8}$ which is

smaller than the function values at other points satisfying the constraint. For instance, $(1, 0)$ satisfies the constraint equation

and $f(1, 0) = 0$. Thus $f\left(\frac{1}{2}, -\frac{1}{4}\right) = -\frac{1}{8}$ is the minimum value subject to the constraint $x - 2y = 1$. (There is no maximum;

we can choose points satisfying the constraint that make $f$ arbitrarily large.)

**3.** $f(x, y) = x^2 + y^2$, $g(x, y) = xy = 1$. We solve the system of equations $f_x = \lambda g_x$ $\Rightarrow$ $2x = \lambda y$, $f_y = \lambda g_y$ $\Rightarrow$

$2y = \lambda x$, and $xy = 1$. From the first equation we have $\lambda = 2x/y$ (note that $y \neq 0$ from the constraint equation) and

substituting into the second equation gives $2y = (2x/y)x$ $\Rightarrow$ $y^2 = x^2$ $\Rightarrow$ $y = \pm x$. If $y = x$, the constraint $xy = 1$

says that $x^2 = 1$ $\Rightarrow$ $x = \pm 1$, so we have possible extreme values at $(1, 1)$ and $(-1, -1)$. If $y = -x$, then the constraint

becomes $-x^2 = 1$ which has no real solution. Evaluating $f$ at these points gives $f(1, 1) = f(-1, -1) = 2$ which is a

minimum value because other points from the constraint, such as $\left(2, \frac{1}{2}\right)$, give larger function values. There is no maximum

value. [The constraint $xy = 1$ allows $x$ or $y$ to become arbitrarily large, and hence $f(x, y) = x^2 + y^2$ can be made arbitrarily

large.]

**5.** $f(x, y) = x^2 y$, $g(x, y) = x^2 + 2y^2 = 6$. We solve the system of equations $f_x = \lambda g_x$ $\Rightarrow$ $2xy = 2\lambda x$ **(1)**,

$f_y = \lambda g_y$ $\Rightarrow$ $x^2 = 4\lambda y$ **(2)**, and $x^2 + 2y^2 = 6$ **(3)**. From (1) we have $2xy - 2\lambda x = 0$ $\Rightarrow$ $2x(y - \lambda) = 0$, so either

$x = 0$ or $y = \lambda$. If $x = 0$, then from (2) we have $0 = 4\lambda y$, so either $\lambda = 0$ or $y = 0$. However, if both $x = 0$ and $y = 0$ then

(3) is not satisfied. So $y \neq 0$ (and $\lambda = 0$) and from (3) we have $0 + 2y^2 = 6$ $\Rightarrow$ $y = \pm\sqrt{3}$. Now we consider the case

where $y = \lambda$: then from (2) we have $x^2 = 4y^2$ and substituting into (3) gives $4y^2 + 2y^2 = 6$ $\Rightarrow$ $y^2 = 1$ $\Rightarrow$ $y = \pm 1$.

Then from $x^2 = 4y^2$ we have $x^2 = 4$ $\Rightarrow$ $x = \pm 2$. Thus $f$ has possible extreme values at the points $\left(0, \pm\sqrt{3}\right)$, $(\pm 2, 1)$,

and $(\pm 2, -1)$. We evaluate $f$ at these six points:

$$f(0, \sqrt{3}) = 0 \qquad f(2, 1) = 4 \qquad f(2, -1) = -4$$
$$f(0, -\sqrt{3}) = 0 \qquad f(-2, 1) = 4 \qquad f(-2, -1) = -4$$

Thus the maximum value of $f$ subject to the constraint is $f(\pm 2, 1) = 4$ and the minimum is $f(\pm 2, -1) = -4$.

**7.** $f(x,y) = x^4 + y^4$, $g(x,y) = x^2 + y^2 = 2$. We solve the equations $f_x = \lambda g_x$ $\Rightarrow$ $4x^3 = 2\lambda x$ **(1)**, $f_y = \lambda g_y$ $\Rightarrow$ $4y^3 = 2\lambda y$ **(2)**, and $x^2 + y^2 = 2$ **(3)**. From (1) we have $4x^3 - 2\lambda x = 0$ $\Rightarrow$ $2x(2x^2 - \lambda) = 0$, so either $x = 0$ or $\lambda = 2x^2$. If $x = 0$, then from (3) we have $y^2 = 2$ $\Rightarrow$ $y = \pm\sqrt{2}$. If $\lambda = 2x^2$ then, substituting into (2), we have $4y^3 = 4x^2 y$ $\Rightarrow$ $4y(y^2 - x^2) = 0$ $\Rightarrow$ $y = 0$ or $y^2 = x^2$. If $y = 0$, (3) gives $x^2 = 2$ $\Rightarrow$ $x = \pm\sqrt{2}$. If $y^2 = x^2$ then (3) gives $2x^2 = 2$ $\Rightarrow$ $x^2 = 1$ $\Rightarrow$ $x = \pm 1$. Since $y^2 = x^2$, $x = 1$ $\Rightarrow$ $y = \pm 1$ and $x = -1$ $\Rightarrow$ $y = \pm 1$. Thus $f$ has possible extreme values at the points $(0, \pm\sqrt{2})$, $(\pm\sqrt{2}, 0)$, $(1, \pm 1)$, and $(-1, \pm 1)$. We evaluate $f$ at these eight points:

$$f(0, \pm\sqrt{2}) = 4 \qquad f(\pm\sqrt{2}, 0) = 4 \qquad f(1, \pm 1) = 2 \qquad f(-1, \pm 1) = 2$$

Thus the maximum value of $f$ subject to the constraint $x^2 + y^2 = 2$ is $f(0, \pm\sqrt{2}) = f(\pm\sqrt{2}, 0) = 4$ and the minimum is $f(1, \pm 1) = f(-1, \pm 1) = 2$.

**9.** $f(x,y,z) = 2x + 6y + 10z$, $g(x,y,z) = x^2 + y^2 + z^2 = 35$. We solve the system of equations $f_x = \lambda g_x$ $\Rightarrow$ $2 = 2\lambda x$, $f_y = \lambda g_y$ $\Rightarrow$ $6 = 2\lambda y$, $f_z = \lambda g_z$ $\Rightarrow$ $10 = 2\lambda z$, and $x^2 + y^2 + z^2 = 35$. From the first three equations we have $x = \dfrac{1}{\lambda}$, $y = \dfrac{3}{\lambda}$, and $z = \dfrac{5}{\lambda}$, and substituting into the constraint gives $\left(\dfrac{1}{\lambda}\right)^2 + \left(\dfrac{3}{\lambda}\right)^2 + \left(\dfrac{5}{\lambda}\right)^2 = 35$ $\Rightarrow$ $\dfrac{35}{\lambda^2} = 35$ $\Rightarrow$ $\lambda^2 = 1$ $\Rightarrow$ $\lambda = \pm 1$. $\lambda = 1$ gives the point $(1, 3, 5)$ and $\lambda = -1$ gives $(-1, -3, -5)$. Evaluating $f$ at these points we see that $f(1, 3, 5) = 70$ is the maximum value of $f$ subject to the constraint, and the minimum is $f(-1, -3, -5) = -70$.

**11.** $f(x,y,z) = 2x + 2y + z$, $g(x,y,z) = x^2 + y^2 + z^2 = 9$. We solve the equations $f_x = \lambda g_x$ $\Rightarrow$ $2 = 2\lambda x$, $f_y = \lambda g_y$ $\Rightarrow$ $2 = 2\lambda y$, $f_z = \lambda g_z$ $\Rightarrow$ $1 = 2\lambda z$, and $x^2 + y^2 + z^2 = 9$. The first three equations imply $x = \dfrac{1}{\lambda}$, $y = \dfrac{1}{\lambda}$, and $z = \dfrac{1}{2\lambda}$. Then substitution into the constraint equation gives $\left(\dfrac{1}{\lambda}\right)^2 + \left(\dfrac{1}{\lambda}\right)^2 + \left(\dfrac{1}{2\lambda}\right)^2 = 9$ $\Rightarrow$ $\dfrac{9}{4\lambda^2} = 9$ $\Rightarrow$ $4\lambda^2 = 1$ $\Rightarrow$ $\lambda = \pm\frac{1}{2}$, giving the points $(2, 2, 1)$ and $(-2, -2, -1)$. The maximum value of $f$ subject to $x^2 + y^2 + z^2 = 9$ is $f(2, 2, 1) = 9$, and the minimum is $f(-2, -2, -1) = -9$.

**13.** Let $x$ and $y$ be the positive numbers. Then we wish to maximize $f(x,y) = xy$ (their product) subject to the constraint $g(x,y) = x + y = 100$ (their sum is 100). We solve the equations $f_x = \lambda g_x$ $\Rightarrow$ $y = \lambda$, $f_y = \lambda g_y$ $\Rightarrow$ $x = \lambda$, and $x + y = 100$. From the first two equations we have $y = x$, and substituting into the constraint gives $x + x = 100$ $\Rightarrow$ $x = 50$, so $y = 50$. We expect a maximum value (not a minimum), and because there is only one solution to the equations, we conclude that $f$ has a maximum value subject to the constraint at $(50, 50)$. [We can also compare values of $f$ at other points satisfying the constraint, such as $f(90, 10) = 900$, and observe that they give smaller values than $f(50, 50) = 2500$.] Thus the numbers are 50 and 50.

**15.** Let $x$, $y$, and $z$ be the positive numbers. Then we wish to maximize $f(x,y,z) = xyz$ subject to the constraint $g(x,y,z) = x + y + z = 100$. We solve the equations $f_x = \lambda g_x$ $\Rightarrow$ $yz = \lambda$, $f_y = \lambda g_y$ $\Rightarrow$ $xz = \lambda$, $f_z = \lambda g_z$ $\Rightarrow$ $xy = \lambda$, and $x + y + z = 100$. From the first three equations we have $yz = \lambda = xz = xy$. Since none of $x$, $y$, $z$ are zero,

$yz = xz$  $\Rightarrow$  $x = y$, and similarly $x = y = z$. Substituting these into the constraint equation gives $x + x + x = 100$  $\Rightarrow$
$x = \frac{100}{3} = y = z$.   $f\left(\frac{100}{3}, \frac{100}{3}, \frac{100}{3}\right)$ is a maximum, as other points from the constraint give smaller values. Thus the
numbers are $\frac{100}{3}$, $\frac{100}{3}$, and $\frac{100}{3}$.

**17.** We wish to maximize $P(x, y) = 8x + 10y$ subject to the constraint $g(x, y) = 16x^2 + 25y^2 = 40{,}000$. We solve the

equations $P_x = \lambda g_x$  $\Rightarrow$  $8 = 32\lambda x$, $P_y = \lambda g_y$  $\Rightarrow$  $10 = 50\lambda y$, and $16x^2 + 25y^2 = 40{,}000$. From the first two

equations we have $\lambda = \dfrac{1}{4x}$ and $\lambda = \dfrac{1}{5y}$, so $\dfrac{1}{4x} = \dfrac{1}{5y}$  $\Rightarrow$  $y = \frac{4}{5}x$, and substituting into the constraint equation gives

$16x^2 + 25\left(\frac{4}{5}x\right)^2 = 40{,}000$  $\Rightarrow$  $16x^2 + 16x^2 = 40{,}000$  $\Rightarrow$  $x^2 = 1250$  $\Rightarrow$  $x = \sqrt{1250} = 25\sqrt{2}$,

$y = \frac{4}{5} \cdot 25\sqrt{2} = 20\sqrt{2}$.   Comparing the values of $f$ at other points from the constraint curve, we see that $P\left(25\sqrt{2}, 20\sqrt{2}\right)$

gives a maximum value, not a minimum. Thus the company should produce $25\sqrt{2} \approx 35.4$ units of A and $20\sqrt{2} \approx 28.3$ units
of B.

**19.** (a) We wish to maximize $P(L, K) = 1200L^{2/3}K^{1/3}$ subject to the constraint $C(L, K) = 100L + 400K = 360{,}000$. We

solve the equations $P_L = \lambda C_L$  $\Rightarrow$  $800L^{-1/3}K^{1/3} = 100\lambda$, $P_K = \lambda C_K$  $\Rightarrow$  $400L^{2/3}K^{-2/3} = 400\lambda$, and

$100L + 400K = 360{,}000$. From the first two equations we have $\lambda = 8L^{-1/3}K^{1/3}$ and $\lambda = L^{2/3}K^{-2/3}$, so

$8L^{-1/3}K^{1/3} = L^{2/3}K^{-2/3}$  $\Rightarrow$  $8K = L$, and substituting into the constraint equation gives

$800K + 400K = 360{,}000$  $\Rightarrow$  $K = 300$, $L = 8(300) = 2400$. We are expecting a maximum (not a minimum), which

must correspond to the only solution to the equations. Thus production is maximized with 2400 units of labor, at a cost of

$240,000, and 300 units of capital, at a cost of $120,000.

(b) From part (a), $\lambda = L^{2/3}K^{-2/3} = (2400)^{2/3}(300)^{-2/3} = 4$. This means that increasing the budget by $1 will increase the
maximum production by 4 units.

**21.** Let the dimensions of the box be $x$ (width), $y$ (length), and $z$ (height). The surface area is $xy$ (base) $+2xz$ (2 sides) $+2yz$ (the

other 2 sides), so wish to minimize $f(x, y, z) = xy + 2xz + 2yz$ subject to the constraint $g(x, y, z) = xyz = 32{,}000$ (the

volume is 32,000 cm). We solve the equations $f_x = \lambda g_x$  $\Rightarrow$  $y + 2z = \lambda yz$, $f_y = \lambda g_y$  $\Rightarrow$  $x + 2z = \lambda xz$,

$f_z = \lambda g_z$  $\Rightarrow$  $2x + 2y = \lambda xy$, and $xyz = 32{,}000$. Solving each of the first three equations for $\lambda$ gives

$\lambda = \dfrac{y + 2z}{yz} = \dfrac{y}{yz} + \dfrac{2z}{yz} = \dfrac{1}{z} + \dfrac{2}{y}$  **(1)**, $\lambda = \dfrac{x + 2z}{xz} = \dfrac{1}{z} + \dfrac{2}{x}$  **(2)**, $\lambda = \dfrac{2x + 2y}{xy} = \dfrac{2}{y} + \dfrac{2}{x}$  **(3)**. Comparing (1) and (2)

we see that $\dfrac{2}{y} = \dfrac{2}{x}$  $\Rightarrow$  $y = x$. Comparing (2) and (3), we have $\dfrac{1}{z} = \dfrac{2}{y}$  $\Rightarrow$  $y = 2z$, so $x = y = 2z$ and then the

constraint equation gives $(2z)(2z)z = 32{,}000$  $\Rightarrow$  $z^3 = 8000$  $\Rightarrow$  $z = 20$ and $x = y = 40$. There is a minimum surface

area for a given volume (not a maximum), and we only have one solution to the equations, so the amount of cardboard is

minimized for a box with a square base with side length 40 cm and height 20 cm.

**23.** Let the dimensions of the box be $x$ (width), $y$ (length), and $z$ (height). We wish to maximize $f(x, y, z) = xyz$ (the volume)

subject to the constraint $g(x, y, z) = 4x + 4y + 4z = 30$ (the sum of the lengths of all 12 edges is 30 ft). We solve the equations $f_x = \lambda g_x \Rightarrow yz = 4\lambda$, $f_y = \lambda g_y \Rightarrow xz = 4\lambda$, $f_z = \lambda g_z \Rightarrow xy = 4\lambda$, and $4x + 4y + 4z = 30$.

Comparing the first two equations we have $yz = xz$, and since $z \neq 0$ we must have $x = y$. Similarly, from the second and

third equations we can conclude that $y = z$, so $x = y = z$ and substituting in the constraint equation gives

$4x + 4x + 4x = 30 \Rightarrow x = \frac{30}{12} = 2.5 = y = z$. There is a maximum volume and we have only one solution to the

equations, so the dimensions of the box are 2.5 ft, 2.5 ft, and 2.5 ft.

## 7    Review

**1.**

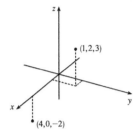

**3.** (a) $g(x, y) = \dfrac{e^{xy}}{x - 1} \Rightarrow g(2, 1) = \dfrac{e^{2 \cdot 1}}{2 - 1} = e^2$

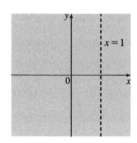

(b) $g(x, y) = \dfrac{e^{xy}}{x - 1}$ is defined whenever $x - 1 \neq 0 \Rightarrow x \neq 1$, so the

domain of $g$ is $\{(x, y) \mid x \neq 1\}$ which is the set of all points not on the line

$x = 1$.

(c) $g_x(x, y) = \dfrac{(x - 1) \cdot e^{xy}(y) - e^{xy} \cdot 1}{(x - 1)^2} = \dfrac{(xy - y - 1)e^{xy}}{(x - 1)^2}$ and $g_y(x, y) = \dfrac{1}{x - 1}e^{xy}(x) = \dfrac{xe^{xy}}{x - 1}$, so

$g_x(2, 1) = \dfrac{(2 - 1 - 1)e^2}{(2 - 1)^2} = 0$ and $g_y(2, 1) = \dfrac{2e^2}{2 - 1} = 2e^2$.

**5.** The level curves of $f(x, y) = y - x^2$ are $y - x^2 = k$ or $y = x^2 + k$, a family

of parabolas centered on the $y$-axis, opening upward, with vertex $(0, k)$.

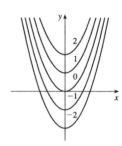

**7.** $f(x, y) = x^2 y^4 - 2xy^5 \Rightarrow f_x(x, y) = 2xy^4 - 2y^5$, $f_y(x, y) = 4x^2 y^3 - 10xy^4$

**9.** $h(x, y) = xe^{2xy} \Rightarrow h_x(x, y) = x \cdot e^{2xy}(2y) + e^{2xy} \cdot 1 = (2xy + 1)e^{2xy}$, $h_y(x, y) = xe^{2xy}(2x) = 2x^2 e^{2xy}$

**11.** $F(\alpha, \beta) = \alpha^2 \ln(\alpha^2 + \beta^2) \Rightarrow F_\alpha = \alpha^2 \cdot \dfrac{1}{\alpha^2 + \beta^2}(2\alpha) + \ln(\alpha^2 + \beta^2) \cdot 2\alpha = \dfrac{2\alpha^3}{\alpha^2 + \beta^2} + 2\alpha \ln(\alpha^2 + \beta^2)$,

$F_\beta = \alpha^2 \cdot \dfrac{1}{\alpha^2 + \beta^2}(2\beta) = \dfrac{2\alpha^2 \beta}{\alpha^2 + \beta^2}$

**13.** $G(x, y, z) = \dfrac{x}{y + 2z} = x(y + 2z)^{-1} \quad \Rightarrow \quad G_x(x, y, z) = \dfrac{1}{y + 2z}, G_y(x, y, z) = -x(y + 2z)^{-2}(1) = -\dfrac{x}{(y + 2z)^2},$

$G_z(x, y, z) = -x(y + 2z)^{-2}(2) = -\dfrac{2x}{(y + 2z)^2}$

**15.** $f(x, y) = 4x^3 - xy^2 \quad \Rightarrow \quad f_x = 12x^2 - y^2, \ f_y = -2xy, \ f_{xx} = 24x, \ f_{yy} = -2x, \ f_{xy} = f_{yx} = -2y$

**17.** $T_x(6, 4) = \lim\limits_{h \to 0} \dfrac{T(6 + h, 4) - T(6, 4)}{h}$, so we can approximate $T_x(6, 4)$ by considering $h = \pm 2$ and using the values given

in the table: $T_x(6, 4) \approx \dfrac{T(8, 4) - T(6, 4)}{2} = \dfrac{86 - 80}{2} = 3, T_x(6, 4) \approx \dfrac{T(4, 4) - T(6, 4)}{-2} = \dfrac{72 - 80}{-2} = 4.$ Averaging

these values, we estimate $T_x(6, 4)$ to be approximately $3.5°\text{C/m}$. Similarly, $T_y(6, 4) = \lim\limits_{h \to 0} \dfrac{T(6, 4 + h) - T(6, 4)}{h}$,

which we can approximate with $h = \pm 2$: $T_y(6, 4) \approx \dfrac{T(6, 6) - T(6, 4)}{2} = \dfrac{75 - 80}{2} = -2.5,$

$T_y(6, 4) \approx \dfrac{T(6, 2) - T(6, 4)}{-2} = \dfrac{87 - 80}{-2} = -3.5.$ Averaging these values, we estimate $T_y(6, 4)$ to be approximately

$-3.0°\text{C/m}$.

**19.** (a) $P(L, K) = 600L^{2/3}K^{1/3} \quad \Rightarrow \quad P_L(L, K) = 600(\frac{2}{3}L^{-1/3})K^{1/3} = 400L^{-1/3}K^{1/3},$

$P_K(L, K) = 600L^{2/3}(\frac{1}{3}K^{-2/3}) = 200L^{2/3}K^{-2/3}$

(b) The marginal productivity of labor is $P_L(100, 80) = 400(100)^{-1/3}(80)^{1/3} \approx 371.3$ and the marginal productivity of

capital is $P_K(100, 80) = 200(100)^{2/3}(80)^{-2/3} \approx 232.1$. This means that if capital investment is held constant at 80 units

and labor is increased from 100 units, then production would increase at a rate of about 371.3 units per unit of labor. If

labor is held constant at 100 units and capital investment is increased from 80 units, then production would increase at a

rate of about 232.1 units per unit of capital.

(c) If we assume that the units of labor and capital have the same monetary value, then the marginal values in part (b) tell us

that an increase in labor results in a larger increase in production than does an increase in capital. If the units of labor and

capital do not have the same monetary value, then we don't have enough information to make a judgment.

**21.** $f(x, y) = x^2 - xy + y^2 + 9x - 6y + 10 \quad \Rightarrow \quad f_x = 2x - y + 9, f_y = -x + 2y - 6, f_{xx} = 2, f_{yy} = 2, f_{xy} = -1.$ Then

$f_x = 0$ implies that $y = 2x + 9$, and substitution into $f_y = -x + 2y - 6 = 0$ gives $-x + 2(2x + 9) - 6 = 0 \quad \Rightarrow$

$3x = -12 \quad \Rightarrow \quad x = -4$, and then $y = 2(-4) + 9 = 1.$ Thus the only critical point is $(-4, 1)$.

$D(-4, 1) = (2)(2) - (-1)^2 = 3 > 0$ and $f_{xx}(-4, 1) = 2 > 0$, so $f(-4, 1) = -11$ is a local minimum.

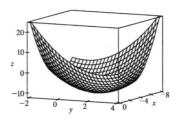

**23.** $f(x,y) = 3xy - x^2y - xy^2 \;\Rightarrow\; f_x = 3y - 2xy - y^2,\; f_y = 3x - x^2 - 2xy,\; f_{xx} = -2y,\; f_{yy} = -2x,$

$f_{xy} = 3 - 2x - 2y$. Then $f_x = 0$ implies $y(3 - 2x - y) = 0$ so $y = 0$ or $3 - 2x - y = 0 \;\Rightarrow\; y = 3 - 2x$. If $y = 0$, then

$f_y = 0$ implies $3x - x^2 = 0 \;\Rightarrow\; x(3 - x) = 0 \;\Rightarrow\; x = 0$ or $x = 3$, giving critical points $(0,0)$ and $(3,0)$. If

$y = 3 - 2x$ then $f_y = 0$ implies $3x - x^2 - 2x(3 - 2x) = 0 \;\Rightarrow\; 3x^2 - 3x = 3x(x - 1) = 0 \;\Rightarrow\; x = 0$ or

$x = 1$. If $x = 0$ then $y = 3$, and if $x = 1$ then $y = 1$, giving critical points $(0,3)$ and $(1,1)$.

$D(x,y) = f_{xx}f_{yy} - f_{xy}^2 = (-2y)(-2x) - (3 - 2x - 2y)^2$, and

$D(0,0) = D(3,0) = D(0,3) = -9 < 0$ so $(0,0)$, $(3,0)$, and $(0,3)$ are

saddle points. $D(1,1) = 3 > 0$ and $f_{xx}(1,1) = -2 < 0$, so $f(1,1) = 1$

is a local maximum.

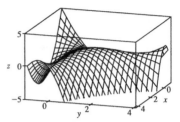

**25.** $f(x,y) = x^2y,\; g(x,y) = x^2 + y^2 = 1$. We solve the system of equations $f_x = \lambda g_x \;\Rightarrow\; 2xy = 2\lambda x$ **(1)**, $f_y = \lambda g_y \;\Rightarrow$

$x^2 = 2\lambda y$ **(2)**, and $x^2 + y^2 = 1$ **(3)**. From (1) we have $2xy - 2\lambda x = 0 \;\Rightarrow\; 2x(y - \lambda) = 0$, so either $x = 0$ or $y = \lambda$.

If $x = 0$, then from (2) we have $0 = 2\lambda y$, so either $\lambda = 0$ or $y = 0$. However, if both $x = 0$ and $y = 0$ then (3) is not satisfied.

So $y \neq 0$ (and $\lambda = 0$) and from (3) we have $0 + y^2 = 1 \;\Rightarrow\; y = \pm 1$. Now we consider the case where $y = \lambda$: then from

(2) we have $x^2 = 2y^2$ and substituting into (3) gives $2y^2 + y^2 = 1 \;\Rightarrow\; y^2 = \frac{1}{3} \;\Rightarrow\; y = \pm\frac{1}{\sqrt{3}}$. Then from $x^2 = 2y^2$ we

have $x^2 = \frac{2}{3} \;\Rightarrow\; x = \pm\sqrt{\frac{2}{3}}$. Thus $f$ has possible extreme values at the points $(0, \pm 1)$, $\left(\pm\sqrt{\frac{2}{3}}, \frac{1}{\sqrt{3}}\right)$, and

$\left(\pm\sqrt{\frac{2}{3}}, -\frac{1}{\sqrt{3}}\right)$. We evaluate $f$ at these six points:

$$f(0, \pm 1) = 0 \qquad f\left(\pm\sqrt{\tfrac{2}{3}}, \tfrac{1}{\sqrt{3}}\right) = \tfrac{2}{3\sqrt{3}} \qquad f\left(\pm\sqrt{\tfrac{2}{3}}, -\tfrac{1}{\sqrt{3}}\right) = -\tfrac{2}{3\sqrt{3}}$$

Thus the maximum value of $f$ subject to the constraint is $f\left(\pm\sqrt{\frac{2}{3}}, \frac{1}{\sqrt{3}}\right) = \frac{2}{3\sqrt{3}}$ and the minimum is

$f\left(\pm\sqrt{\frac{2}{3}}, -\frac{1}{\sqrt{3}}\right) = -\frac{2}{3\sqrt{3}}$.

**27.** $f(x,y,z) = xyz,\; g(x,y,z) = x^2 + y^2 + z^2 = 3$. We solve the equations $f_x = \lambda g_x \;\Rightarrow\; yz = 2\lambda x,\; f_y = \lambda g_y \;\Rightarrow$

$xz = 2\lambda y,\; f_z = \lambda g_z \;\Rightarrow\; xy = 2\lambda z$, and $x^2 + y^2 + z^2 = 3$. If none of $x$, $y$, $z$ are zero then we can solve each of the first

three equations for $\lambda$, giving $\lambda = \frac{yz}{2x} = \frac{xz}{2y} = \frac{xy}{2z}$. Comparing the first two fractions gives $2y^2z = 2x^2z \;\Rightarrow\; y^2 = x^2$, and

similarly the last two give $2xz^2 = 2xy^2 \;\Rightarrow\; z^2 = y^2$. Thus $x^2 = y^2 = z^2$, and substituting into the constraint equation, we

have $x^2 + x^2 + x^2 = 3 \;\Rightarrow\; x^2 = 1 = y^2 = z^2$. Each of $x$, $y$, and $z$ can be $\pm 1$, giving eight possible points: $(1,1,\pm 1)$,

$(1,-1,\pm 1),\; (-1,1,\pm 1),\; (-1,-1,\pm 1)$.

Now we consider the case where one of the variables is zero. If $x = 0$, then from the second equation either $\lambda = 0$ or $y = 0$,

but from the third equation $y = 0$ implies $z = 0$ (if $\lambda \neq 0$), and $x = y = z = 0$ contradicts the constraint equation. Thus if

$x = 0$, the only possibility is $\lambda = 0$. From the first equation we then have $yz = 0$, so either $y = 0$ or $z = 0$. Thus if any of $x$,

$y$, or $z$ is zero, we must have two of the variables equal to zero, and from the constraint we have possible

points $(0, 0, \pm\sqrt{3})$, $(0, \pm\sqrt{3}, 0)$, $(\pm\sqrt{3}, 0, 0)$. $f(x,y,z) = 0$ at all these points, while

$f(1,1,1) = f(1,-1,-1) = f(-1,1,-1) = f(-1,-1,1) = 1$, the maximum value of $f$ subject to the constraint, and

$f(1,1,-1) = f(1,-1,1) = f(-1,1,1) = f(-1,-1,-1) = -1$, the minimum.

# APPENDIXES

## A  Review of Algebra

**1.** $(-6ab)(0.5ac) = (-6)(0.5)(a \cdot abc) = -3a^2bc$

**3.** $2x(x - 5) = 2x \cdot x - 2x \cdot 5 = 2x^2 - 10x$

**5.** $-2(4 - 3a) = -2 \cdot 4 + 2 \cdot 3a = -8 + 6a$

**7.** $4(x^2 - x + 2) - 5(x^2 - 2x + 1) = 4x^2 - 4x + 8 - 5x^2 - 5(-2x) - 5$
$$= 4x^2 - 5x^2 - 4x + 10x + 8 - 5 = -x^2 + 6x + 3$$

**9.** $(4x - 1)(3x + 7) = 4x(3x + 7) - (3x + 7) = 12x^2 + 28x - 3x - 7 = 12x^2 + 25x - 7$

**11.** $(2x - 1)^2 = (2x)^2 - 2(2x)(1) + 1^2 = 4x^2 - 4x + 1$

**13.** $y^4(6 - y)(5 + y) = y^4[6(5 + y) - y(5 + y)] = y^4(30 + 6y - 5y - y^2)$
$$= y^4(30 + y - y^2) = 30y^4 + y^5 - y^6$$

**15.** $(1 + 2x)(x^2 - 3x + 1) = 1(x^2 - 3x + 1) + 2x(x^2 - 3x + 1) = x^2 - 3x + 1 + 2x^3 - 6x^2 + 2x$
$$= 2x^3 - 5x^2 - x + 1$$

**17.** $\dfrac{2 + 8x}{2} = \dfrac{2}{2} + \dfrac{8x}{2} = 1 + 4x$

**19.** $\dfrac{1}{x + 5} + \dfrac{2}{x - 3} = \dfrac{1}{x + 5} \cdot \dfrac{x - 3}{x - 3} + \dfrac{2}{x - 3} \cdot \dfrac{x + 5}{x + 5} = \dfrac{(1)(x - 3) + 2(x + 5)}{(x + 5)(x - 3)}$
$$= \dfrac{x - 3 + 2x + 10}{(x + 5)(x - 3)} = \dfrac{3x + 7}{x^2 + 2x - 15}$$

**21.** $u + 1 + \dfrac{u}{u + 1} = (u + 1) \cdot \dfrac{u + 1}{u + 1} + \dfrac{u}{u + 1} = \dfrac{(u + 1)(u + 1) + u}{u + 1} = \dfrac{u^2 + 2u + 1 + u}{u + 1} = \dfrac{u^2 + 3u + 1}{u + 1}$

**23.** $\dfrac{x/y}{z} = \dfrac{x/y}{z/1} = \dfrac{x}{y} \div \dfrac{z}{1} = \dfrac{x}{y} \cdot \dfrac{1}{z} = \dfrac{x}{yz}$

**25.** $\left(\dfrac{-2r}{s}\right)\left(\dfrac{s^2}{-6t}\right) = \dfrac{-2rs^2}{-6st} = \dfrac{rs}{3t}$

**27.** $\dfrac{1 + \dfrac{1}{c - 1}}{1 - \dfrac{1}{c - 1}} = \dfrac{\dfrac{c - 1 + 1}{c - 1}}{\dfrac{c - 1 - 1}{c - 1}} = \dfrac{\dfrac{c}{c - 1}}{\dfrac{c - 2}{c - 1}} = \dfrac{c}{c - 1} \div \dfrac{c - 2}{c - 1} = \dfrac{c}{c - 1} \cdot \dfrac{c - 1}{c - 2} = \dfrac{c}{c - 2}$

**29.** $2x + 12x^3 = 2x \cdot 1 + 2x \cdot 6x^2 = 2x(1 + 6x^2)$

**31.** The two integers that add to give 7 and multiply to give 6 are 6 and 1. Therefore $x^2 + 7x + 6 = (x + 6)(x + 1)$.

**33.** The two integers that add to give $-2$ and multiply to give $-8$ are $-4$ and $2$. Therefore $x^2 - 2x - 8 = (x-4)(x+2)$.

**35.** $9x^2 - 36 = 9(x^2 - 4) = 9(x-2)(x+2)$   [Equation 3 with $a = x, b = 2$]

**37.** $6x^2 - 5x - 6 = (3x+2)(2x-3)$

**39.** $t^3 + 1 = (t+1)(t^2 - t + 1)$   [Equation 5 with $a = t, b = 1$]

**41.** $4t^2 - 12t + 9 = (2t-3)^2$   [Equation 2 with $a = 2t, b = 3$]

**43.** $x^3 + 2x^2 + x = x(x^2 + 2x + 1) = x(x+1)^2$   [Equation 1 with $a = x, b = 1$]

**45.** $\dfrac{x^2 + x - 2}{x^2 - 3x + 2} = \dfrac{(x+2)(x-1)}{(x-2)(x-1)} = \dfrac{x+2}{x-2}$

**47.** $\dfrac{x^2 - 1}{x^2 - 9x + 8} = \dfrac{(x-1)(x+1)}{(x-8)(x-1)} = \dfrac{x+1}{x-8}$

**49.** $\dfrac{1}{x+3} + \dfrac{1}{x^2 - 9} = \dfrac{1}{x+3} + \dfrac{1}{(x-3)(x+3)} = \dfrac{1(x-3)+1}{(x-3)(x+3)} = \dfrac{x-2}{x^2 - 9}$

**51.** $x^2 + 2x + 5 = [x^2 + 2x] + 5 = [x^2 + 2x + (1)^2 - (1)^2] + 5 = (x+1)^2 + 5 - 1 = (x+1)^2 + 4$

**53.** $x^2 - 5x + 10 = [x^2 - 5x] + 10 = \left[x^2 - 5x + \left(-\frac{5}{2}\right)^2 - \left(-\frac{5}{2}\right)^2\right] + 10 = \left(x - \frac{5}{2}\right)^2 + 10 - \frac{25}{4} = \left(x - \frac{5}{2}\right)^2 + \frac{15}{4}$

**55.** $4x^2 + 4x - 2 = 4[x^2 + x] - 2 = 4\left[x^2 + x + \left(\frac{1}{2}\right)^2 - \left(\frac{1}{2}\right)^2\right] - 2 = 4\left(x + \frac{1}{2}\right)^2 - 2 - 4\left(\frac{1}{4}\right) = 4\left(x + \frac{1}{2}\right)^2 - 3$

**57.** $x^2 - 9x - 10 = 0 \iff (x+10)(x-1) = 0 \iff x + 10 = 0$ or $x - 1 = 0 \iff x = -10$ or $x = 1$.

**59.** Using the quadratic formula, $x^2 + 9x - 1 = 0 \iff x = \dfrac{-9 \pm \sqrt{9^2 - 4(1)(-1)}}{2(1)} = \dfrac{9 \pm \sqrt{85}}{2}$.

**61.** Using the quadratic formula, $3x^2 + 5x + 1 = 0 \iff x = \dfrac{-5 \pm \sqrt{5^2 - 4(3)(1)}}{2(3)} = \dfrac{-5 \pm \sqrt{13}}{6}$.

**63.** $2x^2 + 3x + 4$ is irreducible because its discriminant is negative: $b^2 - 4ac = 9 - 4(2)(4) = -23 < 0$.

**65.** $3x^2 + x - 6$ is not irreducible because its discriminant is nonnegative: $b^2 - 4ac = 1 - 4(3)(-6) = 73 > 0$.

**67.** Using the Binomial Theorem with $k = 6$ we have

$$(a+b)^6 = a^6 + 6a^5 b + \frac{6 \cdot 5}{1 \cdot 2} a^4 b^2 + \frac{6 \cdot 5 \cdot 4}{1 \cdot 2 \cdot 3} a^3 b^3 + \frac{6 \cdot 5 \cdot 4 \cdot 3}{1 \cdot 2 \cdot 3 \cdot 4} a^2 b^4 + 6ab^5 + b^6$$

$$= a^6 + 6a^5 b + 15a^4 b^2 + 20a^3 b^3 + 15a^2 b^4 + 6ab^5 + b^6$$

**69.** Using Equation 9, $\sqrt{32}\sqrt{2} = \sqrt{32 \cdot 2} = \sqrt{64} = 8$.

**71.** $\sqrt{16a^4 b^3} = \sqrt{16a^4 b^2 \cdot b} = \sqrt{16a^4 b^2}\sqrt{b} = 4a^2 b\sqrt{b}$.

**73.** $\sqrt{6p^5r}\sqrt{15r^3} = \sqrt{(6p^5r)\cdot(15r^3)} = \sqrt{90p^5r^4} = \sqrt{(9p^4r^4)(10p)} = \sqrt{9p^4r^4}\sqrt{10p} = 3p^2r^2\sqrt{10p}$

**75.** Using Laws 3 and 1 of Exponents respectively, $3^{10} \times 9^8 = 3^{10} \times (3^2)^8 = 3^{10} \times 3^{2\cdot8} = 3^{10+16} = 3^{26}$.

**77.** Using Laws 4, 1, and 2 of Exponents respectively, $\dfrac{x^9(2x)^4}{x^3} = \dfrac{x^9(2^4)x^4}{x^3} = \dfrac{16x^{9+4}}{x^3} = 16x^{9+4-3} = 16x^{10}$.

**79.** Using Law 2 of Exponents, $\dfrac{a^{-3}b^4}{a^{-5}b^5} = a^{-3-(-5)}b^{4-5} = a^2b^{-1} = \dfrac{a^2}{b}$.

**81.** By definitions 3 and 4 for exponents respectively, $3^{-1/2} = \dfrac{1}{3^{1/2}} = \dfrac{1}{\sqrt{3}}$.

**83.** Using definition 4 for exponents, $125^{2/3} = \left[\sqrt[3]{125}\right]^2 = 5^2 = 25$.

**85.** $(2x^2y^4)^{3/2} = 2^{3/2}(x^2)^{3/2}(y^4)^{3/2} = 2\cdot2^{1/2}\left[\sqrt{x^2}\right]^3\left[\sqrt{y^4}\right]^3 = 2\sqrt{2}\,|x|^3\,(y^2)^3 = 2\sqrt{2}\,|x|^3\,y^6$

**87.** $\dfrac{1}{\left(\sqrt{t}\right)^5} = \dfrac{1}{(t^{1/2})^5} = \dfrac{1}{t^{5/2}} = t^{-5/2}$

**89.** $\dfrac{\sqrt{x}-3}{x-9} = \dfrac{\sqrt{x}-3}{x-9}\cdot\dfrac{\sqrt{x}+3}{\sqrt{x}+3} = \dfrac{(x-9)}{(x-9)\left(\sqrt{x}+3\right)} = \dfrac{1}{\sqrt{x}+3}$

**91.** $\dfrac{x\sqrt{x}-8}{x-4} = \dfrac{x\sqrt{x}-8}{x-4}\cdot\dfrac{x\sqrt{x}+8}{x\sqrt{x}+8} = \dfrac{x^3-64}{(x-4)\left(x\sqrt{x}+8\right)}$

$\qquad = \dfrac{(x-4)(x^2+4x+16)}{(x-4)\left(x\sqrt{x}+8\right)}$   [Equation 4 with $a=x$, $b=4$]   $= \dfrac{x^2+4x+16}{x\sqrt{x}+8}$

**93.** $\dfrac{2}{3-\sqrt{5}} = \dfrac{2}{3-\sqrt{5}}\cdot\dfrac{3+\sqrt{5}}{3+\sqrt{5}} = \dfrac{2\left(3+\sqrt{5}\right)}{9-5} = \dfrac{3+\sqrt{5}}{2}$

**95.** False. See Example 13(b).

**97.** True: $\dfrac{16+a}{16} = \dfrac{16}{16} + \dfrac{a}{16} = 1 + \dfrac{a}{16}$

**99.** False.

**101.** False. Using Law 3 of Exponents, $(x^3)^4 = x^{3\cdot4} = x^{12} \neq x^7$.

**103.** $2x+7 > 3 \;\Leftrightarrow\; 2x > -4 \;\Leftrightarrow\; x > -2$, so the solution is $(-2, \infty)$.

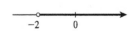

**105.** $1-x \leq 2 \;\Leftrightarrow\; -x \leq 1 \;\Leftrightarrow\; x \geq -1$, so the solution is $[-1, \infty)$.

**107.** $0 \leq 1-x < 1 \;\Leftrightarrow\; -1 \leq -x < 0 \;\Leftrightarrow\; 1 \geq x > 0$, or

equivalently, $0 < x \leq 1$, so the solution is $(0, 1]$.

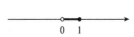

**109.** $(x-1)(x-2) > 0$  *Case 1:* (both factors are positive, so their product is positive)

$$x - 1 > 0 \iff x > 1, \text{ and } x - 2 > 0 \iff x > 2, \text{ so the solution is } (2, \infty).$$

*Case 2:* (both factors are negative, so their product is positive)

$$x - 1 < 0 \iff x < 1, \text{ and } x - 2 < 0 \iff x < 2, \text{ so the solution is } (-\infty, 1).$$

Thus the solution set is $(-\infty, 1) \cup (2, \infty)$.

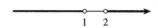

**111.** $x^2 < 3 \iff x^2 - 3 < 0 \iff \left(x - \sqrt{3}\right)\left(x + \sqrt{3}\right) < 0$. *Case 1:* $x > \sqrt{3}$ and $x < -\sqrt{3}$, which is impossible.

*Case 2:* $x < \sqrt{3}$ and $x > -\sqrt{3}$. Thus the solution set is $\left(-\sqrt{3}, \sqrt{3}\right)$.

*Another method:* $x^2 < 3 \iff |x| < \sqrt{3} \iff -\sqrt{3} < x < \sqrt{3}$.

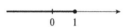

**113.** $x^3 - x^2 \le 0 \iff x^2(x - 1) \le 0$. Since $x^2 \ge 0$ for all $x$, the inequality is satisfied when $x - 1 \le 0 \iff x \le 1$.

Thus the solution set is $(-\infty, 1]$.

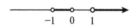

**115.** $x^3 > x \iff x^3 - x > 0 \iff x(x^2 - 1) > 0 \iff x(x-1)(x+1) > 0$. Construct a chart:

| Interval | $x$ | $x - 1$ | $x + 1$ | $x(x-1)(x+1)$ |
|----------|-----|---------|---------|----------------|
| $x < -1$ | $-$ | $-$ | $-$ | $-$ |
| $-1 < x < 0$ | $-$ | $-$ | $+$ | $+$ |
| $0 < x < 1$ | $+$ | $-$ | $+$ | $-$ |
| $x > 1$ | $+$ | $+$ | $+$ | $+$ |

Since $x^3 > x$ when the last column is positive, the solution set is $(-1, 0) \cup (1, \infty)$.

**117.** $C = \frac{5}{9}(F - 32) \implies F = \frac{9}{5}C + 32$. So $50 \le F \le 95 \implies 50 \le \frac{9}{5}C + 32 \le 95 \implies 18 \le \frac{9}{5}C \le 63 \implies$

$\frac{5}{9} \cdot 18 \le \frac{5}{9} \cdot \frac{9}{5}C \le \frac{5}{9} \cdot 63 \implies 10 \le C \le 35$. So the interval is $[10, 35]$.

**119.** (a) Let $T$ represent the temperature in degrees Celsius and $h$ the height in km. $T = 20$ when $h = 0$ and $T$ decreases by $10\,^\circ$C

for every km ($1\,^\circ$C for each 100-meter rise). Thus, $T = 20 - 10h$ when $0 \le h \le 12$.

(b) From part (a), $T = 20 - 10h \implies 10h = 20 - T \implies h = 2 - T/10$. So $0 \le h \le 5 \implies 0 \le 2 - T/10 \le 5 \implies$

$-2 \le -T/10 \le 3 \implies -20 \le -T \le 30 \implies 20 \ge T \ge -30 \implies -30 \le T \le 20$. Thus, the range of

temperatures (in $^\circ$C) to be expected is $[-30, 20]$.

**121.** $|x + 3| = 4 \iff$ either $x + 3 = 4$ or $-4$. In the first case, $x = 1$, and in the second case, $x = -7$. So the solutions are 1

and $-7$.

**123.** $|ab| = \sqrt{(ab)^2} = \sqrt{a^2 b^2} = \sqrt{a^2}\,\sqrt{b^2} = |a|\,|b|$

## B    Coordinate Geometry and Lines

1. Use the distance formula with $P_1(x_1, y_1) = (1, 1)$ and $P_2(x_2, y_2) = (4, 5)$ to get

$$|P_1 P_2| = \sqrt{(4-1)^2 + (5-1)^2} = \sqrt{3^2 + 4^2} = \sqrt{25} = 5$$

3. The distance from $(6, -2)$ to $(-1, 3)$ is $\sqrt{(-1-6)^2 + [3-(-2)]^2} = \sqrt{(-7)^2 + 5^2} = \sqrt{74}$.

5. The distance from $(2, 5)$ to $(4, -7)$ is $\sqrt{(4-2)^2 + (-7-5)^2} = \sqrt{2^2 + (-12)^2} = \sqrt{148} = 2\sqrt{37}$.

7. The slope $m$ of the line through $P(1, 5)$ and $Q(4, 11)$ is $m = \dfrac{11-5}{4-1} = \dfrac{6}{3} = 2$.

9. The slope $m$ of the line through $P(-3, 3)$ and $Q(-1, -6)$ is $m = \dfrac{-6-3}{-1-(-3)} = -\dfrac{9}{2}$.

11. The graph of the equation $x = 3$ is a vertical line with $x$-intercept 3.

    The line does not have a slope.

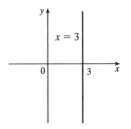

13. By the point-slope form of the equation of a line, an equation of the line through $(1, -8)$ with slope 4 is $y - (-8) = 4(x - 1)$

    or $y + 8 = 4x - 4$ or $y = 4x - 12$.

15. $y - 7 = \frac{2}{3}(x - 1)$ or $y = \frac{2}{3}x + \frac{19}{3}$

17. The slope of the line through $(4, 3)$ and $(3, 8)$ is $m = \dfrac{8-3}{3-4} = \dfrac{5}{-1} = -5$, so an equation of the line is $y - 3 = -5(x - 4)$ or

    $y = -5x + 23$ or $5x + y = 23$.

19. By the slope-intercept form of the equation of a line, an equation of the line is $y = 6x - 4$.

21. Since the $x$-intercept is 2 and the $y$-intercept is $-4$, the line passes through the points $(2, 0)$ and $(0, -4)$, and its slope is

    $m = \dfrac{-4-0}{0-2} = \dfrac{-4}{-2} = 2$. So an equation of the line is $y = 2x - 4$.

23. The line is parallel to the $x$-axis, so it is horizontal and must have the form $y = k$. Since it goes through the point

    $(x, y) = (4, 5)$, the equation is $y = 5$.

25. Putting the line $x + 2y = 6$ into its slope-intercept form gives us $y = -\frac{1}{2}x + 3$, so we see that this line has slope $-\frac{1}{2}$. Thus,

    we want the line of slope $-\frac{1}{2}$ that passes through the point $(1, -6)$: $y - (-6) = -\frac{1}{2}(x - 1)$   $\Leftrightarrow$   $y = -\frac{1}{2}x - \frac{11}{2}$.

27. $2x + 5y + 8 = 0$   $\Leftrightarrow$   $y = -\frac{2}{5}x - \frac{8}{5}$. Since this line has slope $-\frac{2}{5}$, a line perpendicular to it would have slope $\frac{5}{2}$, so the

    required line is $y - (-2) = \frac{5}{2}[x - (-1)]$   $\Leftrightarrow$   $y = \frac{5}{2}x + \frac{1}{2}$.

**29.** $x + 3y = 0 \iff y = -\frac{1}{3}x$, so the slope is $-\frac{1}{3}$ and the $y$-intercept is 0.

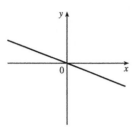

**31.** $y = -2$ is a horizontal line with slope 0 and $y$-intercept $-2$.

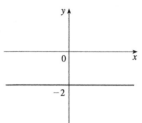

**33.** $3x - 4y = 12 \iff y = \frac{3}{4}x - 3$, so the slope is $\frac{3}{4}$ and the $y$-intercept is $-3$.

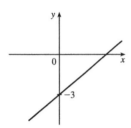

**35.** $\{(x, y) \mid x < 0\}$

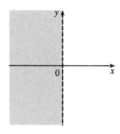

**37.** $\{(x, y) \mid xy < 0\} = \{(x, y) \mid x < 0 \text{ and } y > 0\}$
$\cup \{(x, y) \mid x > 0 \text{ and } y < 0\}$

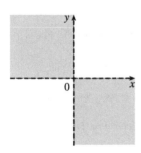

**39.** $\left\{(x, y) \,\middle|\, |x| \le 2\right\} = \{(x, y) \mid -2 \le x \le 2\}$

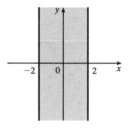

**41.** $\{(x, y) \mid 0 \le y \le 4, x \le 2\}$

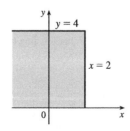

**43.** $\{(x, y) \mid 1 + x \le y \le 1 - 2x\}$

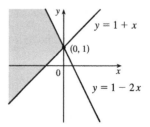

**45.** $2x - y = 4 \iff y = 2x - 4 \implies m_1 = 2$ and $6x - 2y = 10 \iff 2y = 6x - 10 \iff y = 3x - 5 \implies m_2 = 3$.

Since $m_1 \neq m_2$, the two lines are not parallel. To find the point of intersection: $2x - 4 = 3x - 5 \iff x = 1 \implies$

$y = -2$. Thus, the point of intersection is $(1, -2)$.

## C Approximate Integration

**1.** (a) $\Delta x = (b - a)/n = (4 - 0)/2 = 2$

$L_2 = f(x_0) \cdot 2 + f(x_1) \cdot 2 = 2\left[f(0) + f(2)\right] = 2(0.5 + 2.5) = 6$

$R_2 = f(x_1) \cdot 2 + f(x_2) \cdot 2 = 2\left[f(2) + f(4)\right] = 2(2.5 + 3.5) = 12$

$M_2 = f(\overline{x}_1) \cdot 2 + f(\overline{x}_2) \cdot 2 = 2\left[f(1) + f(3)\right] \approx 2(1.6 + 3.2) = 9.6$

(b) 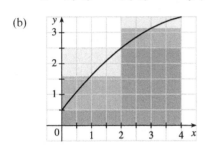 $L_2$ is an underestimate, since the area under the small rectangles is less than the area under the curve, and $R_2$ is an overestimate, since the area under the large rectangles is greater than the area under the curve. It appears that $M_2$ is an overestimate, though it is fairly close to $I = \int_0^4 f(x)\,dx$. See the solution to Exercise 27 for a proof of the fact that if $f$ is concave down on $[a, b]$, then the Midpoint Rule is an overestimate of $\int_a^b f(x)\,dx$.

(c) $T_2 = \left(\tfrac{1}{2}\Delta x\right)[f(x_0) + 2f(x_1) + f(x_2)] = \tfrac{2}{2}[f(0) + 2f(2) + f(4)] = 0.5 + 2(2.5) + 3.5 = 9$.

This approximation is an underestimate, since the graph is concave downward. Thus, $T_2 = 9 < I$. See the solution to Exercise 27 for a general proof of this conclusion.

(d) For any $n$, we will have $L_n < T_n < I < M_n < R_n$.

**3.** (a) $f(x) = \dfrac{x}{1 + x^2}$, $\quad \Delta x = \dfrac{b - a}{n} = \dfrac{2 - 0}{10} = \dfrac{1}{5}$

$M_{10} = \tfrac{1}{5}\left[f\left(\tfrac{1}{10}\right) + f\left(\tfrac{3}{10}\right) + f\left(\tfrac{5}{10}\right) + \cdots + f\left(\tfrac{19}{10}\right)\right] \approx 0.806598$

(b) $S_{10} = \dfrac{1}{5 \cdot 3}\left[f(0) + 4f\left(\tfrac{1}{5}\right) + 2f\left(\tfrac{2}{5}\right) + 4f\left(\tfrac{3}{5}\right) + 2f\left(\tfrac{4}{5}\right) + \cdots + 4f\left(\tfrac{9}{5}\right) + f(2)\right] \approx 0.804779$

*Actual:* $I = \displaystyle\int_0^2 \dfrac{x}{1 + x^2}\,dx = \left[\tfrac{1}{2}\ln\left|1 + x^2\right|\right]_0^2 \qquad [u = 1 + x^2, \, du = 2x\,dx]$

$= \tfrac{1}{2}\ln 5 - \tfrac{1}{2}\ln 1 = \tfrac{1}{2}\ln 5 \approx 0.804719$

*Errors:* $E_M =$ actual $- M_{10} = I - M_{10} \approx -0.001879$

$\phantom{Errors:\ } E_S =$ actual $- S_{10} = I - S_{10} \approx -0.000060$

**5.** $f(x) = \sqrt{x^3 - 1}$, $\Delta x = \dfrac{b - a}{n} = \dfrac{2 - 1}{10} = \dfrac{1}{10}$

(a) $T_{10} = \dfrac{1}{10 \cdot 2}[f(1) + 2f(1.1) + 2f(1.2) + 2f(1.3) + 2f(1.4) + 2f(1.5)$

$\qquad\qquad + 2f(1.6) + 2f(1.7) + 2f(1.8) + 2f(1.9) + f(2)]$

$\approx 1.506361$

(b) $M_{10} = \frac{1}{10}[f(1.05) + f(1.15) + f(1.25) + f(1.35) + f(1.45) + f(1.55) + f(1.65) + f(1.75) + f(1.85) + f(1.95)]$

$\approx 1.518362$

(c) $S_{10} = \frac{1}{10 \cdot 3}[f(1) + 4f(1.1) + 2f(1.2) + 4f(1.3) + 2f(1.4)$

$+ 4f(1.5) + 2f(1.6) + 4f(1.7) + 2f(1.8) + 4f(1.9) + f(2)]$

$\approx 1.511519$

**7.** $f(x) = \dfrac{e^x}{1 + x^2}$, $\Delta x = \dfrac{b - a}{n} = \dfrac{2 - 0}{10} = \dfrac{1}{5}$

(a) $T_{10} = \frac{1}{5 \cdot 2}[f(0) + 2f(0.2) + 2f(0.4) + 2f(0.6) + 2f(0.8) + 2f(1)$

$+ 2f(1.2) + 2f(1.4) + 2f(1.6) + 2f(1.8) + f(2)]$

$\approx 2.660833$

(b) $M_{10} = \frac{1}{5}[f(0.1) + f(0.3) + f(0.5) + f(0.7) + f(0.9) + f(1.1) + f(1.3) + f(1.5) + f(1.7) + f(1.9)]$

$\approx 2.664377$

(c) $S_{10} = \frac{1}{5 \cdot 3}[f(0) + 4f(0.2) + 2f(0.4) + 4f(0.6) + 2f(0.8)$

$+ 4f(1) + 2f(1.2) + 4f(1.4) + 2f(1.6) + 4f(1.8) + f(2)]$

$\approx 2.663244$

**9.** $f(x) = \sqrt{\ln x}$, $\Delta x = \frac{4 - 1}{6} = \frac{1}{2}$

(a) $T_6 = \frac{1}{2 \cdot 2}[f(1) + 2f(1.5) + 2f(2) + 2f(2.5) + 2f(3) + 2f(3.5) + f(4)] \approx 2.591334$

(b) $M_6 = \frac{1}{2}[f(1.25) + f(1.75) + f(2.25) + f(2.75) + f(3.25) + f(3.75)] \approx 2.681046$

(c) $S_6 = \frac{1}{2 \cdot 3}[f(1) + 4f(1.5) + 2f(2) + 4f(2.5) + 2f(3) + 4f(3.5) + f(4)] \approx 2.631976$

**11.** $f(x) = e^{e^x}$, $\Delta x = \frac{1 - (-1)}{10} = \frac{1}{5}$

(a) $T_{10} = \frac{1}{5 \cdot 2}[f(-1) + 2f(-0.8) + 2f(-0.6) + 2f(-0.4) + 2f(-0.2) + 2f(0)$

$+ 2f(0.2) + 2f(0.4) + 2f(0.6) + 2f(0.8) + f(1)]$

$\approx 8.363853$

(b) $M_{10} = \frac{1}{5}[f(-0.9) + f(-0.7) + f(-0.5) + f(-0.3) + f(-0.1) + f(0.1) + f(0.3) + f(0.5) + f(0.7) + f(0.9)]$

$\approx 8.163298$

(c) $S_{10} = \frac{1}{5 \cdot 3}[f(-1) + 4f(-0.8) + 2f(-0.6) + 4f(-0.4) + 2f(-0.2)$

$+ 4f(0) + 2f(0.2) + 4f(0.4) + 2f(0.6) + 4f(0.8) + f(1)]$

$\approx 8.235114$

**13.** (a) $f(t) = \sqrt{t^3 + 2}$, $\Delta t = \frac{2 - 0}{8} = \frac{1}{4}$

(i) $M_8 = \frac{1}{4}\left[f\left(\frac{1}{8}\right) + f\left(\frac{3}{8}\right) + f\left(\frac{5}{8}\right) + f\left(\frac{7}{8}\right) + f\left(\frac{9}{8}\right) + f\left(\frac{11}{8}\right) + f\left(\frac{13}{8}\right) + f\left(\frac{15}{8}\right)\right] \approx 3.85352$

(ii) $T_8 = \frac{1}{4 \cdot 2}\left[f(0) + 2f\left(\frac{1}{4}\right) + 2f\left(\frac{1}{2}\right) + 2f\left(\frac{3}{4}\right) + 2f(1) + 2f\left(\frac{5}{4}\right) + 2f\left(\frac{3}{2}\right) + 2f\left(\frac{7}{4}\right) + f(2)\right] \approx 3.86837$

(iii) $S_8 = \frac{1}{4 \cdot 3}\left[f(0) + 4f\left(\frac{1}{4}\right) + 2f\left(\frac{1}{2}\right) + 4f\left(\frac{3}{4}\right) + 2f(1) + 4f\left(\frac{5}{4}\right) + 2f\left(\frac{3}{2}\right) + 4f\left(\frac{7}{4}\right) + f(2)\right] \approx 3.85842$

(b) Using a graphing calculator, $\int_0^2 \sqrt{t^3 + 2}\, dt \approx 3.85847$. This is extremely close to $S_8$ in part (a).

**15.** $\Delta x = (b - a)/n = (6 - 0)/6 = 1$

  (a) $T_6 = \frac{\Delta x}{2}[f(0) + 2f(1) + 2f(2) + 2f(3) + 2f(4) + 2f(5) + f(6)]$

$\approx \frac{1}{2}[3 + 2(5) + 2(4) + 2(2) + 2(2.8) + 2(4) + 1]$

$= \frac{1}{2}(39.6) = 19.8$

  (b) $M_6 = \Delta x[f(0.5) + f(1.5) + f(2.5) + f(3.5) + f(4.5) + f(5.5)]$

$\approx 1[4.5 + 4.7 + 2.6 + 2.2 + 3.4 + 3.2]$

$= 20.6$

  (c) $S_6 = \frac{\Delta x}{3}[f(0) + 4f(1) + 2f(2) + 4f(3) + 2f(4) + 4f(5) + f(6)]$

$\approx \frac{1}{3}[3 + 4(5) + 2(4) + 4(2) + 2(2.8) + 4(4) + 1]$

$= \frac{1}{3}(61.6) = 20.5\overline{3}$

**17.** $\int_1^5 f(x)\,dx \approx M_4 = \frac{5-1}{4}[f(1.5) + f(2.5) + f(3.5) + f(4.5)] = 1(2.9 + 3.6 + 4.0 + 3.9) = 14.4$

**19.** $T_{\text{ave}} = \frac{1}{24-0}\int_0^{24} T(t)\,dt \approx \frac{1}{24}S_{12} = \frac{1}{24}\frac{24-0}{3(12)}[T(0) + 4T(2) + 2T(4) + 4T(6) + 2T(8) + 4T(10) + 2T(12)$

$+ 4T(14) + 2T(16) + 4T(18) + 2T(20) + 4T(22) + T(24)]$

$\approx \frac{1}{36}[67 + 4(65) + 2(62) + 4(58) + 2(56) + 4(61) + 2(63) + 4(68)$

$+ 2(71) + 4(69) + 2(67) + 4(66) + 64]$

$= \frac{1}{36}(2317) = 64.36\overline{1}°\text{F}.$

The average temperature was about $64.4°\text{F}$.

**21.** By the Net Change Theorem, the increase in velocity is equal to $\int_0^6 a(t)\,dt$. We use Simpson's Rule with $n = 6$ and

$\Delta t = (6 - 0)/6 = 1$ to estimate this integral:

$\int_0^6 a(t)\,dt \approx S_6 = \frac{1}{3}[a(0) + 4a(1) + 2a(2) + 4a(3) + 2a(4) + 4a(5) + a(6)]$

$\approx \frac{1}{3}[0 + 4(0.5) + 2(4.1) + 4(9.8) + 2(12.9) + 4(9.5) + 0] = \frac{1}{3}(113.2) = 37.7\overline{3} \text{ ft/s}$

**23.** By the Net Change Theorem, the energy used is equal to $\int_0^6 P(t)\,dt$. We use Simpson's Rule with $n = 12$ and

$\Delta t = \frac{6-0}{12} = \frac{1}{2}$ to estimate this integral:

$\int_0^6 P(t)\,dt \approx S_{12} = \frac{1/2}{3}[P(0) + 4P(0.5) + 2P(1) + 4P(1.5) + 2P(2) + 4P(2.5) + 2P(3)$

$+ 4P(3.5) + 2P(4) + 4P(4.5) + 2P(5) + 4P(5.5) + P(6)]$

$= \frac{1}{6}[1814 + 4(1735) + 2(1686) + 4(1646) + 2(1637) + 4(1609) + 2(1604)$

$+ 4(1611) + 2(1621) + 4(1666) + 2(1745) + 4(1886) + 2052]$

$= \frac{1}{6}(61{,}064) = 10{,}177.\overline{3} \text{ megawatt-hours}$

**25.** Consider the function $f$ whose graph is shown. The area $\int_0^2 f(x)\,dx$

is close to 2. The Trapezoidal Rule gives

$T_2 = \frac{2-0}{2 \cdot 2}\left[f(0) + 2f(1) + f(2)\right] = \frac{1}{2}\left[1 + 2 \cdot 1 + 1\right] = 2.$

The Midpoint Rule gives $M_2 = \frac{2-0}{2}\left[f(0.5) + f(1.5)\right] = 1[0 + 0] = 0,$

so the Trapezoidal Rule is more accurate.

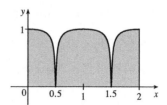

**27.** Since the Trapezoidal and Midpoint approximations on the interval $[a, b]$ are the sums of the Trapezoidal and Midpoint

approximations on the subintervals $[x_{i-1}, x_i]$, $i = 1, 2, \ldots, n$, we can focus our attention on one such interval. The condition

$f''(x) < 0$ for $a \le x \le b$ means that the graph of $f$ is concave downward as in Figure 5. In that figure, $T_n$ is the area of the

trapezoid $AQRD$, $\int_a^b f(x)\,dx$ is the area of the region $AQPRD$, and $M_n$ is the area of the trapezoid $ABCD$, so

$T_n < \int_a^b f(x)\,dx < M_n$. In general, the condition $f'' < 0$ implies that the graph of $f$ on $[a, b]$ lies above the line segment

joining the points $(a, f(a))$ and $(b, f(b))$. Thus, $\int_a^b f(x)\,dx > T_n$. Since $M_n$ is the area under a tangent to the graph, and

since $f'' < 0$ implies that the tangent lies above the graph, we also have $M_n > \int_a^b f(x)\,dx$. Thus, $T_n < \int_a^b f(x)\,dx < M_n$.

**29.** $T_n = \dfrac{\Delta x}{2}\left[f(x_0) + 2\left[f(x_1) + f(x_2) + \cdots + f(x_{n-1})\right] + f(x_n)\right]$ and

$M_n = \Delta x\left[f\left(x_1 - \dfrac{\Delta x}{2}\right) + f\left(x_2 - \dfrac{\Delta x}{2}\right) + \cdots + f\left(x_n - \dfrac{\Delta x}{2}\right)\right]$, so

$\frac{1}{3}T_n + \frac{2}{3}M_n = \frac{1}{3}(T_n + 2M_n) = \dfrac{\Delta x}{3 \cdot 2}\left[f(x_0) + 2\left[f(x_1) + f(x_2) + \cdots + f(x_{n-1})\right] + f(x_n)\right.$

$\left. + 4\left[f\left(x_1 - \dfrac{\Delta x}{2}\right) + f\left(x_2 - \dfrac{\Delta x}{2}\right) + \cdots + f\left(x_n - \dfrac{\Delta x}{2}\right)\right]\right]$

where $\Delta x = \dfrac{b-a}{n}$. Let $\delta x = \dfrac{b-a}{2n}$. Then $\Delta x = 2\delta x$, so

$\frac{1}{3}T_n + \frac{2}{3}M_n = \dfrac{\delta x}{3}\left[f(x_0) + 2\left[f(x_1) + f(x_2) + \cdots + f(x_{n-1})\right] + f(x_n)\right.$

$\left. + 4\left[f(x_1 - \delta x) + f(x_2 - \delta x) + \cdots + f(x_n - \delta x)\right]\right]$

$= \frac{1}{3}\delta x\left[f(x_0) + 4f(x_1 - \delta x) + 2f(x_1) + 4f(x_2 - \delta x)\right.$

$\left. + 2f(x_2) + \cdots + 2f(x_{n-1}) + 4f(x_n - \delta x) + f(x_n)\right]$

Since $x_0, x_1 - \delta x, x_1, x_2 - \delta x, x_2, \ldots, x_{n-1}, x_n - \delta x, x_n$ are the subinterval endpoints for $S_{2n}$, and since $\delta x = \dfrac{b-a}{2n}$ is

the width of the subintervals for $S_{2n}$, the last expression for $\frac{1}{3}T_n + \frac{2}{3}M_n$ is the usual expression for $S_{2n}$. Therefore,

$\frac{1}{3}T_n + \frac{2}{3}M_n = S_{2n}.$

# D   Double Integrals

**1.** $\int_0^5 12x^2 y^3 \, dx = \left[ 12 \dfrac{x^3}{3} y^3 \right]_{x=0}^{x=5} = 4x^3 y^3 \big]_{x=0}^{x=5} = 4(5)^3 \, y^3 - 4(0)^3 \, y^3 = 500y^3,$

$\int_0^1 12x^2 y^3 \, dy = \left[ 12x^2 \dfrac{y^4}{4} \right]_{y=0}^{y=1} = 3x^2 y^4 \big]_{y=0}^{y=1} = 3x^2(1)^4 - 3x^2(0)^4 = 3x^2$

**3.** $\int_1^3 \int_0^1 (1 + 4xy) \, dx \, dy = \int_1^3 \left[ x + 2x^2 y \right]_{x=0}^{x=1} dy = \int_1^3 [(1 + 2y) - (0 + 0)] \, dy = \int_1^3 (1 + 2y) \, dy = \left[ y + y^2 \right]_1^3$

$= (3 + 9) - (1 + 1) = 10$

**5.** $\int_0^2 \int_0^1 (2x + y)^8 dx \, dy = \int_0^2 \left[ \dfrac{1}{2} \dfrac{(2x + y)^9}{9} \right]_{x=0}^{x=1} dy$   [substitute $u = 2x + y \;\Rightarrow\; dx = \tfrac{1}{2} \, du$]

$= \dfrac{1}{18} \int_0^2 [(2 + y)^9 - (0 + y)^9] \, dy = \dfrac{1}{18} \left[ \dfrac{(2 + y)^{10}}{10} - \dfrac{y^{10}}{10} \right]_0^2$

$= \tfrac{1}{180} [(4^{10} - 2^{10}) - (2^{10} - 0^{10})] = \tfrac{1,046,528}{180} = \tfrac{261,632}{45}$

**7.** $\int_1^4 \int_1^2 \left( \dfrac{x}{y} + \dfrac{y}{x} \right) dy \, dx = \int_1^4 \left[ x \ln|y| + \dfrac{1}{x} \cdot \dfrac{1}{2} y^2 \right]_{y=1}^{y=2} dx = \int_1^4 \left( x \ln 2 + \dfrac{2}{x} - 0 - \dfrac{1}{2x} \right) dx = \int_1^4 \left( x \ln 2 + \dfrac{3}{2x} \right) dx$

$= \left[ \tfrac{1}{2} x^2 \ln 2 + \tfrac{3}{2} \ln|x| \right]_1^4 = 8 \ln 2 + \tfrac{3}{2} \ln 4 - \tfrac{1}{2} \ln 2 = \tfrac{15}{2} \ln 2 + 3 \ln 4^{1/2} = \tfrac{21}{2} \ln 2$

**9.** $\int_0^4 \int_0^{\sqrt{y}} xy^2 \, dx \, dy = \int_0^4 \left[ \tfrac{1}{2} x^2 y^2 \right]_{x=0}^{x=\sqrt{y}} dy = \int_0^4 \tfrac{1}{2} y^2 [(\sqrt{y})^2 - 0^2] dy = \tfrac{1}{2} \int_0^4 y^3 \, dy = \tfrac{1}{2} \left[ \tfrac{1}{4} y^4 \right]_0^4 = \tfrac{1}{2}(64 - 0) = 32$

**11.** $\int_0^1 \int_{x^2}^x (1 + 2y) \, dy \, dx = \int_0^1 \left[ y + y^2 \right]_{y=x^2}^{y=x} dx = \int_0^1 \left[ x + x^2 - x^2 - (x^2)^2 \right] dx = \int_0^1 (x - x^4) dx$

$= \left[ \tfrac{1}{2} x^2 - \tfrac{1}{5} x^5 \right]_0^1 = \tfrac{1}{2} - \tfrac{1}{5} - 0 + 0 = \tfrac{3}{10}$

**13.** $\iint_R (6x^2 y^3 - 5y^4) \, dA = \int_0^3 \int_0^1 (6x^2 y^3 - 5y^4) \, dy \, dx = \int_0^3 \left[ \tfrac{3}{2} x^2 y^4 - y^5 \right]_{y=0}^{y=1} dx = \int_0^3 \left( \tfrac{3}{2} x^2 - 1 \right) dx$

$= \left[ \tfrac{1}{2} x^3 - x \right]_0^3 = \tfrac{27}{2} - 3 = \tfrac{21}{2}$

**15.** $\iint_R xye^{x^2 y} \, dA = \int_0^2 \int_0^1 xye^{x^2 y} \, dx \, dy = \int_0^2 \left[ \tfrac{1}{2} e^{x^2 y} \right]_{x=0}^{x=1} dy$   [let $u = x^2 y \;\Rightarrow\; du = 2xy \, dx$]

$= \tfrac{1}{2} \int_0^2 (e^y - 1) \, dy = \tfrac{1}{2} \left[ e^y - y \right]_0^2 = \tfrac{1}{2} [(e^2 - 2) - (1 - 0)] = \tfrac{1}{2} (e^2 - 3)$

**17.**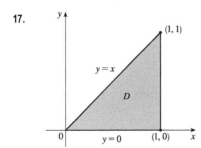

$D$ is the region below the line $y = x$ and above the line $y = 0$ for

$0 \le x \le 1$, so $D = \{(x, y) \mid 0 \le x \le 1, 0 \le y \le x\}$ and

$\displaystyle\iint_D xy^2 \, dA = \int_0^1 \int_0^x xy^2 \, dy \, dx = \int_0^1 x \left[ \tfrac{1}{3} y^3 \right]_{y=0}^{y=x} dx$

$= \int_0^1 \tfrac{1}{3} x^4 \, dx = \left[ \tfrac{1}{3} \cdot \tfrac{1}{5} x^5 \right]_0^1 = \tfrac{1}{15} - 0 = \tfrac{1}{15}$

**19.** $\iint_D x^3 \, dA = \int_1^e \int_0^{\ln x} x^3 \, dy \, dx = \int_1^e \left[ x^3 y \right]_{y=0}^{y=\ln x} dx = \int_1^e x^3 \ln x \, dx$   $\left[ \begin{array}{l} \text{integrate by parts} \\ \text{with } u = \ln x, dv = x^3 \, dx \end{array} \right]$

$= \left[ \tfrac{1}{4} x^4 \ln x - \tfrac{1}{16} x^4 \right]_1^e = \tfrac{1}{4} e^4 - \tfrac{1}{16} e^4 - 0 + \tfrac{1}{16} = \tfrac{3}{16} e^4 + \tfrac{1}{16}$

**21.** The plane $3x + 2y + z = 12$ is the function $z = 12 - 3x - 2y$, so the volume of the solid is

$$V = \iint_R (12 - 3x - 2y)\, dA = \int_{-2}^3 \int_0^1 (12 - 3x - 2y)\, dx\, dy = \int_{-2}^3 \left[ 12x - \tfrac{3}{2}x^2 - 2xy \right]_{x=0}^{x=1} dy$$

$$= \int_{-2}^3 \left( \tfrac{21}{2} - 2y \right) dy = \left[ \tfrac{21}{2}y - y^2 \right]_{-2}^3 = \tfrac{95}{2}$$

**23.** The region of integration $D$ is the region below the line $x + 3y = 7 \iff y = (7 - x)/3$ and above the line $y = 1$ for

$1 \le x \le 4$. Thus

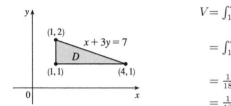

$$V = \int_1^4 \int_1^{(7-x)/3} xy\, dy\, dx = \int_1^4 \left[ \tfrac{1}{2}xy^2 \right]_{y=1}^{y=(7-x)/3} dx$$

$$= \int_1^4 \tfrac{1}{2}x \left[ \tfrac{1}{9}(7 - x)^2 - 1 \right] dx = \tfrac{1}{2} \int_1^4 \tfrac{1}{9}x[(7 - x)^2 - 9]\, dx$$

$$= \tfrac{1}{18} \int_1^4 (x^3 - 14x^2 + 40x)\, dx = \tfrac{1}{18} \left[ \tfrac{1}{4}x^4 - \tfrac{14}{3}x^3 + 20x^2 \right]_1^4$$

$$= \tfrac{1}{18} \left( \tfrac{256}{3} - \tfrac{187}{12} \right) = \tfrac{31}{8}$$

**25.** $R$ is the rectangle $[-1, 1] \times [0, 5]$. Thus, $A(R) = 2 \cdot 5 = 10$ and

$$f_{\text{ave}} = \frac{1}{A(R)} \iint_R f(x, y)\, dA = \tfrac{1}{10} \int_0^5 \int_{-1}^1 x^2 y\, dx\, dy = \tfrac{1}{10} \int_0^5 \left[ \tfrac{1}{3}x^3 y \right]_{x=-1}^{x=1} dy = \tfrac{1}{10} \int_0^5 \tfrac{2}{3}y\, dy = \tfrac{1}{10} \left[ \tfrac{1}{3}y^2 \right]_0^5 = \tfrac{5}{6}.$$

**27.** The average value of a function $f$ of two variables defined on a rectangle $R$ was defined as $f_{\text{ave}} = \frac{1}{A(R)} \iint_R f(x, y)\, dA$, where

$A(R)$ is the area of the region of integration $R$. Extending this definition to general regions $D$, we have

$f_{\text{ave}} = \frac{1}{A(D)} \iint_D f(x, y)\, dA$.

Here $D = \{(x, y) \mid 0 \le x \le 1, 0 \le y \le 3x\}$, so $A(D) = \tfrac{1}{2}(1)(3) = \tfrac{3}{2}$

and

$$f_{\text{ave}} = \frac{1}{A(D)} \iint_D f(x, y)\, dA = \tfrac{1}{3/2} \int_0^1 \int_0^{3x} xy\, dy\, dx$$

$$= \tfrac{2}{3} \int_0^1 \left[ \tfrac{1}{2}xy^2 \right]_{y=0}^{y=3x} dx = \tfrac{1}{3} \int_0^1 9x^3\, dx = \tfrac{3}{4}x^4 \Big]_0^1 = \tfrac{3}{4}$$

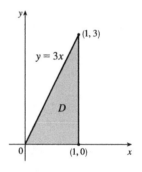

**29.** If we place the origin at the southwest corner of the state, then the region is described by the rectangle

$R = \{(x, y) \mid 0 \le x \le 388, 0 \le y \le 276\}$. The area of $R$ is $388 \cdot 276 = 107{,}088$, and the average snowfall was

$$f_{\text{ave}} = \frac{1}{A(R)} \iint_R f(x, y)\, dA = \frac{1}{107{,}088} \int_0^{388} \int_0^{276} (4.6 + 0.02x - 0.01y + 0.0001xy)\, dy\, dx$$

$$= \frac{1}{107{,}088} \int_0^{388} \left[ 4.6y + 0.02xy - 0.005y^2 + 0.00005xy^2 \right]_{y=0}^{y=276} dx$$

$$= \frac{1}{107{,}088} \int_0^{388} (1269.6 + 5.52x - 380.88 + 3.8088x)\, dx = \frac{1}{107{,}088} \int_0^{388} (888.72 + 9.3288x)\, dx$$

$$= \frac{1}{107{,}088} \left[ 888.72x + 4.6644x^2 \right]_0^{388} = \frac{1}{107{,}088}(1{,}047{,}020.794) \approx 9.77 \text{ in}$$

Made in the USA
Las Vegas, NV
11 August 2021